Handbook of Experimental Pharmacology

Volume 289

The *Handbook of Experimental Pharmacology* is one of the most authoritative and influential book series in pharmacology. It provides critical and comprehensive discussions of the most significant areas of pharmacological research, written by leading international authorities. Each volume in the series represents the most informative and contemporary account of its subject available, making it an unrivalled reference source.

HEP is indexed in PubMed and Scopus.

Erica L. Bradshaw • Mary E. Spilker
Editors

Quantitative Systems Pharmacology

Springer

Editors
Erica L. Bradshaw
Biomosaic Consulting
Carlsbad, CA, USA

Mary E. Spilker
Pharmacokinetics, Dynamics and Metabolism
Pfizer Inc.
La Jolla, CA, USA

ISSN 0171-2004 ISSN 1865-0325 (electronic)
Handbook of Experimental Pharmacology
ISBN 978-3-032-02691-0 ISBN 978-3-032-02692-7 (eBook)
https://doi.org/10.1007/978-3-032-02692-7

This Springer imprint is published by the registered company Springer Nature Switzerland AG
The registered company address is: Gewerbestrasse 11, 6330 Cham, Switzerland

Preface

Modeling and simulation has progressively become a critical pillar of modern drug discovery and development, accelerating and improving the efficiency, safety, and successful delivery of therapeutics. This advancement stems from the successful application of techniques such as physiologically based pharmacokinetics (PBPK), toxicology, population pharmacokinetic (PK), and PK-pharmacodynamic (PD) modeling.

Quantitative systems pharmacology (QSP) is a relatively new technique formally defined herein as: "a modeling discipline that integrates mathematical representations of physiological and pathophysiological processes with pharmacology to support an improved mechanistic understanding of how drugs interact with biology and the systemic implications of drug administration." QSP has emerged as a transformative discipline at the intersection of systems biology, pharmacometrics, and mechanistic modeling. It serves as a powerful tool and framework that integrates quantitative methods with biological knowledge to inform drug discovery, development, and clinical decision-making. Since the publication of the National Institutes of Health (NIH) Workshop White Paper (Sorger et al. 2011) on QSP in 2011, the field has rapidly gained interest and traction across academia, industry, and regulatory agencies. Notably, more than a third of all QSP publications have emerged in the past 5 years. The need for rigorous, mechanistically grounded, and predictive modeling has never been more urgent—especially as the complexity of therapeutic modalities, disease mechanisms, and data landscapes continues to grow. This is highlighted in part by recent communication from regulatory bodies, such as the Food and Drug Administration (FDA) and the European Medicines Agency (EMA), which have not only embraced but now actively encourage the use of modeling and simulation across a range of activities, including to justify dose selection, aid clinical trial design, enable extrapolation across populations, inform on regulatory decision-making, and support label claims and post-market adjustments (ICH M15 General Principles for Model-Informed Drug Development (https://database.ich.org/sites/default/files/ICH_M15_EWG_Step2_DraftGuideline_2024_1031.pdf)).

This volume of the Handbook of Experimental Pharmacology series was conceived to provide a "go-to" reference by offering a structured and comprehensive introduction to QSP, intended for a wide audience: from newcomers seeking foundational knowledge, to experienced practitioners and decision-makers looking for clear guidance and illustrative applications while also reflecting the current state and future potential of the field. To achieve this objective, the book is organized into three sections. The first section of the book, *Fundamental Modeling*, introduces the historical context and conceptual underpinnings of QSP, core pharmacological (PK and PD) modeling principles, practical considerations for model execution, and a commentary on the importance of the datasets QSP models leverage. This section also comments on the evolving integration of machine learning with QSP. The second section focuses on *Model Development and Application.* This section presents a diverse set of real-world case studies spanning neuroscience, rare disease, oncology, and toxicology. These chapters illustrate how QSP is currently being used to generate insights, enable decision-making, and bridge preclinical and clinical domains. A dedicated chapter also explores the collaborative landscape of QSP model development across academia, industry, and contract research organizations. The final section and chapter of the book, *Current Challenges and Future Directions*, presents a forward-looking view on QSP; envisioning how QSP can evolve by leveraging emerging technologies, diverse data, and computational advancements. It also highlights near and long-term opportunities and research directions that will shape the next generation of QSP innovation.

The editors would like to thank all chapter authors as well as the reviewers for their contributions. Each chapter has been authored and reviewed by renowned experts and thought leaders in the field, many of whom have played pivotal roles in the advancement of QSP. We would also like to thank the Handbook of Experimental Pharmacology book series Editor, Dr. James E Barrett, for the ideation of this book volume and extending this editorial invitation to us, as well as Susanne Dathe and Alamelu Damodharan from Springer for their continued support to publish this volume.

We believe this volume will serve as both a foundational textbook and a practical resource. It is intended to foster greater understanding, dialogue, and progress in QSP that will facilitate the use of more predictive science and rational decision-making to ultimately deliver effective and safe therapies to patients. We hope that students, scientists, educators, and stakeholders across biomedicine will find value in this resource, and that it contributes meaningfully to the growth and evolution of QSP.

Carlsbad, CA, USA Erica L. Bradshaw
La Jolla, CA, USA Mary E. Spilker

Reference

Sorger PK et al (2011) Quantitative and systems pharmacology in the post-genomic era: new approaches to discovering drugs and understanding therapeutic mechanisms. An NIH white paper by the QSP workshop group. NIH, Bethesda, pp 1–48

Contents

Part I

Fundamental Modeling

What Is QSP and Why Does It Exist?: A Brief History

Christina Friedrich

Contents

C. Friedrich (✉)
Rosa & Co, LLC, San Carlos, CA, USA
e-mail: cfriedrich@rosaandco.com

E. L. Bradshaw, M. E. Spilker (eds.), *Quantitative Systems Pharmacology*,
Handbook of Experimental Pharmacology 289, https://doi.org/10.1007/164_2024_733

Abstract

Quantitative systems pharmacology (QSP) is a modeling approach employed in drug research and development that combines mechanistic representations of biological processes with drug pharmacology to deepen biological understanding and predict the responses to novel drugs or protocols. QSP has evolved from and is related to other modeling approaches, but has a number of unique attributes and applications. Here, we clarify the definition of QSP and its key features, trace its evolution, briefly compare it to other approaches, and explain why and how it can be used to reduce risk and improve efficiency in drug research and development.

Keywords

Drug development · Mechanistic modeling · Model-informed drug development · Pharmacology · Physiological modeling · QSP modeling · Quantitative systems pharmacology · Systems biology

1 Introduction

Quantitative systems pharmacology (QSP) is a relatively new and quickly evolving modeling discipline with applications throughout the drug research and development process. In this chapter, we define QSP and some of the key concepts involved in this type of modeling. We also tackle the question, "Why does QSP exist?" from two angles: (1) what is the nature of the problems that QSP evolved to address; and (2) what confluence of factors led to QSP evolving when and in the way that it has? We end by briefly touching on the current state and future directions of QSP. This chapter serves as an initial introduction and context for the more detailed discussions and case studies in the following chapters.

Definition and key features of QSP models	
Definition	QSP is a modeling discipline that integrates mathematical representations of physiological and pathophysiological processes with pharmacology to support an improved mechanistic understanding of how drugs interact with biology and the systemic implications of drug administration
Key features	• Includes mechanistic representation of biology independent of drug effects • Includes drug pharmacokinetics and pharmacodynamics • Integrates results from reductionist datasets with the appropriate biological context to explore system-level behaviors

2 What Is QSP?

QSP is a modeling discipline that integrates mathematical representations of physiological and pathophysiological processes with pharmacology to support an improved mechanistic understanding of how drugs interact with biology and the systemic implications of drug administration.

Consider an illustration that summarizes and integrates findings from various sources to propose a set of biological connections and signaling cascades involved in a disease process; for example, a schematic showing key cells and mediators involved in atopic dermatitis (AD) pathophysiology and the interactions between them (Fig. 1). This is a common example of a *qualitative* mechanistic model of biology. This type of model can enhance our understanding of the system as well as illustrate knowledge gaps and support the generation of testable hypotheses. However, it does not reproduce quantitative data or support quantitative predictions.

QSP modeling brings the qualitative mechanistic model to life. It asks and answers questions such as: can this conceptualization of biology reproduce key behaviors of interest? How would outcomes change if we interfered with some part of the biology?

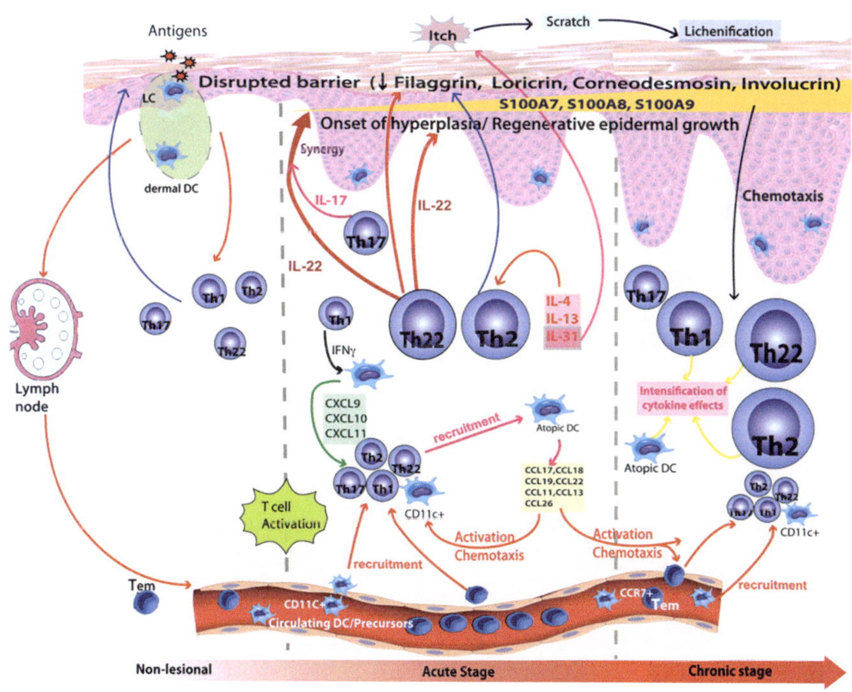

Fig. 1 A schematic illustration of the initiation of acute atopic dermatitis (AD) and progression to chronic skin lesions is an example of a qualitative mechanistic model. Reproduced from Gittler et al. (2012)

To achieve this, QSP first makes the mechanistic conceptualization of the biology precise and quantitative by defining each interaction in the diagram with a mathematical representation. For example, the recruitment of T cells, interactions with dendritic cells, T cell differentiation, activation, and production of cytokines, the effects of those cytokines on skin cells, etc., are all characterized by mathematical relationships. An example of an atopic dermatitis QSP model diagram is shown in Fig. 2.

The model can now run simulations that recreate the conditions of test protocols, such as the administration of existing atopic dermatitis therapies, to dynamically reproduce the observed results. For example, response to dupilumab, a monoclonal antibody against interleukin-4 receptor α can be tested by administering a virtual dose to the model. The drug inhibits the binding of IL-4 and IL-13 in the model and thereby reduces the effects of these cytokines on the biology represented (including keratinocyte (KC) activation), ultimately leading to a reduction in the Eczema Area and Severity Index (EASI) score that can be compared to data (Fig. 3).

Finally, by representing the mechanism of action (MoA) of new drugs in the context of the modeled biology, the QSP model can answer the predictive question, how will this new type of drug affect outcomes? Many specific applications of QSP modeling in drug discovery and development build on this foundation of representing biology in a simulation-capable mathematical model, as further elucidated later in this introduction and in the application chapters (Chapters "QSP Model Development and Application in Neuroscience", "Rare Disease (Niemann-Pick Disease); Highlight Extrapolation from Adults to Pediatrics & FDA Interactions", "Applications of Quantitative Systems Pharmacology in Cancer Drug Development: Target Identification, Dose Optimization, and Efficacy Prediction", and "Toxicology").

Most commonly QSP models are implemented as systems of ordinary differential equations (ODEs) that depict the dynamical interactions between components of the biological system as well as drug pharmacokinetics (PK) and pharmacodynamics (PD). Other approaches, such as agent-based modeling or partial differential equations are also possible; the classification as QSP does not depend on the mathematical approach.

3 Features of QSP Models

3.1 QSP Models Represent Biology

QSP models represent aspects of biology that exist independent of the drug(s) being applied. The scope of biology included may range from a complete representation of known disease processes comprising multiple organ systems and hundreds of components to a much more focused exploration of a specific component of disease biology, such as a ligand-receptor interaction. In the absence of a drug, QSP model simulations recapitulate the dynamic interactions between biological entities that are involved in producing an outcome of interest. For example, the model may reproduce the untreated cancer cell proliferation dynamics reported in literature, or the glucose and insulin dynamics of a patient with diabetes. The mechanistic

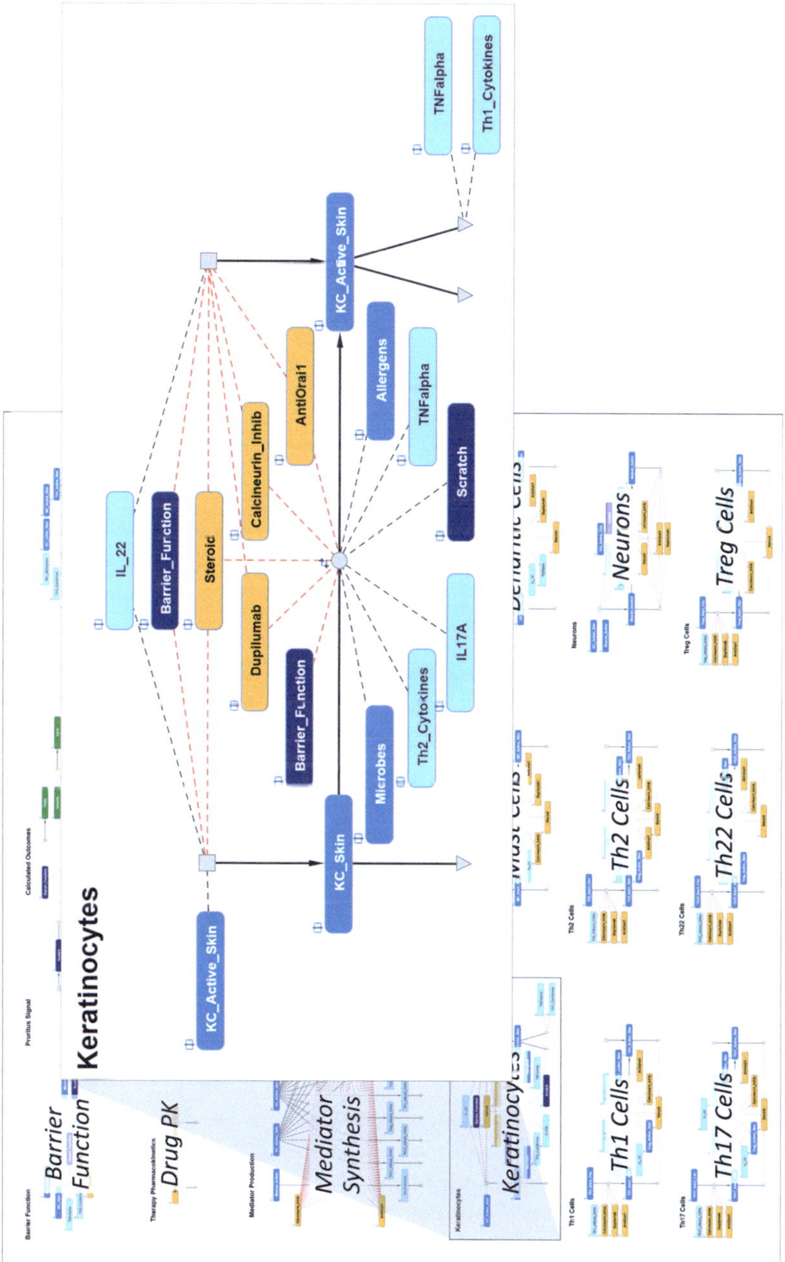

Fig. 2 An Atopic Dermatitis PhysioMap®, the graphical component of a QSP model of atopic dermatitis, from Friedrich et al. (2019). The zoomed-in module shows skin cell (keratinocyte, KC) production, clearance, and activation. Every line indicates a reaction that is defined mathematically. Solid lines represent processes involving mass transport, dashed lines represent regulation of processes in which the regulator is not consumed. Model developed by Rosa & Co, implemented in MATLAB® SimBiology®

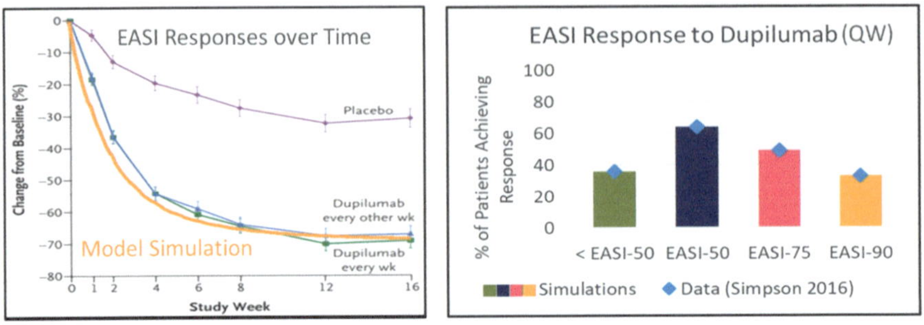

Fig. 3 Simulations (300 mg QW) vs. clinical trial data for dupilumab. Simulation results from Friedrich et al. (2019), data from Simpson et al. (2016), left figure modified from Simpson et al. (2016)

representation of biology independent of drug effects is a defining feature of QSP models. It is this inclusion of biological mechanisms that enables QSP models to be used for the prediction of novel scenarios beyond currently available data.

3.2 QSP Models Represent Pharmacology

As the name suggests, QSP models also include pharmacokinetics and pharmacodynamics of drugs of interest. In the context of QSP modeling research, the drugs of interest may include not only the novel compound under investigation but also any relevant existing drugs in the therapeutic area of interest. Importantly, drug effects are represented mechanistically, i.e., they perturb a part of the represented biological system, and the model is then expected to reproduce the observed pharmacological outcomes. If multiple drugs are included in the model, each drug is represented by separate drug-specific PK and MoA parameters, but a single set of *biological* parameters must be able to produce reasonable responses to all of the drugs administered. Existing drugs with known mechanisms and available clinical data provide excellent system-level constraints on the behavior of a QSP model, increasing confidence in predictions of novel scenarios.

3.3 QSP Models Explore System-Level Behaviors

Empirical biological research is inherently reductionist, focused on identifying components of a system that may be associated with clinical or subclinical phenomena of interest, e.g., cell types, genes, or proteins that are upregulated in a disease state or in response to a stimulus. To isolate the component or effect of interest experimentally, other layers of variability or complexity are stripped away, which allows the experimentalist to draw more confident conclusions, but also removes the system from its in vivo context, thus removing some of the relevance. This approach produces the building blocks that QSP models then put back together and animate via system-level dynamic simulations. In the process, QSP modeling can lead to novel insights about emergent system properties, namely, the system-level behaviors that only become apparent when the parts of the system are combined and their relationships explored.

3.4 QSP Models Use Relevant Data, Knowledge, and Hypotheses

The representations of biology in a QSP model are informed not only by data, but also by knowledge of the biology and hypotheses. To this end, the scope of biology included and the level of detail of the mechanistic representation are decided by the modeling team (which should ideally include life science experts, clinicians, and other persons with relevant domain knowledge) and should be guided by the research context – what questions does the model need to address, by when, and what data are available (Friedrich 2016; Ramanujan et al. 2019; Kuemmel et al. 2020)? During the scoping and model design phase, modeling teams must consider prior knowledge of disease processes, including clinical and non-clinical evidence that sheds light on the pathways involved in disease pathophysiology and their connection to outcomes. Because of the significant complexity of biological systems, QSP models will leave out more than they include; the task of the QSP modeling team is to make informed choices about what essential elements must be included to achieve the research objective. Broader biological knowledge and judgment come into play during this stage, as modelers design a simplified version of the real biological system that nonetheless can both recapitulate relevant known behaviors and have sufficient detail to support the prediction of novel scenarios.

During model building and calibration, modelers take a deep dive into literature and proprietary sources detailing the quantitative interactions between biological components. Here again, QSP modeling is a deeply scientific endeavor. The emphasis during this stage of the project shifts to what is commonly referred to as "the data" (i.e., specific, quantitative data points illustrating cause and effect in in vitro, animal, or clinical studies). A particular strength of QSP modeling is that it utilizes, integrates, and contextualizes many different types of data; however, the challenge is that appropriate interpretation of data can be tricky (see Chapter "QSP Model Execution" for a more detailed discussion of data used in QSP modeling). In the future, Artificial Intelligence (AI) tools may be helpful in quickly identifying and evaluating relevant data for QSP modeling.

3.5 QSP Models Predict Novel Scenarios

In drug development, empirical statistical modeling approaches, such as those traditionally applied in pharmacometrics, are essential for analyzing the data generated, and identifying trends, covariates, etc. Empirical modeling is also used for the analysis of -omics data, which may yield useful insights into biomarker patterns associated with outcomes of interest. While some degree of interpolation and extrapolation is possible; for example, analysis of existing data for two dose levels can yield a confident prediction of response to another dose level, the strength of these methods is a detailed analysis of the data in hand.

Unlike empirical statistical models, QSP models' structure is not wholly inferred from any one set or type of data, and their biological parameters are not fitted to a single data set, but instead should be biologically plausible. QSP models are explicitly built to represent biological mechanisms, and the model's performance must be consistent with a range of data. These attributes support their ability to predict novel scenarios well beyond the capabilities of statistical approaches. Indeed, QSP models are often utilized in contexts where clinical data are not yet available to better anticipate trial outcomes, optimize trial designs, ensure optimal use of biomarkers, and generally stay a step ahead of the clinical data.

3.6 QSP Models Help Interrogate Biological Uncertainty

In the context of QSP modeling, biological uncertainty refers to the inherent knowledge gaps in human understanding of biology. The systematic analysis of literature data that is necessary to confidently use the data in QSP modeling routinely exposes inconsistencies (e.g., same assay, different results) and data gaps (e.g., quantities are unmeasurable but important for disease dynamics). It can also lead to new insights and hypotheses. It is common for QSP modeling to inspire additional research to resolve uncertainties.

Where necessary data are not available, QSP models must incorporate assumptions. A standard part of the QSP modeling process must be to document these uncertainties and assumptions (Friedrich 2016; Gadkar et al. 2016), so that they may later be examined in the context of the entire biological system represented in the QSP model. A vital function of a QSP model is to support investigations of which uncertainties have a significant impact on outcomes (i.e., sensitivity analysis) and which hypotheses are consistent with existing data. By this process, QSP modeling can help reduce uncertainty and identify factors that may pose a risk to the clinical success of the asset under development.

QSP skeptics may point to biological uncertainty as a reason not to develop QSP models, asking how it is possible to build mechanistic models when aspects of the biological mechanisms are unknown. This is a misunderstanding of the role of QSP modeling. If everything were known, a model would no longer be necessary. Engaging in the process of representing the relevant biology, to the best of our current knowledge, and interrogating the impact of remaining uncertainty is a powerful way to further our understanding of the biology.

3.7 QSP Models Connect Mechanistic and Outcome Variability

In the context of QSP modeling, variability refers to the differences in biological components, mechanisms, and outcomes between subjects or between observations.

Because of the complexity of biological systems, including feedback and redundancy, the dynamic and systemic consequences of changes in some parts of the system are often not obvious. Thus, another excellent use of QSP models is to investigate what differences in biological mechanisms may be responsible for observed clinical variability in the untreated or treated state. For example, what are all the biologically reasonable ways in which a patient may be a poor responder to a given therapy? Would the same patient respond better to a different therapy? How would different types of patients respond to combination therapy?

These types of investigations are typically conducted in QSP models using a range of virtual patients (VPs), i.e., distinct instances of the same structural model with different parameter values representing mechanistic variations (Bangs 2005). When the model is simulated, alternative parameters lead to different predicted outcomes. This facilitates both the recapitulation of observed clinical variability and the prediction of variability in novel scenarios. It should be noted that investigation of uncertainty can also be conducted using VPs because the typical VP contains some parameter values that are known to vary between patients, and some parameter values that represent an assumption about an uncertainty.

The investigation of the connection between biological mechanisms and outcomes goes in both directions. Starting from the mechanism, the QSP model can be used to illustrate the impact of perturbing different parts of the model on the expected outcomes, e.g., to assess the likely impact of a novel therapy. The QSP model can also be used to address an inverse problem, i.e., given the range of responses reported for a given drug, what parts of the system are likely to be different between patients? VPs should be developed to expressly investigate both mechanistic variability and outcome variability. In the process, the connection between mechanisms and outcomes at the population level comes into focus. A collection of VPs that jointly reproduce the range and distribution of clinical responses observed in data is known as a virtual population (VPop).

4 Why Does QSP Exist?

4.1 What Is the Nature of the Problems QSP Evolved to Address?

4.1.1 Tackling Diseases with Mechanistic Complexity

Since Louis Pasteur introduced the world to germ theory in 1861, medical and pharmaceutical innovation has been astounding. Global life expectancy has more than doubled in the past century, in no small part due to medical advances. The diseases with the greatest remaining unmet medical needs tend to be biologically highly complex (e.g., auto-immune diseases), difficult to study clinically (e.g., neurodegenerative diseases, rare diseases), and/or dynamically evolving (e.g., cancer). In addition, many current treatment approaches are not tailored to the individual's pathophysiology and may be associated with toxicity burden and/or fail to fully address the underlying causes of disease progression (e.g., heart failure).

QSP modeling can be helpful in these challenging scenarios. For example, for biologically complex diseases, QSP can integrate what is known, including mechanistic evidence, to better understand the pathophysiology. For diseases that are difficult to study clinically, QSP modeling may anticipate clinical outcomes so that costly and lengthy clinical trials are optimized. For dynamically evolving diseases such as cancer, QSP modeling can be used to evaluate protocols that account for dynamic complexities, for example to minimize mutation risk, or to test response over time to different drug combinations. In all cases, a better mechanistic understanding of variability helps inform personalized treatment regimens. QSP modeling may not completely eliminate uncertainty, but the discipline of integrating multiple lines of evidence and types of data helps improve biological understanding, clarifies system-level behaviors, and reduces uncertainty and risk when predicting novel scenarios.

While not a QSP model per se (no pharmacology included), one of the earliest examples of a mathematical model that was developed to understand a complex biological system is the Guyton model of circulation, published in 1972 (Guyton et al. 1972). The field of physiology yielded other early examples of systems models used in scientific research as highlighted in (Hester et al. 2011). Early commercial development of QSP models focused on large models of complex diseases, such as the diabetes, asthma, rheumatoid arthritis, and cardiovascular models developed by Entelos (Rullmann et al. 2005; Young et al. 2006; Waters et al. 2009; Shoda et al. 2010).

4.1.2 Enabling Mechanistic Inference

On the other end of the model complexity spectrum, Bergman et al. published the minimal model of glucose utilization in 1979 (Bergman et al. 1979; Bergman 2020). Consisting of just two quasi-linear equations, the model was able to recapitulate the dynamic signatures of glucose and insulin in response to glucose administration and led to a number of fundamental insights that would not have been possible without a dynamic model. The authors found that there must be a delay between insulin release and its effects, a finding that was later verified experimentally. This shows that insights can be gained from very small, focused models and is an excellent early illustration of mechanistic models used to explain unusual dynamic phenomena and support mechanistic inference. When a disease is better understood, better treatment approaches can emerge.

4.1.3 Avoiding Predictable Failures

An improved understanding of the biological mechanisms of diseases and the impact that drugs have on them helps to avoid costly failures. Not every clinical failure is predictable, but many are with the right tools and approaches. Predictable failures may be due to inherent flaws in the hypothesis of how the drug's target is involved in disease pathophysiology or how modulating that target is likely to affect outcomes. There may also be sub-populations for whom the drug will be more or less efficacious due to differences in underlying disease drivers. Exploring a novel drug's MoA in a QSP model before moving into clinical development reduces the risk of moving forward with an asset that is unlikely to be efficacious and offers opportunities to refine clinical programs to maximize the probability of success.

4.1.4 Identifying Sub-Populations and Biomarkers

Patients differ from each other, and their differences can lead to different responses to drugs. If there was a way to capture each individual's biology in a model, we could test drugs in the model instead of in the person. Unnecessary exposure to drugs unlikely to work could be avoided, dosing could be optimized for each patient, and the right combination could be identified. This is the promise of personalized medicine, of which QSP could be an integral part.

One reason to pursue mechanistic approaches like QSP modeling is that they bring us a step closer to that vision of personalized medicine. While QSP models do not capture every aspect of every patient's biology, they do capture components of biology associated with, for example, response and non-response to therapy. QSP models can be systematically interrogated to identify the biological factors that distinguish strong responders from weak responders. Some of these factors may be clinically measurable and could serve as inclusion criteria or become measured biomarkers to ensure the selection of a clinical population most likely to show efficacy in response to treatment. Because they connect mechanisms with outcomes, QSP models help elucidate which of each patient's many unique factors may be most associated with predicted response.

4.2 What Confluence of Factors Led to QSP Evolving When It Did?

4.2.1 Evolution of and from Related Disciplines

The term QSP came into wide use following the publication of an NIH White Paper in 2011 (Sorger et al. 2011) that summarized two workshops held at the NIH in 2008 and late 2010 with participation from academia, industry, and government. QSP's origins can be traced back to physiological modeling, pharmacology modeling, and systems biology (Fig. 4). Admittedly, some models could be fairly described as fitting into multiple categories, and the goal of the brief descriptions here is to give the reader a sense of the features of typical models in each discipline.

4.2.2 Physiological Modeling

Physiological modeling draws on disciplines including mathematics, physics, engineering, biology, chemistry, and biochemistry to develop models that capture and explain biological phenomena of interest. Mathematical modeling of natural phenomena goes back centuries, and some of the basic building blocks of QSP models, such as equations describing biochemical reactions, were derived many decades ago (Beard et al. 2005). Physiological models are often mathematically sophisticated, derived from first principles, and focused on elucidating the fundamental mechanisms of a specific sub-system of interest (e.g., Beard and Bassingthwaighte 2001; Kavdia et al. 2002). QSP models may incorporate aspects of physiological models, but may also include more empirical or semi-mechanistic components.

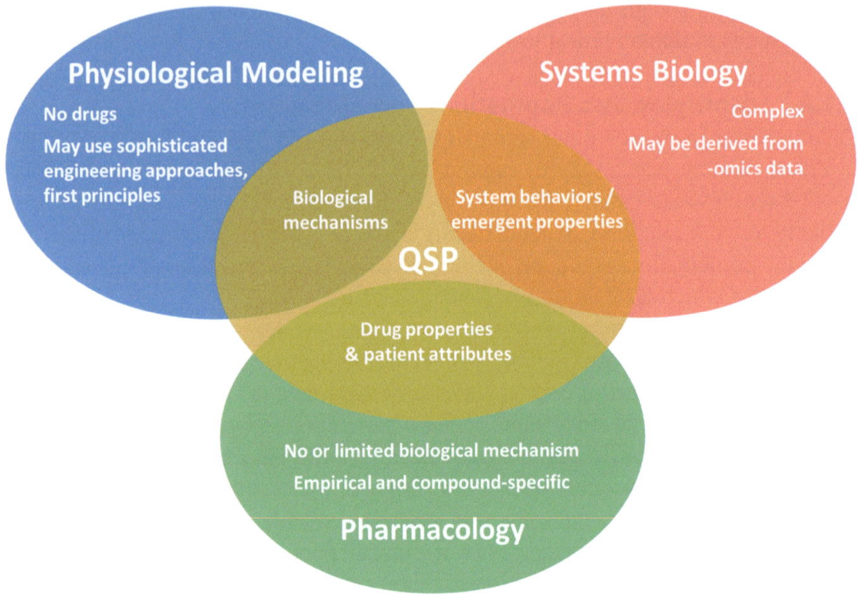

Fig. 4 Overlap and distinctions between modeling disciplines. The figure describes typical model attributes, exceptions may exist within each category

Typical QSP models are broader in scope to capture system-level behaviors, and they must include pharmacology, by definition.

4.2.3 Pharmacology Modeling

Pharmacology is the study of drugs, and pharmacology modeling is used to describe both the effects of the body on the drug (PK) and the effects of the drug on the body (PD). PK modeling mathematically describes the drug's kinetic properties, while population PK modeling further identifies co-variates that affect a patient's exposure. PD modeling links exposure to an observed biomarker or endpoint ("response"). Pharmacology modeling is essential and routine for every drug in development. Most pharmacology modeling is empirical, fitting to a single or small number of data sets specific for the compound under development to identify best-fit parameters which can then be used for prediction of the next stages of development.

Pharmacology modeling is not replaced by QSP, rather, QSP was in part inspired by and often incorporates pharmacological modeling concepts and constructs and combines them with biological mechanisms. The use of virtual patients and virtual clinical trials to simulate outcomes that account for inter-patient variability is shared between the disciplines. A key difference is that QSP models capture biological mechanisms and QSP virtual patients therefore include mechanistic biological variability that classical pharmacology modeling approaches do not capture. Fundamentals of pharmacology modeling are discussed in more detail in Chapter "Breadth of Pharmacology Modeling".

Note that physiologically-based PK (PBPK) modeling is a different type of PK modeling that is more mechanistic in nature, including, for example, transport and fluid-dynamic equations derived from prior knowledge of the processes affecting drugs in the body. PBPK models are particularly useful for predicting the distribution of drugs into tissues of interest. As such, PBPK modeling is itself a hybrid at the intersection of physiological modeling and pharmacology modeling, but unlike QSP, PBPK modeling does not typically include pharmacodynamics, i.e., the response or downstream effects of what the drug does to the body.

4.2.4 Systems Biology

Systems biology modeling emphasizes the investigation of the emergent behaviors of complex systems using a range of mathematical approaches. These models emerged around the turn of the twenty-first century in part to make sense of vast new amounts of information at the genomic and proteomic levels. Systems biology models often include networks at the cellular and intracellular level and tend to be quite complex (e.g., Janes et al. 2005; Hwang et al. 2009). Systems biology shares with QSP a focus on improving understanding of the biological system as a whole, and the role that individual components play in the system-level behavior. A difference is that systems biology models may not include pharmacology and may have more empirically inferred structural components.

4.2.5 Data Availability

It would be difficult to overstate the degree to which the acquisition of biological data has been transformed in the last 50 years or so. From the development of transgenic mice to high-throughput screening of compound libraries, to genomics and proteomics, to new imaging methods, the number of techniques and amount of data available to support biological research and drug discovery and development is vast, which creates a new set of issues. One, how do we make sense of the large amounts of data, and two, how do we contextualize the information to turn it into learning? The use of data for QSP modeling is discussed in further detail in Chapter "QSP Model Execution", and QSP's integration with big data and machine learning is examined in Chapter "Data to Inform Models".

The increasing availability of data supported the evolution of QSP as a discipline and continues to support the development of QSP models. In turn, QSP models add value to the drug discovery and development process by contextualizing data so that it is used to further our understanding, i.e., turning information into knowledge. Further, by utilizing systematic sensitivity analysis and VPs, QSP modeling can help identify the uncertainties that have the greatest potential impact on predicted outcomes and can therefore help prioritize additional empirical research.

4.2.6 Computing Resources

No less astounding than the advances in biological data acquisition in the last 50 years are the advances in computing resources. Sixty years ago, Guyton's earliest models were built as analog circuits and required highly specialized equipment and operators to perform simulations. Advances in digital computing, processing power, networking, and modeling software have transformed modeling and simulation from an academic research exercise supported by specialized equipment and expertise to a routine activity feasible to support the pace and needs of drug discovery and development.

4.2.7 Structural Conditions

Investor Financing

Improved access to biological data, computing resources, and advances in related disciplines led to pioneering companies such as Entelos being able to raise capital in the late 1990s and early 2000s. Investor-financed commercialization of the QSP modeling approach helped the field jump the gap from academic labs, in which many modeling techniques originated, to commercial drug development. Many pharmaceutical and biotech firms now have internal QSP functions (Nijsen et al. 2018; Cucurull-Sanchez 2024). In addition, contract research organizations (CROs) with expertise in QSP are a vital part of the marketplace infrastructure that supplies additional capacity and expertise.

Regulatory Support for Modeling

With PK/PD modeling approaches paving the way, regulatory agencies including the Food and Drug Associate (FDA) in the US and the European Medicines Agency (EMA) in Europe are increasingly supportive of model-informed drug development (MIDD). The role that QSP modeling can and should play in regulatory submissions is currently an area of active research and discussion (Bai et al. 2021a). FDA engagement raises QSP's profile, adds legitimacy, and supports arguments for company investment in QSP as a discipline.

Academic Programs

The growth of academic programs in computational and systems biology, pharmacology, bioinformatics, and other QSP-related fields is both a reflection of the aforementioned trends and a necessary condition for the continued growth of the discipline. While graduates of these programs can and have gone on to become successful QSP modelers, academic programs specifically focused on QSP are still lacking. Such programs would help to further cement the role of QSP in drug development and would reduce the need for on-the-job training that can be a deterrent to hiring talented newcomers to the field.

Increasing Costs of Drug Development

The ever-increasing costs of drug development have been well-documented (Wouters et al. 2020) and explain the need for greater efficiency, including through MIDD. QSP models can help investigate the scientific case for advancing assets, anticipate trial outcomes years in advance, and avoid costly mistakes, thus de-risking the process and increasing the probability of success.

5 The Current State of QSP

5.1 What Do QSP Models Look Like?

While early QSP models tended to be large in scope and rich in detail, this is no longer necessarily the case. Today's QSP models can range in size from several hundred nodes to less than ten. Different-sized models can be appropriate in different research contexts, depending on the questions asked, the data available, and the intended application of the modeling results. Several QSP examples can be found in Chapters "QSP Model Development and Application in Neuroscience", "Rare Disease (Niemann-Pick Disease); Highlight Extrapolation from Adults to Pediatrics & FDA Interactions", "Applications of Quantitative Systems Pharmacology in Cancer Drug Development: Target Identification, Dose Optimization, and Efficacy Prediction", and "Toxicology".

5.2 How Are QSP Models Used?

Recent surveys shed some light on the current use of QSP in pharmaceutical research and development. In brief, not all companies currently use QSP. In those that do, QSP modeling applications range from target validation to patient stratification and span a wide range of disease areas (Nijsen et al. 2018; Bradshaw et al. 2019; Lemaire et al. 2023; Cucurull-Sanchez 2024). Applications build on the foundation of elucidating pathophysiology and drug mechanism of action. For example, QSP has proved useful in supporting pre-clinical to clinical translation or translation between populations, where attributes such as body weight, metabolic differences, or disease status can be explicitly accounted for in the model to yield a more confident prediction of effects in the clinical population of interest. Dose setting and exploration of a range of dosing regimens is also facilitated by QSP modeling, as is exploration of a much broader range of combination therapies than would ever be practical to test clinically. By the same token, differentiation from competitors or standards of care via virtual head-to-head trials is commonly tested and can be crucial for go/no-go decisions. In addition, the use of QSP to predict toxicity (sometimes referred to as quantitative systems toxicology (QST); covered in Chapter "Toxicology") can provide crucial insight into the compound's therapeutic window and likely clinical utility. Analysis of the drivers of response to treatment in a mechanistically diverse QSP VPop has also been used to support the identification of biomarkers and optimal inclusion/exclusion criteria. More recently, QSP modeling has been successfully employed to clarify the complex dose-response dynamics of novel treatment modalities, such as bi-specific antibodies, T-cell engagers, or cell-based therapies, in which the lines between PK and PD become increasingly blurred. The case examples in Chapters "QSP Model Development and Application in Neuroscience", "Rare Disease (Niemann-Pick Disease); Highlight Extrapolation from Adults to Pediatrics & FDA Interactions", "Applications of Quantitative Systems Pharmacology in Cancer Drug Development: Target

Identification, Dose Optimization, and Efficacy Prediction", and "Toxicology" illustrate some of these applications.

5.3 Challenges and Optimal Conditions

Some of QSP's strengths – its flexible nature, the integration of biological science and mathematics, the use of disparate types of data – also pose challenges for its broader adoption. Its flexible nature makes it difficult to prescribe a step-by-step qualification or validation approach (Friedrich 2016; Ramanujan et al. 2019; Bai et al. 2021b; Chan et al. 2022). The integration of biology and mathematics and the use of disparate types of data challenge practitioners to reach across departmental silos and develop a holistic understanding of the biological process. The models can be complex and their value non-obvious without significant investment in understanding not only the model but the scientific judgments and assumptions that went into it. Documentation is a challenge because it must include not only what is in the model, but also what is not and why. Sometimes there is great value to be derived from a wrong model prediction because it can indicate that the understanding of biology represented in the model is incomplete, leading to novel questions and more targeted empirical research. It is the scientific method in action, but it can be jarring to an audience used to thinking of models as prediction engines.

Communication is key. Stakeholders must have clarity on the purpose of the QSP modeling effort and appropriate expectations. The decision-making context must be considered, and an appropriate qualification approach should be set out in advance so that the goalposts do not move. Ideally, modelers should have ready access to relevant experts and stakeholders, and modeling decisions get sign-off and buy-in to create a sense of shared ownership of the model and a commitment to follow through.

5.4 Where Are QSP Models Going?

QSP practice is evolving in a number of exciting directions (see Chapter "Current Challenges & Future Direction"). The integration of machine learning/AI and QSP modeling holds a lot of promise, as further discussed in Chapters "Data to Inform Models" and "Current Challenges & Future Direction". The push toward personalized medicine will continue to provide momentum to QSP modeling efforts. There is increasing interest in model sharing and reuse, which could reduce development timelines when done properly. Standards around model qualification continue to evolve and may in time include more prescriptive criteria for specific common contexts of use. Regulatory agencies may broaden the acceptance of QSP results and provide an impetus for continued investment. And of course, the costs of drug development will continue to rise, making a potent argument for improving efficiency through wise use of QSP modeling in drug discovery and development.

References

Bai JPF, Earp JC, Florian J, Madabushi R, Strauss DG, Wang Y, Zhu H (2021a) Quantitative systems pharmacology: landscape analysis of regulatory submissions to the US Food and Drug Administration. CPT Pharmacometrics Syst Pharmacol 10:1479–1484

Bai JPF, Schmidt BJ, Gadkar KG, Damian V, Earp JC, Friedrich C, van der Graaf PH, Madabushi R, Musante CJ, Naik K, Rogge M, Zhu H (2021b) FDA-industry scientific exchange on assessing quantitative systems pharmacology models in clinical drug development: a meeting report, summary of challenges/gaps, and future perspective. AAPS J 23:60

Bangs A (2005) Predictive biosimulation and virtual patients in pharmaceutical R and D. Stud Health Technol Inform 111:37–42

Beard DA, Bassingthwaighte JB (2001) Modeling advection and diffusion of oxygen in complex vascular networks. Ann Biomed Eng 29:298–310

Beard DA, Bassingthwaighte JB, Greene AS (2005) Computational modeling of physiological systems. Physiol Genomics 23:1–3. discussion 4

Bergman RN (2020) Origins and history of the minimal model of glucose regulation. Front Endocrinol (Lausanne) 11:583016

Bergman RN, Ider YZ, Bowden CR, Cobelli C (1979) Quantitative estimation of insulin sensitivity. Am J Phys 236:E667–E677

Bradshaw EL, Spilker ME, Zang R, Bansal L, He H, Jones RDO, Le K, Penney M, Schuck E, Topp B, Tsai A, Xu C, Nijsen M, Chan JR (2019) Applications of quantitative systems pharmacology in model-informed drug discovery: perspective on impact and opportunities. CPT Pharmacometrics Syst Pharmacol 8:777–791

Chan JR, Allen R, Boras B, Cabal A, Damian V, Gibbons FD, Gulati A, Hosseini I, Kearns JD, Saito R, Cucurull-Sanchez L, Selimkhanov J, Stein AM, Umehara K, Wang G, Wang W, Neves-Zaph S (2022) Current practices for QSP model assessment: an IQ consortium survey. J Pharmacokinet Pharmacodyn:1–13

Cucurull-Sanchez L (2024) An industry perspective on current QSP trends in drug development. J Pharmacokinet Pharmacodyn. https://doi.org/10.1007/s10928-024-09905-y

Friedrich CM (2016) A model qualification method for mechanistic physiological QSP models to support model-informed drug development. CPT Pharmacometrics Syst Pharmacol 5:43–53

Friedrich C, Ito T, Kudrycki K, Pryor M, Hurez V, Yamada S, Kiyosawa N, Nishimura M, Atsumi R, Morimoto K (2019) A simulation study for clinical efficacy of an anti-ORAI1 antibody (DS-2741a) on atopic dermatitis using quantitative systems pharmacology (QSP) modeling for preclinical-to-clinical translation. In: European Academy of Dermatology & Venerology (EADV) Congress. Madrid, Spain

Gadkar K, Kirouac DC, Mager DE, van der Graaf PH, Ramanujan S (2016) A six-stage workflow for robust application of systems pharmacology. CPT Pharmacometrics Syst Pharmacol 5:235–249

Gittler JK, Shemer A, Suarez-Farinas M, Fuentes-Duculan J, Gulewicz KJ, Wang CQ, Mitsui H, Cardinale I, de Guzman Strong C, Krueger JG, Guttman-Yassky E (2012) Progressive activation of T(H)2/T(H)22 cytokines and selective epidermal proteins characterizes acute and chronic atopic dermatitis. J Allergy Clin Immunol 130:1344–1354

Guyton AC, Coleman TG, Granger HJ (1972) Circulation: overall regulation. Annu Rev Physiol 34:13–46

Hester RL, Iliescu R, Summers R, Coleman TG (2011) Systems biology and integrative physiological modelling. J Physiol 589:1053–1060

Hwang D, Lee IY, Yoo H, Gehlenborg N, Cho JH, Petritis B, Baxter D, Pitstick R, Young R, Spicer D, Price ND, Hohmann JG, Dearmond SJ, Carlson GA, Hood LE (2009) A systems approach to prion disease. Mol Syst Biol 5:252

Janes KA, Albeck JG, Gaudet S, Sorger PK, Lauffenburger DA, Yaffe MB (2005) A systems model of signaling identifies a molecular basis set for cytokine-induced apoptosis. Science 310:1646–1653

Kavdia M, Tsoukias NM, Popel AS (2002) Model of nitric oxide diffusion in an arteriole: impact of hemoglobin-based blood substitutes. Am J Physiol Heart Circ Physiol 282:H2245–H2253

Kuemmel C, Yang Y, Zhang X, Florian J, Zhu H, Tegenge M, Huang SM, Wang Y, Morrison T, Zineh I (2020) Consideration of a credibility assessment framework in model-informed drug development: potential application to physiologically-based pharmacokinetic modeling and simulation. CPT Pharmacometrics Syst Pharmacol 9:21–28

Lemaire V, Bassen D, Reed M, Song R, Khalili S, Lien YTK, Huang L, Singh AP, Stamatelos S, Bottino D, Hua F (2023) From cold to hot: changing perceptions and future opportunities for quantitative systems pharmacology modeling in cancer immunotherapy. Clin Pharmacol Ther 113:963–972

Nijsen M, Wu F, Bansal L, Bradshaw-Pierce E, Chan JR, Liederer BM, Mettetal JT, Schroeder P, Schuck E, Tsai A, Xu C, Chimalakonda A, Le K, Penney M, Topp B, Yamada A, Spilker ME (2018) Preclinical QSP modeling in the pharmaceutical industry: an IQ consortium survey examining the current landscape. CPT Pharmacometrics Syst Pharmacol 7:135–146

Ramanujan S, Chan JR, Friedrich CM, Thalhauser CJ (2019) A flexible approach for context-dependent assessment of quantitative systems pharmacology models. CPT Pharmacometrics Syst Pharmacol 8:340–343

Rullmann JA, Struemper H, Defranoux NA, Ramanujan S, Meeuwisse CM, van Elsas A (2005) Systems biology for battling rheumatoid arthritis: application of the Entelos PhysioLab platform. Syst Biol (Stevenage) 152:256–262

Shoda L, Kreuwel H, Gadkar K, Zheng Y, Whiting C, Atkinson M, Bluestone J, Mathis D, Young D, Ramanujan S (2010) The Type 1 diabetes PhysioLab platform: a validated physiologically based mathematical model of pathogenesis in the non-obese diabetic mouse. Clin Exp Immunol 161:250–267

Simpson EL, Bieber T, Guttman-Yassky E, Beck LA, Blauvelt A, Cork MJ, Silverberg JI, Deleuran M, Kataoka Y, Lacour JP, Kingo K, Worm M, Poulin Y, Wollenberg A, Soo Y, Graham NM, Pirozzi G, Akinlade B, Staudinger H, Mastey V, Eckert L, Gadkari A, Stahl N, Yancopoulos GD, Ardeleanu M, Solo, and Solo Investigators. (2016) Two phase 3 trials of Dupilumab versus placebo in atopic dermatitis. N Engl J Med 375:2335–2348

Sorger PK, Allerheiligen SR, Abernethy DR, Altman RB, Brouwer KL, Califano A, D'Argenio DZ, Iyengar R, Jusko WJ, Lalonde R (2011) An NIH white paper by the QSP workshop group. Quantitative and systems pharmacology in the post-genomic era: new approaches to discovering drugs and understanding therapeutic mechanisms

Waters SB, Topp BG, Siler SQ, Alexander CM (2009) Treatment with sitagliptin or metformin does not increase body weight despite predicted reductions in urinary glucose excretion. J Diabetes Sci Technol 3:68–82

Wouters OJ, McKee M, Luyten J (2020) Estimated Research and Development investment needed to bring a new medicine to market, 2009-2018. JAMA 323:844–853

Young DL, Ramanujan S, Kreuwel HT, Whiting CC, Gadkar KG, Shoda LK (2006) Mechanisms mediating anti-CD3 antibody efficacy: insights from a mathematical model of Type 1 diabetes. Ann N Y Acad Sci 1079:369–373

Breadth of Pharmacology Modeling: Fundamentals of Pharmacokinetics, Pharmacodynamics, and Mechanistic Modeling

Donald E. Mager

Contents

Abstract

The basic principles of pharmacokinetics and pharmacodynamics represent the foundational knowledge base upon which complex quantitative systems

D. E. Mager (✉)
Division of Pharmacokinetics-Pharmacodynamics and Systems Pharmacology, Department of Pharmaceutical Sciences, University at Buffalo, SUNY, Buffalo, NY, USA

Enhanced Pharmacodynamics, LLC, Buffalo, NY, USA
e-mail: dmager@buffalo.edu

E. L. Bradshaw, M. E. Spilker (eds.), *Quantitative Systems Pharmacology*, Handbook of Experimental Pharmacology 289, https://doi.org/10.1007/164_2025_746

pharmacology models of drug action are built. This chapter provides a high-level overview of fundamental factors that determine the disposition and physiological responses to drugs and the application of compartmental models to characterize the time-course of drug exposure and pharmacological effects. Many of these processes are subject to capacity-limitation, which is defined by a nonlinear function containing a driving substrate concentration and parameters representing the capacity of the process and a substrate affinity constant. Most contemporary mechanism-based pharmacodynamic models are developed by integrating an appropriate drug exposure forcing function, a mathematical model of the interaction between the drug and its target (i.e., binding and transduction), and the physiological turnover (or production and loss) of the biomarker of drug response. Numerous complexities can be introduced to basic models, such as homeostatic feedback, tolerance mechanisms, disease progression, drug interactions, circadian rhythms, and many others. These basic and advanced models can be viewed as the groundwork for the development of comprehensive quantitative systems pharmacology models that are applicable across biological spatiotemporal scales.

Keywords

Mathematical modeling · Pharmacodynamics · Pharmacokinetics · Systems pharmacology · Target-binding · Turnover

1 Introduction

Mathematical modeling in pharmacology has formed the basis for understanding how the body processes drugs, which determines their time-course and net exposure (pharmacokinetics or PK), and how that exposure is related to the time-course and intensity of their pharmacological effects (pharmacodynamics or PD). Such models provide a structural framework for identifying sources of variability in patient responses to drugs and have become essential for the discovery, development, and utilization of drugs to treat disease. The complexity of PK/PD models ranges substantially, from empirical data-driven models with a relatively small number of variables and identifiable parameters to large multi-scale system level models featuring numerous physiological constraints, signal transduction process, state variables, and parameters, which are often not readily identifiable from routine clinical data. Regardless of complexity, mathematical models of drug action should be based on an appreciation of underlying biochemical processes, and as predicted by Levy, PK/PD models have evolved "from abstract number crunching to a physiologic mechanism-oriented endeavor" (Levy 1994a).

The three main pillars of mechanism-based PK/PD models are the PK of the drug, its pharmacology, and the (patho)physiological constraints and turnover processes of the system (Jusko 2013, 2016; Mager et al. 2003b). In the schematic shown in Fig. 1, the PK component is meant to represent the temporal pattern of drug concentrations

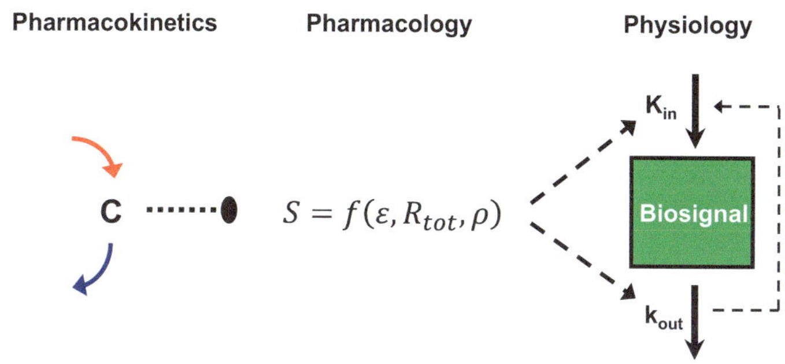

Fig. 1 The main pillars of mechanism-based pharmacodynamic models. Drug concentrations (C) are at the site of action, and the function defining the pharmacological signal (S) is defined in the text. Rate constants K_{in} and k_{out} control the production and loss of the biosignal, and the feedback arrow represents pathophysiological or homeostatic mechanisms and functional adaptation processes that can affect responses to drugs

at the site of action or biophase (C). Most PK studies will include drug concentrations in plasma to assess systemic exposure, and the use of plasma PK models as forcing functions typically assumes that drug distribution to the site of action does not influence its systemic exposure and that plasma drug concentrations are proportional to concentrations at the biophase. The direct measurement (and modeling) of drug concentrations as close to the site of action as possible is most desirable to avoid any disconnect between plasma and biophase drug concentrations; however, this is often precluded owing to logistical and ethical issues in obtaining specific tissue drug concentrations. When drug distribution to the biophase is responsible for a short delay in the pharmacological effect, a link compartment can be specified to mimic this distributional delay (see Sect. 2.2.1) or more mechanistically described using a physiologically based PK (PBPK) model (see Sect. 2.1.5). The pharmacology component in Fig. 1 depicts a simple expression defining a pharmacological signal (S). Stephenson suggested that the drug effect could be defined by any mathematical function, $f(S)$, which is driven by the product of drug efficacy (e) and receptor occupancy (ρ) (Stephenson 1956). Furchgott expanded the definition of the efficacy parameter to include total receptor concentration (R_{tot}) and the intrinsic efficacy of the drug (ϵ) (Kenakin 1997). This theoretical expression is phenomenological in nature, and numerous mathematical models can be derived to describe the drug-target interaction and subsequent signal transduction based on available data and assumptions regarding the influence and constraints of various system-specific properties (see Sect. 2.2). For rapid direct drug effects, the PK of the drug is rate limiting, and the PD may be defined simply by an explicit function that is directly proportional to ρ. However, the onset and time to peak drug effects are often delayed relative to drug exposure owing to drugs acting on turnover (i.e., production and loss) processes controlled by (patho)physiological mechanisms (Fig. 1).

There are a host of complexities that can influence all three pillars, including (among others) PK and PD drug interactions, disease progression, tolerance mechanisms, maturation or aging, environmental factors, and genetic polymorphisms and splice variants that can give rise to changes in protein expression or activity, thereby altering drug disposition, drug-target interactions, and homeostatic feedback. Thus, mathematical models in pharmacology can span from simple expressions to large systems level models depending on the pre-stated goals and objectives of the analysis, the nature and quality of the available data, and the allotted time for model building and validation. This chapter will introduce fundamental PK and PD concepts, define basic and more advanced mathematical models that characterize drug exposure-response relationships, and highlight how these primary models can be used to construct increasing complex multi-scale models such as PBPK and quantitative systems pharmacology (QSP) models.

2 Fundamentals of Pharmacokinetic and Pharmacodynamic Modeling

One of the most ubiquitous concepts in pharmacology modeling is the *law of mass action* and its consequence of capacity-limited processes. As an example, the law of mass action applied to drug binding to its pharmacological target is often expressed as:

$$C + R \underset{k_{\mathrm{off}}}{\overset{k_{\mathrm{on}}}{\rightleftharpoons}} RC$$

with R and RC representing the concentrations of the free receptor and drug-receptor complex, k_{on} is a second-order association rate constant, and k_{off} is a first-order dissociation rate constant. In this case, C is defined as the free drug concentration at the biophase. An ordinary differential equation defining the rate of change of RC can be written as:

$$\frac{dRC}{dt} = k_{\mathrm{on}} \cdot C \cdot R - k_{\mathrm{off}} \cdot RC$$

Under equilibrium conditions and assuming constant total receptor concentration (R_{tot}), this equation can be rearranged to produce the Hill equation of target occupancy (Hill 1910):

$$\frac{RC}{R_{\mathrm{tot}}} = \rho = \frac{C}{K_D + C}$$

with K_D as the equilibrium dissociation constant ($K_D = k_{\mathrm{off}}/k_{\mathrm{on}}$), which is the free drug concentration producing 50% target occupancy and ρ defined to be receptor occupancy (RC/R_{tot}). This equation is analogous to the Michaelis-Menten equation for the rate of enzymatic metabolism, in which the capacity parameter is the zero-order rate constant of maximum reaction velocity ($V_{\max}$), and the affinity term is the

Table 1 Pharmacological processes described by common capacity-limited functions

Process	Equation	Substrate	Capacity	Affinity
Metabolism	$\frac{dM}{dt} = \frac{V_{\max} \cdot C}{K_m + C}$	Concentration at enzymatic site	$V_{\max}$	K_m
Transport	$\frac{dA_T}{dt} = \frac{V_{\max} \cdot C}{K_m + C}$	Concentration at transporter site	$V_{\max}$	K_m
Protein binding	$D_b = \frac{n \cdot P_T \cdot D_f}{1/K_A + D_f}$	Free drug concentration	n: Number of protein binding sites P_T: Total protein concentration	K_A: Equilibrium association constant
Receptor binding	$RC = \frac{R_{\text{tot}} \cdot C}{K_D + C}$	Concentration at receptor site	R_{tot}	K_D: Equilibrium dissociation constant
Direct effects	$E = \frac{E_{\max} \cdot C^{\gamma}}{EC_{50}^{\gamma} + C^{\gamma}}$	Concentration at site of action	$E_{\max}$	EC_{50}

Adapted from Jusko (1989, 2016)

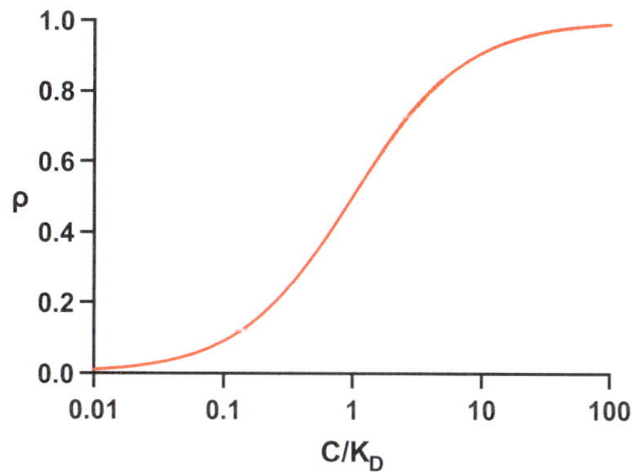

Fig. 2 Hypothetical receptor occupancy (ρ) as a function of drug concentration normalized by the equilibrium dissociation constant (K_D)

Michaelis-Menten constant (K_M), which is the drug concentration at which the reaction is half-maximal. These and other examples of pharmacological processes described by capacity-limitation are listed in Table 1, including facilitated drug transport, protein binding, organ clearance, and simple direct drug effects (Jusko 1989, 2016).

The equations listed in Table 1 all exhibit a similar sigmoidal relationship, and a normalized example for ρ is shown in Fig. 2. It is critical to recognize that the capacity and affinity parameters that define this curve are highly correlated and that a sufficient range of substrate concentrations are needed to numerically resolve or estimate both terms. This caveat was made clear by Dutta and colleagues who conducted simulations of the direct effect equation in Table 1 and then attempted to re-estimate the parameters from a simulated dataset with varying degrees of truncation (Dutta et al. 1996). They showed that drug concentrations producing up to 95% of the maximal effect were needed to re-estimate the capacity and affinity parameters that were used to create the original dataset with good precision. It is now commonplace that PK and PD studies are conducted over a sufficient range of dose levels to determine whether such nonlinear relationships exist and, if so, to ensure that the data will allow for the precise estimation of the capacity and affinity terms when a structurally identifiable PK or PD model is applied. Multiple-dosing regimens are also needed to assess whether the system is stationary (i.e., PK and/or PD parameters stay constant with time) or if additional phenomena (e.g., autoinduction of drug metabolism and functional adaptation leading to drug tolerance) can alter the PK and/or PD properties over time.

2.1 Pharmacokinetic Concepts and Compartmental Modeling

This section will introduce the primary physiological parameters of interest in PK and the most common types of mathematical models that are used to describe the time-course of drug exposure, calculate primary and secondary PK parameters, and provide a suitable forcing function for modeling biomarkers of drug response. The three primary PK parameters of interest are drug absorption, the steady-state volume of distribution, and the clearance of drugs. A major challenge in PK analysis is that these properties are not directly measured but inferred or calculated from temporal plasma concentration-time profiles. Although preclinical studies provide an opportunity to (among others) directly measure tissue drug concentrations, administer drug directly into various anatomical locations, and isolate clearance organs for perfusion studies, which can provide further insights into the determinants of drug absorption and disposition; most clinical PK studies involve the measurement of drug concentrations in relatively easily accessible biological fluids such as plasma, saliva, or urine. Thus, elements of the study design, analytical drug assay, and assumptions of the mathematical model can all influence the ability to characterize the PK of drugs (i.e., identify the best mathematical model and estimate or calculate PK parameters with good precision). There are numerous textbooks available that provide a foundational basis of pharmacokinetics, which is beyond the scope of this chapter, and the reader is referred to several (among many others) cited here (Gibaldi and Perrier 1982; Jusko 1995; Rowland and Tozer 2010).

2.1.1 Systemic Clearance

Total systemic clearance (CL), an important primary PK parameter, is defined as the volume of blood that is cleared of drug per unit of time. One way to conceptualize CL after an intravenous (IV) bolus dose (D_{IV}) is:

$$\mathrm{CL} = \frac{dA_e/dt}{C_p(t)}$$

with dA_e/dt representing the overall rate of elimination of the drug, which is normalized by the plasma drug concentration (C_p) at any time (t). By multiplying both sides of this equation by $C_p(t)$ and integrating both sides, one can derive one of the most important equations in PK:

$$\mathrm{CL} = \frac{D_{\mathrm{IV}}}{\mathrm{AUC}}$$

with AUC as the area under the plasma concentration-time curve. Here one can observe that under linear conditions, the dose and CL determine the net drug exposure (AUC), with AUC inversely related to CL. For oral doses (D_{PO}), the equation includes overall bioavailability (F), which is the fraction of the dose that reaches the systemic circulation after the loss of drug as it passes, for the first time, through organs of elimination (so-called first-pass effect):

$$\mathrm{CL} - \frac{F \cdot D_{PO}}{\mathrm{AUC}}$$

Total systemic clearance is also a primary determinant of the steady-state plasma drug concentration (C_{ss}) after about five terminal elimination half-lives following continuous IV infusion:

$$C_{ss} = \frac{K_0}{\mathrm{CL}}$$

with K_0 as a zero-order infusion rate constant (in units of amount/time), or after multiple bolus dosing events (Wagner et al. 1965):

$$C_{ss} = \frac{F \cdot \mathrm{Dose}}{\mathrm{CL} \cdot \tau}$$

with τ representing the dosing interval for either IV or oral doses. Thus, the C_{ss} value is proportional to dose levels and inversely proportional to dosing intervals, and C_{ss} is independent of the volume of distribution. Interestingly, the AUC after a single dose is equal to the AUC during a dosing interval at steady-state under linear and stationary conditions, and therefore, the C_{ss} after multiple dosing can be predicted from single-dose concentration-time data, which could and should be verified experimentally.

Total systemic clearance is equal to the linear sum of all organ clearance processes:

$$\mathrm{CL} = \mathrm{CL}_{\mathrm{H}} + \mathrm{CL}_{\mathrm{R}} + \mathrm{CL}_{\mathrm{other}}$$

with subscripts H and R representing hepatic and renal clearance. Rowland introduced organ clearance concepts, based on Fick's first law (Fick 1855), to gain further insights into the physiological determinants of these clearance processes (Rowland 1972). A general expression of Fick's first law applied to the movement or flux of the amount of drug (A) is:

$$\frac{dA}{dt} = -D \cdot \Delta C$$

with D representing blood flow (Q for perfusion) or a permeability-surface area coefficient (PS for diffusion), and ΔC is a concentration gradient. Consider an eliminating organ such as the liver or kidney in which organ blood flow (Q_{organ}) delivers drug to the organ and biotransformation (e.g., hepatic metabolism) or excretion processes (e.g., renal filtration) result in drug removal:

$$\mathrm{CL}_{\mathrm{organ}} = \frac{Q_{\mathrm{organ}} \cdot (C_{\mathrm{in}} - C_{\mathrm{out}})}{C_{\mathrm{in}}} = Q_{\mathrm{organ}} \cdot \mathrm{ER}$$

with C_{in} and C_{out} representing the input and output blood concentrations, and ER is the organ extraction ratio, which can be defined by mathematical models. For the well-stirred model (WSM) of hepatic clearance, which assumes that the liver is a single well-mixed compartment (with rapid mixing of portal venous and hepatic arterial blood) and uniform blood concentration throughout the liver, hepatic clearance can be defined as (Rowland et al. 1973):

$$\mathrm{CL}_{\mathrm{H}} = Q_h \cdot \mathrm{ER} = \frac{Q_{\mathrm{H}} \cdot f_{ub} \cdot \mathrm{CL}_{\mathrm{int,WSM}}}{Q_{\mathrm{H}} + f_{ub} \cdot \mathrm{CL}_{\mathrm{int,WSM}}}$$

with f_{ub} as the free fraction of drug in the blood, and $\mathrm{CL}_{\mathrm{int,WSM}}$ is the intrinsic clearance, which represents the intrinsic ability of an eliminating organ to remove free drug from the system. This equation thus defines the upper limit of CL_{H} as either hepatic blood flow, when $Q_{\mathrm{H}} \ll f_{ub} \cdot \mathrm{CL}_{\mathrm{int,\ WSM}}$, or $f_{ub} \cdot \mathrm{CL}_{\mathrm{int,\ WSM}}$ when the opposite is true. Other common models to define the ER include the parallel tube model (Pang and Rowland 1977) and the dispersion model (Roberts and Rowland 1986). The choice of which model to apply is still debatable and can be dependent on drug properties; however, the WSM remains the most used model despite its physiological limitations (Pang et al. 2019).

The establishment of these clearance concepts laid the groundwork for in vitro–in vivo extrapolation (IVIVE) techniques in which bioassays (e.g., microsomal and hepatocyte assays) are used to calculate an in vitro intrinsic clearance for small molecule drugs that can be scaled to predict the overall in vivo hepatic clearance (Houston 1994; Rane et al. 1977). A major limitation of the early clearance concepts is that they do not account for the role of drug transporters and the influence of uptake and efflux processes. Later models of hepatic clearance include influx and efflux clearances and are referred to as extended clearance models (Pang et al. 2019). These models can address complex clearance characteristics for drugs that are uptake-limited, where the overall or net intrinsic clearance will approach the influx clearance. The extended clearance concept and other (semi-) mechanistic models may help to improve the predictive performance of IVIVE coupled with physiological modeling for anticipating in vivo hepatic clearance (Sugiyama and Aoki 2023).

Renal clearance is another major organ of drug elimination that can also be understood based on clearance concepts. An overall equation that incorporates the primary processes affecting CL_{R} is:

$$CL_R = Q_{RP} \cdot ER = f_{up}(GFR + ATS) \cdot (1 - R_F)$$

with Q_{RP} as the effective renal plasma flow (about 600 mL/min), f_{up} is the free fraction in plasma, GFR is the glomerular filtration rate (about 110 mL/min), ATS represents active tubular secretion, and R_F is the fraction of drug reabsorbed in the tubules. In the simplest case, if a drug is only filtered, and all filtered drug is eliminated in the urine, then the rate of excretion will equal the rate of filtration, and $CL_R = f_{up} \cdot GFR$. Some drugs, such as amoxicillin, are actively secreted into the tubules, and a common operative equation is:

$$ATS = \frac{Q_{RP} \cdot CL_{int}^{S}}{Q_{RP} + f_{up} \cdot CL_{int}^{S}}$$

with CL_{int}^{S} as the intrinsic secretory clearance (note that f_{up} in the numerator was factored out in the prior equation for CL_R). This equation is analogous to the WSM and has the similar limiting conditions of Q_{RP} and $f_{up} \cdot CL_{int}^{S}$ depending on their relative values. Active secretion is one mechanism by which renal clearance can be greater than *GFR*, and this process can also be subject to capacity limitation (see transport equation in Table 1) and drug-drug interactions. For example, probenecid has been shown to inhibit the tubular secretion of amoxicillin and increase net exposure (Staniforth et al. 1983).

The passive reabsorption of drugs in the renal tubules can depend upon lipid solubility, pKa, urine pH, and urine flow rate. Examples of drugs exhibiting these dependencies are references for further review: (Beckett and Rowland 1965; Hinderling et al. 1984; Linton et al. 1967). Although the general CL_R equation above integrates the physiological processes involved, it does not lend itself to calculating renal clearance, which is more commonly calculated from collecting urine to (1) determine the absolute cumulative amount of drug excreted $\left(A_e^{\infty}\right)$ and dividing by AUC $\left(CL_R = A_e^{\infty}/AUC\right)$ or (2) generate a plot of urinary excretion rate as a function of plasma drug concentration (obtained during the mid-point of urine collection intervals) and calculating the resulting positive slope (Tucker 1981).

2.1.2 Volume of Distribution

The volume of distribution is a theoretical proportionality constant that relates the amount of drug in body with measured plasma drug concentrations. One often thinks of a central physiological volume, such as the volume of plasma (V_p, about 3.5 L in a 70 kg adult) (Brown et al. 1997; Davies and Morris 1993) and a peripheral volume outside of the central space reflecting drug distribution to tissues. The volume of drug distribution at steady-state (V_{ss}) can be generally expressed as the sum of V_p and the apparent volume of tissue distribution:

$$V_{ss} = V_p + \sum K_{p,i} \cdot V_i$$

with $K_{p,i}$ and V_i as the tissue:plasma partition coefficient and volume of the ith tissue. In this context, *steady-state* refers to the point in time following an acute bolus dose when the early distributive phase of drug disposition ends and a distribution equilibrium is attained between the central and tissue concentrations. Another way to define V_{ss} is to start with a simple description of the amount of drug in the body (A_B):

$$A_{\mathrm{B}} = V_{\mathrm{p}} \cdot C_{\mathrm{p}} + V_{\mathrm{T}} \cdot C_{\mathrm{T}}$$

with V_{T} and C_{T} as the distribution volume and drug concentration in extravascular tissue. At distribution equilibrium, the free drug concentration in the plasma and tissue compartments are equal:

$$C_{\mathrm{p}} \cdot f_{up} = C_{\mathrm{T}} \cdot f_{uT}$$

with f_{uT} representing the free fraction in tissue. This expression can be solved for C_{T} and substituted back into the previous equation to derive the Gillette equation for V_{ss} (Gillette 1971):

$$\frac{A_{\mathrm{B,ss}}}{C_{\mathrm{p,ss}}} = V_{\mathrm{ss}} = V_{\mathrm{p}} + \frac{f_{up}}{f_{uT}} V_{\mathrm{T}}$$

The Gillette equation can provide a simple basis for understanding calculated values of V_{ss} from either noncompartmental analysis or mathematical modeling. This general expression provides a convenient way to conceptualize the extent of drug distribution and explain why the calculated values for small molecule drugs can be relatively small for weak acids, which bind extensively to albumin (e.g., 7 L for ibuprofen), but supraphysiological for weak bases (e.g., 300 L for propranolol), which often bind moderately to alpha-1-acid glycoprotein and extensively to tissue proteins. Of course, the influence of drug transporters and other complexities (e.g., lysosomal trapping, disease conditions, maturation or development, and physicochemical properties) are not considered and better addressed with physiological models.

The rate or kinetics of distribution may be influenced by blood flow/perfusion, diffusion/permeability, transporters, and binding kinetics. A common model parameter used to represent distribution kinetics is CL_D, or so-called distribution clearance, which can be confusing as this "clearance" process does not result in removal of drug from the body, and the volume of distribution and clearance are independent parameters. Blood flow and drug permeability across capillaries are major determinants of the CL_D (Renkin 1955):

$$\mathrm{CL}_D = Q\left(1 - e^{-PS/Q}\right)$$

For drugs with high PS values, Q will become the rate-limiting process for CL_D, whereas CL_D can be limited by diffusion/permeability when the PS term is low. This expression only considers passive diffusion, and active transport processes are often defined by capacity-limited transport equations (Table 1).

A noncompartmental, model-independent, approach to calculating V_{ss}, again assuming linear and stationary conditions, is to multiply drug clearance by the mean residence time (MRT) after an IV bolus dose (Benet and Galeazzi 1979):

$$V_{ss} = \mathrm{CL} \cdot \mathrm{MRT_{IV}}$$

MRT is defined as the average amount of time a drug molecule will remain in the system before being permanently cleared and can be readily calculated according to the ratio of the area under the first-moment curve (AUMC) and the AUC (Gibaldi and Perrier 1982).

As stated above, V_{ss} and CL are independent PK parameters, which can be a source of confusion on considering the noncompartmental equation for V_{ss}, which is only useful as a means of calculation and does not specify a fundamental pharmacological relationship. Instead, consider the fact that these two independent parameters together influence the half-life ($T_{1/2}$) of the drug (i.e., the time it takes for the concentration of a drug to decrease by 50%). For example, the $T_{1/2}$ for a drug exhibiting monoexponential disposition can be defined as:

$$T_{1/2} = \frac{\ln(2) \cdot V}{\mathrm{CL}}$$

with V as the apparent volume of distribution (NB: monoexponential kinetics assumes an instantaneous distribution and V is used instead of V_{ss}). This is a fundamental relationship that describes how changes or differences in the distribution volume or clearance can impact the $T_{1/2}$ of drugs and should not be used or rearranged to imply a pharmacological relationship between CL and V_{ss}.

2.1.3 Oral Absorption

Oral drug absorption is complex and results from a dynamic interplay among drug-specific properties and physiological processes, which include drug release kinetics from rapid or slow-release formulations, dissolution, gastrointestinal (GI) motility, food effects, changes in pH along the GI tract, aqueous solubility and permeability, active transporters, and first-pass metabolism (Cummins et al. 2002; Martinez and Amidon 2002). Several pharmacokinetic parameters that are typically used to characterize the absorption of small molecule drugs following oral administration include overall oral bioavailability (F), the rate of absorption, and the maximum or peak drug concentration ($C_{\max}$) and the time at which it is achieved ($T_{\max}$). The latter two terms are easily obtained from an assessment of the PK profile; however, care must be taken to ensure a reasonably appropriate blood sampling scheme to accurately determine the values without the use of modeling.

Oral bioavailability can be conceptually defined by the following equation:

$$F = f_a \cdot f_g \cdot f_h$$

with f_a as the fraction of the dose absorbed, f_g is the fraction of the dose that passes through the gut wall without being metabolized, and f_h is the fraction of the dose that escapes first-pass hepatic metabolism. However, it is most commonly calculated using a ratio of AUC values following oral and IV administration:

$$F = \frac{D_{\text{IV}} \cdot \text{AUC}_{\text{PO}}}{D_{\text{PO}} \cdot \text{AUC}_{\text{IV}}}$$

From these relationships, it becomes clear that overall bioavailability is dependent on both absorption (f_a and f_g) and clearance mechanisms (f_h).

For many oral dosage forms, the rate of drug absorption can be characterized by a first-order rate constant (k_a). Fick's first law can be applied to describe the rate of change in drug concentration in the GI fluid (C_g) from the apparent unidirectional transfer of passively absorbed drugs (Wagner 1971):

$$\frac{dC_g}{dt} = -\frac{D \cdot \text{SA} \cdot P}{V_g \cdot h} \Delta C$$

with D as the diffusion coefficient, SA is the surface area, P is the drug partition coefficient between aqueous and lipid phases, V_g is the volume of the GI fluid, and h is the effective thickness of the GI wall. Under sink conditions, $\Delta \text{C} \rightarrow C_g$, and therefore, the right-hand side of the equation can be reduced to: $-k_a \cdot C_g$, which explains and supports why oral drug absorption can often be described with a first-order rate constant that is partially associated with drug lipophilicity. The absorption process can also reflect zero-order input (K_0), particularly for controlled release drug products. In practice, a number of techniques are available to estimate k_a and K_0 (along with more complex absorption rate profiles) from PK data that range from deconvolution to the application of mathematical models (Jusko 1995).

2.1.4 Protein Therapeutics

Protein drugs represent a diverse class of therapeutic molecules that can be grouped into several primary groups based on their structure and mechanisms of action including (1) proteins that replace deficient or abnormal concentrations, (2) proteins that augment activity through existing signaling pathways, (3) proteins with novel function or activity, (4) immunotherapy, such as monoclonal antibodies (mAbs), antibody-drug conjugates (ADCs), and other antibody-based constructs, and (5) gene and cell-based therapies (Leader et al. 2008; Shah et al. 2022). The PK of macromolecules present with unique complexities as compared with small molecule drugs, which result from the wide spectrum of molecular weights, distinct interactions with physiological processes, and the physicochemical properties of the drugs (Diao and Meibohm 2013; Shah 2015; Wang et al. 2008). Some of the major differences between small and large molecule drugs are listed in Table 2 (Mahmood and Green 2007), and this section will highlight these in the context of

Table 2 Differences between macromolecular and small molecule drugs

Properties	Protein drugs	Small-molecule drugs
Molecular weight	$>$1,000 Da	$<$1,000 Da
Route of administration	Parenteral (IV, SC, IM)	All routes
Impurities	Difficult to standardize	Purity standards well established
Plasma protein binding	Of negligible importance	May be important
Volume of distribution	Mainly limited to blood volume	Dictated by hydrophobicity and plasma and tissue protein binding
Half-life	Wide range (from minutes to hours for peptides and proteins to weeks for mAb	Short (usually in hours)
Cell-surface receptor interactions	Plays a significant role and may result in nonlinear PK	May be important in some cases
Elimination	Phagocytosis, endocytosis, proteolysis, and renal (MW $<$60–70 kDa)	Biotransformation (oxidation and conjugation), renal
Immunogenicity	May play a significant role	Not important

Adapted from Mahmood and Green (2007) and Shah et al. (2022)

the major PK parameters described in the previous sections. It is important to recognize that several of the assumptions for noncompartmental analysis are often not valid for protein drugs, such as linear conditions and rapid equilibrium with the site of elimination, which can render such analyses misleading or inappropriate and necessitate the application of computational modeling to best delineate dose-exposure (PK) and exposure-response (PK/PD) relationships.

The clearance of protein therapeutics in general is primarily through three major pathways: (1) renal excretion, (2) non-specific catabolic pathways like proteolysis in the extracellular environment or inside cells following uptake via pinocytosis or convection, and (3) specific pathways such as receptor-mediated endocytosis. Peptides and proteins that are $<$60–70 kDa are subject to renal filtration and excretion. For example, IL-10 is a relatively small cytokine (MW = 18 kDa) and its total systemic clearance is positively correlated with renal function as determined by creatinine clearance (Andersen et al. 1999; Meibohm and Zhou 2012). In contrast, mAbs and antibody constructs ($>$70 kDa) are too large to be filtered, and renal excretion is not a major mechanism of elimination. Whereas renal clearance depends on a molecular size cutoff, tissue catabolism following non-specific cellular uptake (i.e., pinocytosis or convection) is a common clearance mechanism for macromolecules. Therefore, it is not surprising that body weight is frequently identified as a significant covariate on total systemic CL in population PK studies, particularly for mAbs (Bajaj et al. 2019). A major difference between smaller proteins and mAbs is that the typical half-lives of the former are on the order of minutes to hours as compared to weeks for mAbs, which are protected from proteolysis by binding to FcRn in endothelial cells. The FcRn receptor is a widely expressed glycoprotein, which is located in the endosomes of endothelial cells and

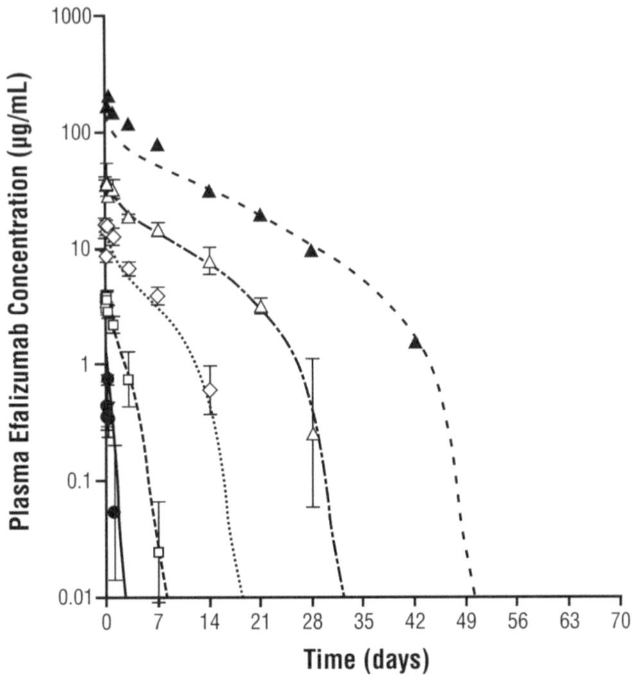

Fig. 3 Pharmacokinetic profiles of efalizumab in psoriasis patients following a range of single IV bolus doses. Efalizumab exhibits target-mediated drug disposition (TMDD), in which drug binding to the pharmacological target influences the PK properties of the drug. Profiles show steep distributive phases at low doses, which become saturated with increasing dose levels. Drugs exhibiting TMDD typically display an extended terminal elimination phase, but this can be obscured by assays with insufficient quantification limits. Lines represent fitted profiles using a Michaelis-Menten approximation model of TMDD. (Partial reprint from Joshi et al. (2006) with permission)

binds to the Fc domain of antibodies in a pH-dependent manner. It does not bind to the Fc domain at physiological pH, but FcRn strongly binds to this region at the acidic pH of endosomes (6.0–6.5) and bound Fc containing proteins are protected from endosomal catabolism and trafficked back to the systemic circulation.

The binding of many protein therapeutics to their pharmacological target can result in receptor-mediated endocytosis and lysosomal catabolism and influence both their apparent distribution and clearance (Diao and Meibohm 2013; Wang et al. 2008). This phenomenon is referred to as target-mediated drug disposition (TMDD), which was originally coined by Levy in describing the unique PK properties of a few small molecule drugs that showed target specific, saturable binding in tissues that affected their systemic disposition (Levy 1994b). This process can become saturated as drug concentrations exceed target capacity and manifest as dose-dependent distribution and/or elimination (Fig. 3). The time-course of drug concentrations at low dose levels often show steep distributive phases but with extended terminal elimination half-lives. These processes become saturated at high doses, and the drug may follow linear kinetics until concentrations begin to decrease toward non-saturating concentrations. TMDD can be further complicated by additional processes that can alter target expression, such as disease, the PD effect of the drug, feedback adaptation, and drug interactions, and such dynamic and nonlinear behavior can complicate the analysis of PK/PD relationships for small and macromolecular drugs (An 2020; Mager 2006).

The generation of anti-drug antibodies (ADAs) is an immunogenic reaction that has the potential to alter the PK and/or PD of therapeutic proteins (Chirmule et al. 2012; Dingman and Balu-Iyer 2019). The hierarchical structure of the protein, its aggregation state, and the drug product formulation can contribute to the risk of developing ADA. Other factors related to the dosing regimen such as dose, frequency, and route of administration can also influence their immunogenic potential. The development of ADAs is often delayed in time and associated with accelerating the clearance of the drug. Such curves can sometimes resemble the signature profile of TMDD, and care must be taken to distinguish ADA effects from TMDD behavior. Whereas binding ADAs can alter the PK properties of protein drugs, neutralizing ADAs also will interfere with the molecular recognition of the drug by its pharmacological target and subsequent pharmacodynamics.

The volume of distribution of macromolecules is limited by their larger molecular size as compared to small molecule drugs, and most proteins and antibodies will exhibit a V_{ss} that is one to two-times blood volume or about 15 L for a 70 kg person (Diao and Meibohm 2013; Wang et al. 2008). This is essentially the extracellular volume, whereas the central volume of distribution is approximately equal to the plasma volume. Of the major pathways for tissue uptake (convection, diffusion, and transcytosis), convection is the primary mechanism by which antibodies extravasate into tissues, which is driven by the blood–tissue hydrostatic pressure gradient. The efficiency of this uptake will vary depending on the relative size of the protein and the pores in tissue capillaries. In another biological application of Fick's law, the influx of the amount of drug (A) owing to convection can be defined as:

$$\frac{dA}{dt} = L \cdot (1 - \sigma) \cdot C$$

with L representing lymph flow, about 0.2–4% of cardiac output, C is the driving drug concentration (e.g., arterial concentration), and σ is the vascular reflection coefficient, which will depend on the capillary structure in each organ. Molecular diffusion may represent a significant tissue distribution pathway for relatively small proteins but not large mAbs or antibody-constructs. The relationship between plasma and tissue concentrations of mAbs tends to be linear, defined by distribution coefficients, regardless of the absolute plasma concentration and species (Shah and Betts 2013).

Protein drugs are generally not bioavailable following oral dosing, owing to their instability at gastric pH, and are usually administered parenterally (i.e., IV, SC, and IM), with the SC route being most preferred. On injection into the skin hypodermis, lymphatic drainage from the SC site of injection is a major mechanism by which protein drugs gain access to the systemic circulation (McLennan et al. 2005; Richter et al. 2012; Supersaxo et al. 1990). The percentage of dose of biologics recovered in the lymphatics appears to be positively correlated with molecular weight and approaches 100% for proteins greater than 40 kDa. The role of lymphatic recovery in the uptake of protein drugs after SC dosing can vary substantially across species and is understudied in humans, but there is consensus that lymphatic circulation is a major pathway for the absorption of biologics following this route of administration. Smaller proteins and peptides injected subcutaneously can achieve rapid uptake of a small fraction of the dose into the systemic circulation followed by a slower absorption phase through lymphatic delivery (Radwanski et al. 1998). In contrast, mAbs simply show slow absorption primarily through lymph flow (T_{max} of 2–8 days) because the reflection coefficient is far greater for capillaries at the site of injection relative to lymphatic drainage (Diao and Meibohm 2013; Wang et al. 2008). The extent of SC absorption is also highly variable and often incomplete ($F < 1$), and major factors influencing the bioavailability of protein drugs include anatomic injection sites, injection site proteolysis and/or TMDD, differences across species, molecular size, FcRn binding affinity of mAbs, and co-administered formulation excipients (Kagan 2014; Kagan and Mager 2013; Richter and Jacobsen 2014).

Overall, there are unique differences in the physiological processes that regulate the PK of protein therapeutics as compared to small molecule drugs (Table 2). In addition, there is a greater likelihood that these processes can show capacity limitation and plasma drug concentrations are often not in rapid equilibrium with the site of drug elimination, which significantly limits the use of simple noncompartmental analysis to characterize PK properties. Mathematical modeling is therefore essential for characterizing the PK of biologics, which can then be leveraged to establish PK-PD relationships that can inform the drug development process and provide a means for individualizing pharmacotherapy.

2.1.5 Compartmental Modeling

The use of simple compartments to describe the time-course of drug concentrations in relevant biological fluids has been a major component of the discipline since its foundational establishment (Gibaldi and Perrier 1982; Wagner 1971). Ordinary differential equations (ODEs) are specified that define the rate of change of the amount or concentration of drug in each compartment as influenced by input and processes of disposition covered in the previous section. In the early days of PK modeling, considerable efforts were made to solve such equations explicitly to derive analytical solutions to improve computational speed, address potential instabilities associated with stiff differential equations, and derive further equations that relate key summary PK properties to the model parameters. However, the continued advancement in computer hardware and software/algorithms has largely addressed the concern for long computational run times and the need for robust differential equation solvers. Most importantly, capacity-limited or nonlinear processes make explicit analytical equations nearly impossible. Thus, most contemporary PK models are simply left in the format of ODEs. Five of the most common compartmental models are shown in Fig. 4 and are described further in this section.

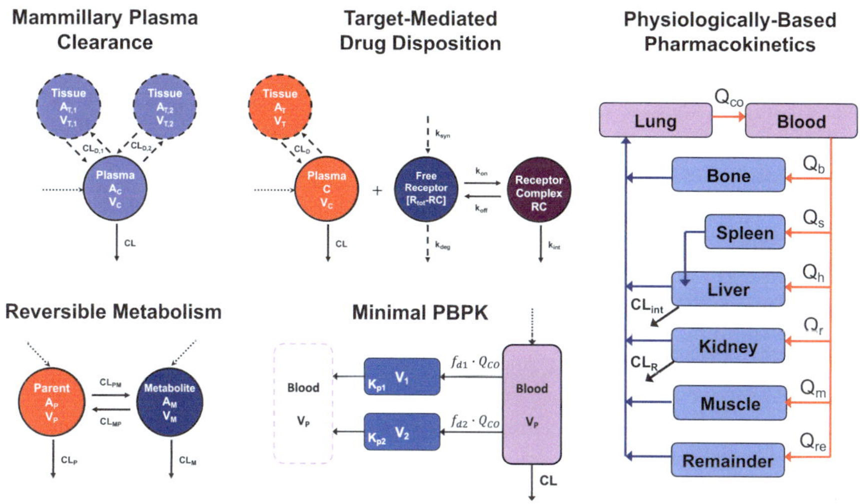

Fig. 4 Common compartmental model structures for pharmacokinetic analyses

A review of the art of modeling, standards and best practices, and hierarchical statistical components of mixed effects models are beyond the scope of this chapter and have been covered extensively elsewhere (Bonate 2006; Ette and Williams 2007; Gabrielsson and Weiner 2016).

Mamillary Plasma Clearance Models

Of the compartmental models shown in Fig. 4, the mamillary plasma clearance models are by far the most common. These are nested models consisting of a central compartment (representing plasma concentration) and a number of additional tissue compartments depending on the number of exponentials that can be supported by the data. A drug that exhibits monoexponential disposition can be described with just a central compartment (instantaneous distribution in a well-mixed compartment), biexponential kinetics will require two compartments (one central and one tissue), and so forth. A general equation for the amount of drug in the central compartment (A_C) after an IV bolus dose (D_{IV}) can be defined as:

$$\frac{dA_C}{dt} = -\left(\frac{CL}{V_C} + \sum_{i=1}^{n} \frac{CL_{D,n}}{V_C}\right) \cdot A_C + \sum_{i=0}^{n} \frac{CL_{D,n}}{V_{T,n}} \cdot A_{T,n}$$

with n as the number of peripheral tissue compartments, the initial condition is: $A_C(0) = D_{\mathrm{IV}}$, and the observed plasma concentration is: $C = A_C/V_c$. The rate of change of the drug amount in any of the tissue compartments is:

$$\frac{dA_{T,n}}{dt} = \frac{CL_{D,n}}{V_C} \cdot A_C - \frac{CL_{D,n}}{V_{T,n}} \cdot A_{T,n}$$

and the initial condition is zero. These equations can also be written in terms of first-order rate constants, in which $k_{10} = CL/V_C$, $k_{1n} = CL_{D,\,n}/V_C$, and $k_{n1} = CL_{D,\,n}/V_{T,\,n}$. Models using rate constants or CL and V terms are mathematically equivalent; however, it is often preferable to use a CL and V parameterization to improve interpretability of parameter values and especially if population or nonlinear mixed effects modeling is being applied to assess potential covariate relationships with model parameters. Should any of the distribution or elimination clearances show capacity-limitation, a Michaelis-Menten equation can be substituted for the above concentration-independent parameters (Table 1). The choices in specifying the structural model should be based on a careful balance among known physiological and pharmacological mechanisms of disposition for the study drug and standard model fitting criteria, such as (among others) visual inspection of observed and fitted data, distribution of residuals, model convergence, and the precision of parameter estimates (Bonate 2006; Gabrielsson and Weiner 2016).

Mamillary plasma clearance models can also be readily extended to accommodate simple and complex absorption kinetics. Consider a one-compartment model with the oral absorption process from an absorption site (A_a) defined by a first-order rate constant (k_a):

$$\frac{dA_a}{dt} = -k_a \cdot A_a, \quad A_a(0) = F \cdot D_{PO}$$

$$\frac{dA_C}{dt} = k_a \cdot A_a - \frac{\mathrm{CL}}{V_C} \cdot A_C, \quad A_c(0) = 0$$

In this case, PK data following both IV and oral administration would be required to identify and estimate the bioavailability (F). If the absorption process is delayed, the model can be modified to shift the dosing event by an estimated t_{lag} parameter, in which concentrations prior to t_{lag} are set to zero, and the dosing event is shifted in time to occur at t_{lag}. This mathematical delay simply shifts the expected curve to the right and does not characterize a PK curve exhibiting a gradual absorption process. Blood sampling in the early phases of absorption are typically sparse, and so the simple t_{lag} approach can work well, but this method will not describe a gradual increase in drug concentrations. Savic and colleagues have popularized the use of a series of transit compartments to better define the gradual absorption of drugs (Savic et al. 2007). Instead of specifying a number of compartments by trial and error, the series of transit compartments were described by the gamma distribution function, which was integrated into the drug absorption site equation:

$$\frac{dA_a}{dt} = D_{PO} \cdot F \cdot k_{tr} \cdot \frac{(k_{tr} \cdot t)^n \cdot e^{-k_{tr} \cdot t}}{\sqrt{2\pi} \cdot n^{n+0.5} \cdot e^{-n}} - k_a \cdot A_a, \quad A_a(0) = 0$$

The gamma distribution function is shown above multiplied by $D_{PO} \cdot F$ (with the Sterling approximation for $n!$ in the denominator), in which k_{tr} is a first-order rate constant between transit compartments, and n is the number of transit compartments (estimated as a real number and not an integer).

For small protein drugs that gain early access to the systemic circulation, such as IL-10, a fraction of the dose can be estimated to be delivered by a rapid zero-order process, with the remaining absorbable fraction following a slower absorption phase presumably through the lymphatics (Radwanski et al. 1998). Modifications of such a dual input model have been described for other protein drugs such as erythropoietin and interferon-beta (Mager et al. 2003a; Woo et al. 2006), both of which showed increased bioavailability with dose. In contrast, mAbs tend to follow simple linear first-order absorption in humans, and interestingly, age was identified as a covariate on k_a for the absorption of canakinumab, which was interpreted as decreasing k_a with increasing age owing to decreased lymphatic flow (Chakraborty et al. 2013). Drug formulations can alter such simple behavior, and a combination of modeling approaches can be used to describe complex absorption profiles for long-acting SC injectable drug products (Cirincione et al. 2017).

Reversible Metabolism

Some small molecule drugs can undergo reversible metabolism, in which drug metabolites can be further metabolized and biotransformed back to the parent drug (Cheng and Jusko 1993). Traditional noncompartmental analysis methods of such systems will yield erroneous results. Separate experiments must be conducted by administering the parent compound $D_{\mathrm{IV},\,P}$ and synthesized metabolite $D_{\mathrm{IV},\,M}$, the plasma parent and metabolite concentrations should be measured following the administration of either compound, and the combined PK data should be modeled simultaneously (see basic model in Fig. 4) to identify all PK parameters (Ebling et al. 1985). This framework can be further expanded to characterize the PK of drugs showing enterohepatic circulation (Li et al. 2014).

Target-Mediated Drug Disposition

A general compartmental model of TMDD was developed to incorporate the binding of small molecule drugs and biologics to their pharmacological target as part of their disposition (Mager and Jusko 2001a). Assuming no peripheral distribution, the rate of change of drug in the central compartment after an IV bolus dose is defined as:

$$\frac{dA_C}{dt} = -\left(\frac{CL}{V_C} + \frac{k_{\mathrm{on}}}{V_C} \cdot R_f\right) \cdot A_C + k_{\mathrm{off}} \cdot RC, \quad A_C(0) = D_{\mathrm{IV}}$$

with R_f as the amount of free receptor, which can be described with a turnover model governed by zero-order synthesis and first-order degradation rate constants (K_{syn} and k_{deg}):

$$\frac{dR_f}{dt} = K_{\text{syn}} - \left(k_{\text{deg}} + \frac{k_{\text{on}}}{V_C} \cdot A_C\right) \cdot R_f + k_{\text{off}} \cdot \text{RC}, \quad R_f(0) = R_{\text{tot}} = \frac{K_{\text{syn}}}{k_{\text{deg}}}$$

The initial condition is equal to the total amount of available receptor in the absence of drug (R_{tot}), which is determined by the ratio of K_{syn} to k_{deg}. The rate of change of the drug-target complex is defined as:

$$\frac{dRC}{dt} = \frac{k_{\text{on}}}{V_C} \cdot R_f \cdot A_C - (k_{\text{off}} + k_{\text{int}}) \cdot \text{RC}, \quad \text{RC}(0) = 0$$

with k_{int} representing a first-order internalization rate constant. This system is flexible, and changes can be made to accommodate (among others) peripheral tissue distribution, the lack of *RC* internalization (by setting $k_{\text{int}} = 0$), and any type of drug input function (e.g., absorption and short- and long-term IV infusions), depending on known or suspected pharmacological mechanisms, model-fitting criteria, and route of drug administration. The model can also be further simplified if the rate constants of free receptor degradation and internalization of the drug-target complex are equal (as determined by model fitting and parameter estimation). Under such conditions, the total amount of receptor becomes constant over time, and R_f can be simply expressed as: $R_f = R_{\text{tot}} - \text{RC}$, eliminating the need for a differential equation to define R_f. When $k_{\text{deg}} \neq k_{\text{int}}$, the total amount of receptor will either increase ($k_{\text{deg}} > k_{\text{int}}$) or decrease ($k_{\text{deg}} < k_{\text{int}}$) over time before returning to baseline conditions as drug is removed from the system (Mager and Krzyzanski 2005). Further details on the mathematical properties of the general TMDD model and which parameters influence the various phases of disposition for low and high doses have been assessed (Peletier and Gabrielsson 2012).

The TMDD model is a priori structurally identifiable (Eudy et al. 2015); however, study design considerations (e.g., range of dose levels, number and timing of blood samples, and limits of assay quantification) can generate numerical identifiability concerns that can limit the ability to estimate model parameters with good precision, especially drug-target binding micro-constants (i.e., k_{on} and k_{off}). One remedy is to apply an approximation model that re-parameterizes the full TMDD model in terms of total drug and total receptor concentrations, assuming rapid binding or quasi-steady-state conditions, and calculating drug concentrations not associated with the drug-target complex through a solution of the quadratic formula (Gibiansky et al. 2008; Mager and Krzyzanski 2005). These approximation models also show that under certain limiting conditions, the Michaelis-Menten function (Table 1) is mathematically equivalent to the approximation models and can improve numerical identifiability (albeit with the loss of tracking components of target occupancy). The evolution of compartmental TMDD models has been reviewed (Dua et al. 2015), and further modifications can be made to account for additional complexities, such as (among many others) the PD effect of the drug altering target expression (Jin and Krzyzanski 2004; Krzyzanski et al. 2016), bispecific binding compounds (Schropp et al. 2019), and disease-dependent effects on target expression (Jonsson et al. 2005). Importantly, by their very nature, TMDD models include target-occupancy, which can be extended and linked to mechanism-based pharmacodynamic models.

Physiologically Based Pharmacokinetic (PBPK) Models

Physiologically based models of PK is a systems level approach that provides a general whole-body framework for integrating biological and pharmacological mechanisms of drug absorption and disposition. In his landmark article, Teorell was the first to describe a physiological basis for the kinetics of administered substances in the body and included physiological compartments with tissue volumes, partitioning, and rate constants for distribution and elimination processes (Teorell 1937). However, without the availability of computational resources, the first model that one would recognize as a conceptual PBPK model would not be published until 1960 by Jacquez et al. (Jacquez et al. 1960). Later, Bischoff and Dedrick would become known as the fathers of computational PBPK modeling and popularized the approach using methotrexate and thiopental as test drugs of interest (Bischoff and Dedrick 1968; Bischoff et al. 1971). The major components and requirements of PBPK models include a mathematical structure that features primary tissue compartments anatomically connected through circulating blood (Fig. 4), physiological parameters and processes (e.g., tissue volumes and blood flows, GFR, lymph flow, and others), and relevant pharmacological properties and mechanisms (e.g., transport processes, intrinsic clearance, tissue partitioning, protein and tissue binding, and others).

Reviewing comprehensive details of a traditional PBPK model is beyond the scope of this section (Nestorov 2003), and only a few common elements of basic models are defined here. Following on Fick's Law, a simple equation can be specified for the rate of change of small molecule drug concentration in a non-eliminating tissue (C_T):

$$\frac{dC_T}{dt} = \frac{Q_T}{V_T} \cdot \left(C_a - \frac{C_T}{K_p} \right)$$

with Q and V as blood flow and volume of tissue (T), C_a is the arterial drug concentration, and K_p is the tissue:plasma partition coefficient. For preclinical studies, the K_p values can be calculated from continuous infusion experiments ($K_p = C_{ss,\ T}/C_{ss}$), single dose studies ($K_p = \text{AUC}_T/\text{AUC}$), or in vitro homogenate binding assays (Lin et al. 1982), or the values can be estimated during model fitting of tissue level data. The K_p values can also be approximated in silico from explicit equations based on drug lipophilicity and tissue-specific properties, such as water, neutral lipid, and phospholipid content (Graham et al. 2012). Physiological parameter values are typically fixed to reported values (Brown et al. 1997; Davies and Morris 1993); however, Xu et al. also showed how individual organ parameters (i.e., blood flows and tissue volumes) could be estimated using prior population information and a Maximum a Posteriori Bayesian estimation algorithm within a dual PBPK model for 17-(allylamino)-17-demethoxygeldanamycin and its active metabolite in tumor-bearing mice (Xu et al. 2003). The equation above is considered perfusion-limited and assumes that permeability or diffusion into the tissue is not rate-limiting. For drugs that feature permeability-limited uptake, tissue compartments are further divided into interstitial and tissue spaces, and drug distribution between these compartments is a function of the permeability-surface area product for each tissue. Active transport processes can also be added to specific tissue sites, which are often expressed using traditional Michaelis-Menten equations (Table 1). For drug eliminating organs, a clearance process can be added:

$$\frac{dC_T}{dt} = \frac{Q_T}{V_T} \cdot \left(C_a - \frac{C_T}{K_p}\right) - CL_{\text{int}} \cdot \frac{C_T}{K_p}$$

However, the mathematical functions defining clearance will depend on many factors, such as capacity-limitation, protein binding, and appropriate driving concentrations of the clearance process (e.g., renal filtration would be driven by free plasma concentrations). As mentioned in Sect. 2.1.1, intrinsic clearance values may be extrapolated from in vitro bioassays but can also be estimated from retrograde model fitting. For the liver, models should also feature incoming blood flow from both the hepatic artery and the portal vein, which carries venous blood from gastrointestinal organs (such as the stomach, spleen, pancreas, and intestine). Additional complexities in base models can include (among many others): linear and nonlinear plasma and tissue binding, TMDD, enterohepatic recirculation, and distribution to red blood cells. Models must also be substantially modified for proteins and antibodies. Although some base models for small proteins can follow a similar structure as defined for small molecule drugs (Sato et al. 1987), early PBPK models of antibodies were designed to account for convection and sieving coefficients, lymphatic distribution, tissue sub-compartments for interstitial and cell-associated spaces, and a two-pore theory of tissue uptake (Baxter et al. 1994; Covell et al. 1986; Rippe and Haraldsson 1994). Garg and Balthasar introduced an endosomal compartment for each tissue that incorporates antibody binding to FcRn, the trafficking of bound drug to the vascular and interstitial spaces, and the degradation of unbound antibody (Garg and Balthasar 2007). Base platform models have since emerged for monoclonal antibodies (Shah and Betts 2012) and endogenous and exogenous proteins of varying size (Liu et al. 2024). Finally, whereas full PBPK models provide a comprehensive whole-body framework, the opportunity exists to lump together tissues exhibiting similar kinetic properties to form reduced models (Nestorov et al. 1998) or to utilize minimal PBPK models (Cao et al. 2013; Cao and Jusko 2012) for focused fit-for-purpose modeling of limited specific tissues while preserving physiological structures and meaningful system-specific parameter values.

Many advances and applications of PBPK modeling during the twentieth century were made in toxicological sciences and non-cancer risk assessment of exposure to chemicals or drugs (Clewell and Andersen 1996). The interest in the development and application of PBPK models for drug discovery, development, and therapeutics has increased exponentially owing in large part to the encouragement and acceptance of PBPK model-informed studies by regulatory agencies, the impact such models have had in clinical drug development and regulatory decision-making, and the availability of sophisticated PBPK software packages, such as Simcyp (Certara, Sheffield, UK), GastroPlus® (Simulations Plus, Lancaster, CA), and PK-Sim (Open Systems Pharmacology) (Grimstein et al. 2019; Huang et al. 2013; Rowland et al. 2015; Rowland et al. 2011). Major advancements have been made in the use of PBPK for (among others) predicting PK drug-drug interactions, drug exposures in specific patient populations (e.g., pediatrics and oncology), drug absorption and the role of formulations, drug exposures in patients with organ impairment, first-in-human dose selection, and the role of genetic polymorphisms (Jamei 2016; Yoshida et al. 2017). Rostami-Hodjegan and Toon provide an excellent perspective on the evolution of PBPK within the context of model-informed drug development, and interdisciplinary collaborative team science will be needed to realize the full potential of PBPK modeling with new applications such as personalized pharmacotherapy becoming possible (Rostami-Hodjegan and Toon 2020). PBPK modeling is the state-of-the art in PK systems analysis and explains sources of inter-individual drug exposures in the systemic circulation and sites of drug action in a holistic manner. Predicted concentration-time profiles can be used to drive empirical and mechanistic pharmacodynamic and systems pharmacology models that relate drug exposures to the intensity and time-course of pharmacological effects.

2.2 Pharmacodynamic Concepts and Basic Models

Small molecule drugs are typically categorized at a high level according to how they interact with their pharmacological target (e.g., agonist, antagonist, partial agonist, and inverse agonist) (Kenakin 1997). In turn, drug targets are represented by a broad range of structural motifs, including (among others) G-protein coupled receptors, enzymes, ion channels, calcium release receptors, transporters, and receptors regulating nuclear transcription (Brunton and Knollmann 2023). In this section, the terms receptor and target will be used generically regardless of their structural basis, and drug effect (E) or response (R) may be used interchangeably. Drugs can bind to their target in a reversible or irreversible manner, and the former can be described by the target-binding and receptor occupancy equations introduced in the beginning of this chapter. Drug-target interactions are subsequently transduced through signaling pathways or networks, and biomarkers of drug response can be classified on the basis of whether they inform on drug-target interactions or mechanisms of drug action. Mechanism-based biomarkers are often downstream mediators within the causal pathway of clinical (or surrogate) drug effects, which would include genes, proteins, cells, and components of biochemical and physiological pathways. Pharmacodynamic models are used to integrate physiological turnover processes regulating such biomarkers with PK driving functions and pharmacological drug-target interactions to test and establish drug exposure-response relationships, identify sources of inter-subject variability, and guide the design of optimal dosing regimens.

The time-course of pharmacological effects are regulated by a vast array of mechanisms, which can complicate the identification of a suitable base model as well as modifications that can be made to accommodate various drug- or system-specific complexities. Most drug effects will return toward baseline conditions in a zero-order manner, and Levy was the first to identify the determinants of the linear decay of drug responses (Levy 1964). He recognized for rapid direct effects that the monoexponential decrease in plasma drug concentrations could be mathematically associated with the linear slope (m) between drug effects and the log of drug concentrations (see Fig. 2 between 20% and 80% occupancy) to derive a simple equation describing the zero-order loss of drug effect over time:

$$E = E_0 - \frac{k_{10} \cdot m}{2.303} \cdot t$$

with E_0 as a theoretical intercept. This expression is not used to model data owing to its limited application to just the linear phase of the loss of drug effect for rapid direct drug effects; however, it explained that such drug effects dissipate at a rate defined by the $k_{10} \cdot m$ product, could be used to understand why the duration of drug effects were directly proportional the half-life and the log of the administered dose, and ushered in the era of mathematically describing the time-course of drug effects.

Whenever possible, PD model selection should be based on mechanisms of drug action and the primary rate-limiting steps in the biological system. Study design can also have an impact on model selection and the ability to resolve drug and system parameters, and good practices should be followed such as conducting studies over a suitable range of dose-levels to resolve capacity and sensitivity parameters, the inclusion of sampling times that can facilitate an assessment of baseline (or placebo response) conditions and cover a sufficient wash-out period to follow the nature of the biomarker return to baseline, and conducting multiple-dose or continuous infusion experiments to test for PD stationarity. In many cases, a careful evaluation of the PK-PD data within the context of mechanisms of action and basic PD model expectations (or signature profiles) can provide insights into a suitable base model and a means for calculating initial parameter estimates.

2.2.1 Simple Direct Effects and Biophase Modeling

The term *simple direct effect* is used to refer to a rapid drug response that is rate-limited by the PK of the drug, such as circulating enzyme activity, the activity of circulating cells not dependent on turnover (e.g., platelet aggregation), and other biomarkers with very fast turnover (e.g., EKG, EEG, and muscle activity). The time-course of simple direct effects can be readily described by the E_{max} model (Table 1) (Wagner 1968), with the capacity and sensitivity terms defined by the maximum drug effect (E_{max}) and the drug concentration resulting in 50% of E_{max} (EC_{50}). The E_{max} equation assumes that drug effects are directly proportional to target occupancy (or linear transduction) (Ariëns 1954; Clark 1926) and can also be modified to incorporate baseline biomarker values in the absence of drug (E_0):

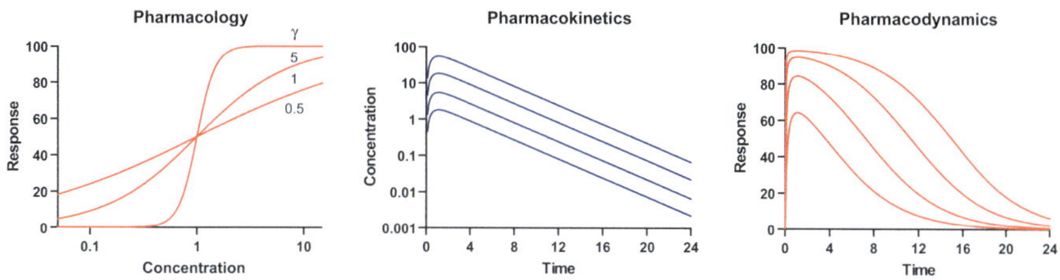

Fig. 5 Simulated profiles of drug concentration-effect relationships (left panel), pharmacokinetics (middle panel), and time-course of simple direct effects (right panel). Concentration-effect curves were simulated with the sigmoidal E_{max} model ($E_{max} = 100$ AU, $EC_{50} = 1$ AU) with 3 different values of the Hill coefficient (γ). Pharmacokinetic profiles were simulated using a one-compartment model with first-order absorption and elimination ($k_a = 2\ h^{-1}$, CL = 3 AU/h, $V_c = 10$ AU) and for doses of 25, 75, 250, and 750 AU. Pharmacodynamic curves are driven by the PK profiles (middle panel) and the concentration-effect relationship shown in the left panel (γ = 1)

$$E = E_0 \pm \frac{E_{\max} \cdot C^{\gamma}}{EC_{50}^{\gamma} + C^{\gamma}}$$

For antagonist or inhibitory effects, $E_{\max}$ and EC_{50} are often replaced by $I_{\max}$ and IC_{50}. The time-course of plasma drug concentrations (C) can be defined by any PK model, and the Hill coefficient (γ) can be added, which influences the steepness of the concentration-effect relationship (Fig. 5). Simulations of the time-course of effects defined by the E_{max} model, driven by a one-compartment PK model with first-order absorption and elimination rate constants, reveal a signature profile in which (1) the time to peak concentration (T_{max}) is the same as the time to peak response (T_{Rmax}), (2) high doses (or high concentrations relative to the EC_{50}) show a plateau phase near peak response, and (3) a zero-order decrease in drug effects that is linear, parallel, and independent of dose (in accordance with the Levy $k_{10} \cdot m$ concept) (Fig. 5).

The distribution of drugs to the site of action (or biophase) could represent a rate-limiting step in producing their pharmacological effects. Under such conditions, the time to peak response is delayed relative to $T_{\max}$ in plasma, and plots of effect as a function of log drug concentration will no longer show a Hill-type relationship but instead will exhibit a hysteresis loop (Fig. 6, left panel). Although measuring drug concentrations near or at the site of action would be most desirable, and biophase concentrations could be used to drive the $E_{\max}$ model (Minematsu et al. 2001), it is rarely feasible or impossible to do so (particularly for clinical studies). To address this temporal delay, Sheiner et al. utilized a hypothetical effect compartment to link plasma concentrations of d-tubocurarine to its pharmacological effect (Sheiner et al. 1979). Like Fick's law, the rate of change of drug concentrations in the effect compartment (C_e) was originally defined as:

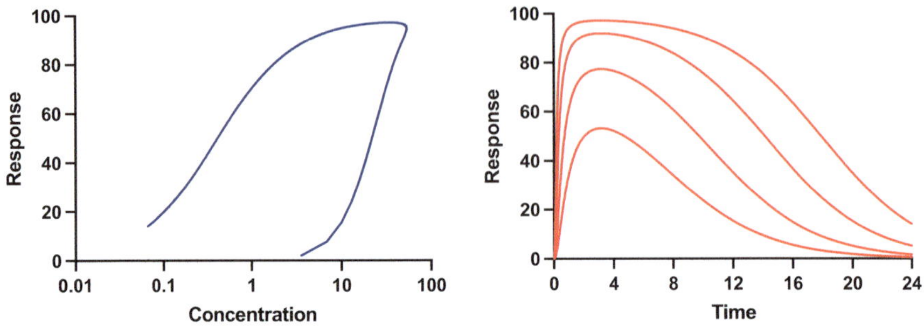

Fig. 6 Simulated hysteresis curve (left panel) and pharmacodynamic profiles according to the biophase model (right panel). The PK curves in Fig. 5 were used to drive the PD profiles ($k_{e0} = 0.5$, $E_{max} = 100$ AU, and $EC_{50} = 1$ AU). The hysteresis curve (left panel) shows the apparent concentration-effect relationship for the highest dose level

$$\frac{dC_e}{dt} = k_{1e} \cdot C - k_{e0} \cdot C_e, \quad C_e(0) = 0$$

with k_{1e} and k_{e0} as first-order rate constants for drug entering and exiting the effect compartment. These rate constants are often not numerically identifiable, and they are set equal to each other ($k_{1e} = k_{e0}$) and the right-hand side of the equation simplifies to: $k_{e0} \cdot (C - C_e)$. The plasma drug concentrations in the E_{max} equation above are replaced by effect compartment concentrations, which will produce a later $T_{R\max}$ relative to $T_{\max}$ (Fig. 6, right panel). The amount of drug distributing to the effect compartment is assumed to be negligible and does not affect the plasma drug concentrations, which can be described by any PK model. This model is highly useful for describing short delays in the pharmacological effect owing to drug distribution (typically on the order of minutes). The signature profile of this model shows that $T_{R\max}$ is dose independent (Fig. 6) because drug distribution to the effect compartment is a linear process, and peak concentrations in that compartment will therefore occur at the same time. Other details of the signature profile are shared with the E_{max} model such as plateaued effects near $T_{R\max}$ for high doses and parallel and linear decreases in effect over time when the effect is between 20% and 80% maximal.

2.2.2 Irreversible Effects and Turnover

Some drugs, such as chemotherapeutic agents, antimicrobials, and enzyme inhibitors, will elicit irreversible inactivation of enzymes, cells, and other proteins. Jusko introduced the use of a second-order rate constant (k) for the irreversible cell killing effects of phase-nonspecific chemotherapeutic agents (Jusko 1971):

$$\frac{dR}{dt} = -k \cdot C \cdot R$$

with R representing the biomarker of interest (in this case number of tumor cells), and the initial condition or $R(0)$ is the observed initial value of R (or R_0). This simple irreversible biomolecular interaction model produces a linear relationship between the log of the cellular survival fraction and the dose of the drug, and the product of k and AUC of the drug is the major determinant of net effects. This model was further modified to include a peripheral compartment of quiescent cells to describe chemotherapeutic drugs that are cell-cycle specific (Jusko 1973). Zhi et al. expanded the phase-nonspecific model to describe the pharmacodynamics of piperacillin on P. aeruginosa in neutropenic mice by including a first-order growth rate constant for the bacteria (Zhi et al. 1988):

$$\frac{dR}{dt} = k_{ng} \cdot R - k \cdot C \cdot R, \quad R(0) = R_0$$

with k_{ng} representing a first-order net growth rate constant. Here the k_{ng} term can be interpreted as the difference between a first-order growth and first-order death rate constants. This model is able to capture the exponential growth of bacteria in the absence of drug (control), whereas the bacterial survival fractions initially decrease in an exponential manner with treatment, which level off and bacteria grow again in an exponential manner as drug is removed from the system. The growth component of the model can also be modified to account for an upper limit of bacterial growth (R_{ss}), with production defined as: $k_{ng} \cdot R \cdot (1 - R/R_{ss})$.

The general turnover of a response variable (R) can be defined by a simple differential equation:

$$\frac{dR}{dt} = K_{in} - k_{out} \cdot R$$

with K_{in} as a zero-order production rate constant, and k_{out} is a first-order removal rate constant. If the value of R is constant, then the initial condition can be defined by the ratio of the production and loss rate constants: $R(0) = R_0 = K_{in}/k_{out}$. Yamamoto et al. combined this baseline equation with the irreversible inactivation equation from Jusko to describe the time-course of the antiplatelet effects of aspirin (Yamamoto et al. 1996):

$$\frac{dR}{dt} = K_{in} - (k_{out} + k \cdot C) \cdot R$$

Simulations of this model show a rapid decrease in R after drug is administered, and the time to maximum inhibition shifts to the left with greater drug concentrations owing to the second-order nature of the equation (Fig. 7). This model, in its original or modified forms, has been applied to other enzyme systems including 5α-reductase (Gisleskog et al. 1998) and H^+,K^+-ATPase (Katashima et al. 1998).

2.2.3 Indirect Response Models

One of the first accounts of indirect drug responses was described by Ariëns, who noted that some drug effects are not reflective of direct receptor interactions but are rather mediated by such interactions influencing the protection or release of

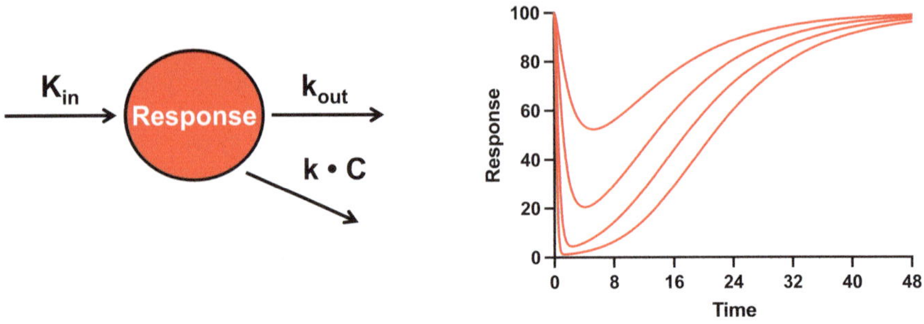

Fig. 7 Model diagram and signature profile for a turnover model with irreversible inactivation. PK curves in Fig. 5 were used to drive the PD profiles (baseline or $R_0 = 100$ AU, $k_{out} = 0.02\ h^{-1}$, $k = 0.15\ AU^{-1}\ h^{-1}$, and $K_{in} = R_0 \cdot k_{out}$ or 2 AU/h). Abbreviations are defined in the text

endogenous substances (Ariëns 1964). The first PD model to incorporate an indirect mechanism of action was published by Nagashima et al. for the inhibition of prothrombin complex activity production by warfarin (Nagashima et al. 1969). Later, the landmark paper by Dayneka et al. introduced a family of four basic indirect responses models (Dayneka et al. 1993) that has shown broad applicability to many drugs that alter the production or loss of biomarkers of drug effects (Jusko and Ko 1994). A general differential equation for the basic indirect response models can be written as:

$$\frac{dR}{dt} = K_{\text{in}} \cdot \{1 \pm H_1(C)\} - k_{\text{out}} \cdot \{1 \pm H_2(C)\} \cdot R$$

with $H_n(C)$ representing the E_{max} (or I_{max}) model. In this context, the capacity parameters represent fold-change in the production or loss of the biomarker, and I_{max} will range between 0 (no effect) and 1 (complete inhibition at high drug concentrations) and E_{max} can be any positive number. The four models are defined by the following conditions: Model 1 (inhibition of production) in which $H_1(C)$ is subtracted and $H_2(C) = 0$; Model 2 (inhibition of loss) in which $H_1(C) = 0$ is and $H_2(C)$ is subtracted; Model 3 (stimulation of production) in which $H_1(C)$ is added and $H_2(C) = 0$; and Model 4 (stimulation of loss) in which $H_1(C) = 0$ is and $H_2(C)$ is added. Figure 8 features the signature profiles of all four models that show gradual increases or decreases in the response and a delayed time to T_{Rmax} that shifts to the right in a dose-dependent manner. Drug responses reach a maximum (or minimum) value and then dissipate in a zero-order manner with linear recession slopes. The dose-dependent time to peak response for these models distinguishes them from the biophase model and can be explained by drug concentrations (after high dose levels) remaining above the EC_{50} (or IC_{50}) for longer durations and maintaining inhibitory or stimulatory conditions. Additional mathematical properties of this important class of models have been reviewed (Krzyzanski and Jusko 1998; Sharma and Jusko 1998).

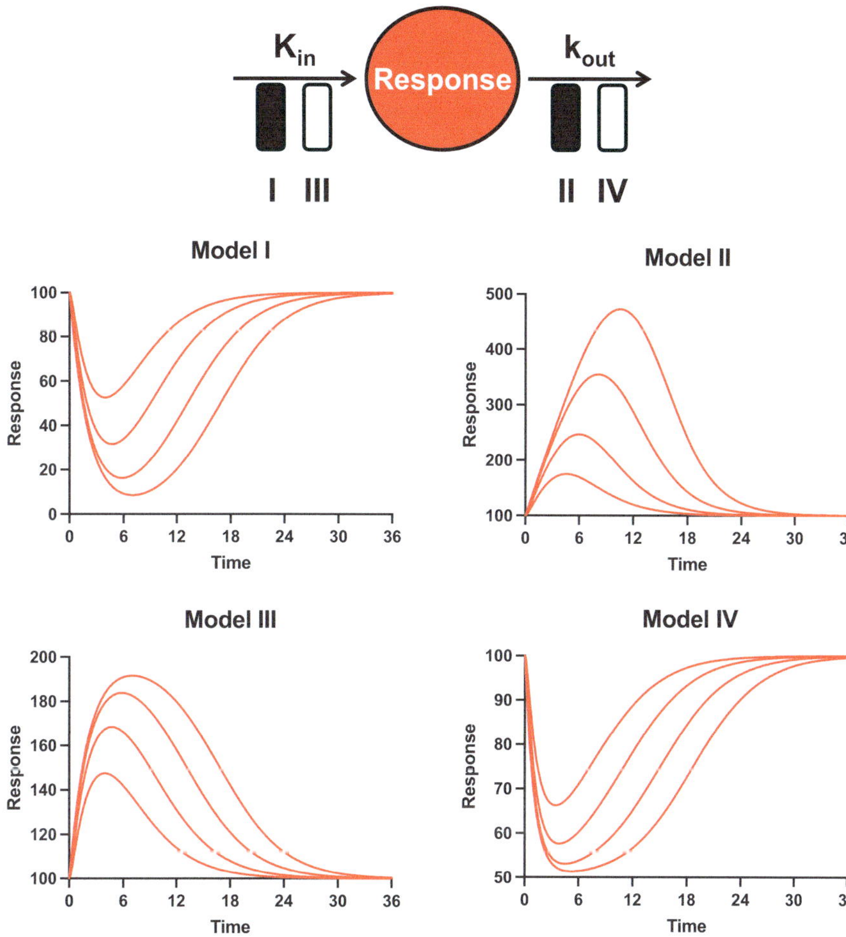

Fig. 8 Family of four basic indirect response models. Models I and II are for drugs that inhibit the production or loss of the biomarker (closed rectangles), and models III and IV are defined by drug stimulation of production or loss of the biomarker (open rectangles). PK curves in Fig. 5 were used to drive the PD profiles (baseline or $R_0 = 100$ AU, $k_{out} = 0.5\ h^{-1}$, $E_{max} = 1$ AU, $EC_{50} = 1$ AU, and $K_{in} = R_0 \cdot k_{out}$ or 50 AU/h). Abbreviations are defined in the text

Indirect response models are also common for characterizing the exposure-response relationships for protein therapeutics. Nieforth et al. characterized the interferon-α2a stimulation of MX protein following SC injection in healthy volunteers using Model 3 (Nieforth et al. 1996). The PD model developed by Bauer et al. for an anti-CD11a mAb is mathematically equivalent to Model 4 driven by a Michaelis-Menten PK model (likely approximating TMDD) in both chimpanzees and humans (Bauer et al. 1999). A response variable could also be regulated by an endogenous substance, such as the free PCSK9-mediated inhibition of LDL cholesterol (Gibbs et al. 2017):

$$\frac{dR}{dt} = K_{\text{in}} - k_{\text{out}} \cdot \left(1 - \frac{I_{\max} \cdot \text{PCSK9}}{IC_{50} + \text{PCSK9}}\right) \cdot R$$

with R representing LDL cholesterol, and IC_{50} is the free PCSK9 concentration producing 50% of $I_{\max}$. The PK of an anti-PCSK9 mAb was described using the quasi-steady-state approximation of the general TMDD model (Gibiansky et al. 2008), and the decrease in free PCSK9 concentrations translated into less inhibition of cholesterol removal (or accelerated clearance). The data were well captured, and the final model provided key insights into various dosing regimen designs. Turnover processes and basic and extended indirect response models are a major component of many contemporary PD models for protein-based drugs.

2.2.4 Nonlinear and Time-Dependent Transduction

The $E_{\max}$ model for simple direct effects (Table 1) can be viewed as a rapid, linear transduction process in which the effect is directly proportional to target occupancy (ρ) (Ariëns 1954; Clark 1926). However, transduction steps subsequent to drug-target binding could also be nonlinear (or any function of ρ). As described in the *Introduction* of this chapter, Stephenson introduced a modification to receptor theory, such that the effect of a drug could be represented by any mathematical function (f) of a signal (S), which is driven by drug efficacy (e) and target occupancy (Stephenson 1956). The efficacy term was further expanded by Furchgott to include total receptor concentration (R_{tot}) and the intrinsic efficacy of the drug (ϵ) (Kenakin 1997):

$$E = f(S) = f(e \cdot \rho) = f(\epsilon \cdot R_{\text{tot}} \cdot \rho)$$

These concepts laid the groundwork for Black and Leff to propose the operational model of agonism, which assumes that the drug effect can be nonlinearly related to RC based on the following *transducer function*:

$$E = \frac{E_{\max} \cdot RC}{K_E + RC}$$

with K_E as the RC concentration producing 50% of the system maximum. Substituting $RC = R_{\text{tot}} \cdot \rho$ and rearranging terms yields:

$$E = \frac{E_{\max} \cdot \tau \cdot C}{K_D + (\tau + 1) \cdot C}$$

with τ defined as an efficacy parameter or *transducer ratio*: $\tau = R_{\text{tot}}/K_E$. This equation can account for partial agonism, and simulations in Fig. 9 show the role of τ in determining the apparent capacity and sensitivity of the concentration-effect relationship. If a Hill-type coefficient is included in the transducer function, then the effect equation becomes:

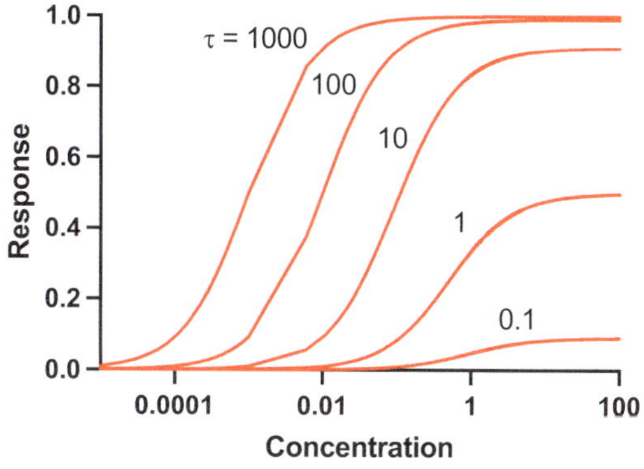

Fig. 9 Simulated concentration-effect relationship as defined by the operational model of agonism for different values of the transducer ratio (τ). Other model parameters were fixed at: $E_{max} = 1$ AU, and $K_D = 1$ AU. Large values of τ allow the system to approach a curve associated with a full agonist, whereas decreasing values alter the apparent capacity and sensitivity of the profiles (partial agonism)

$$E = \frac{E_{\max} \cdot \tau^{\gamma} \cdot C^{\gamma}}{(K_D + C)^{\gamma} + \tau^{\gamma} \cdot C^{\gamma}}$$

This equation uniquely identifies system-specific ($E_{\max}$ and γ) and drug-specific (K_D and τ) parameters and was successfully used to characterize the in vivo PD for a series of adenosine A_1 receptor agonists (Van Der Graaf et al. 1997). For numerical purposes, the true $E_{\max}$ needs to be resolved from experiments that include a full agonist. Interestingly, the K_D and τ parameters for the adenosine A1 agonists were highly correlated with measurements from in vitro bioassays, which suggested that the in vivo PD of future analogs could be predicted. In the event of hysteresis owing to drug distribution to the biophase, the drug concentration in the equation can be replaced by that in a hypothetical effect compartment (Cox et al. 1998). Furthermore, nonlinear transduction processes need not be constrained by the Black and Leff transducer function. For example, Visser et al. derived a general stimulus-response model for $GABA_A$ receptor modulators, which was informed by a series of benzodiazepines and neuroactive steroids, using an asymmetrical parabolic transducer function (Visser et al. 2003).

Transduction processes can also be time-dependent, enough so that the PD response variable might remain unchanged at earlier times following drug administration despite the existence of high systemic drug concentrations. Such substantial delays in the onset of effect are not consistent with the signature profiles of the PD models described thus far. In a theoretical study, Sun and Jusko proposed a series of transit compartments (M_n) to link drug exposure to delayed responses:

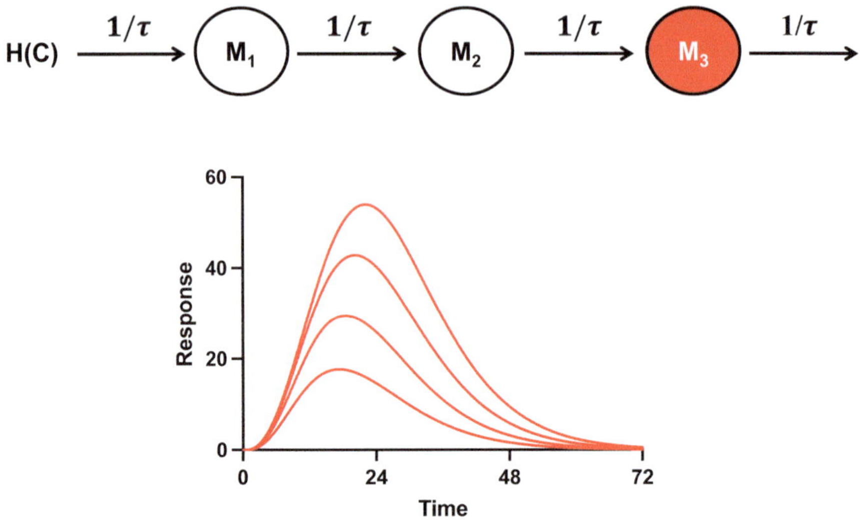

Fig. 10 Model diagram and signature profile of a 3-compartment transduction model. A Hill function serves as the input signal, driven by the PK profiles in Fig. 5, and the response is set to the value of the M_3 compartment. Here, τ is defined as the mean transit time in any one compartment (M_n). Parameter values were set to: baseline or $R_0 = 0$ AU, $\tau = 6$ h, $E_{max} = 100$ AU, and $EC_{50} = 1$ AU. As described in the text, the response can be set equal to the last compartment (M_3 in this case), a baseline ± the last compartment, the sum of all transit compartments, or the last compartment could be used as a signal acting on a biomarker

$$\frac{dM_1}{dt} = \frac{1}{\tau} \cdot (RC - M_1) \cdots \frac{dM_n}{dt} = \frac{1}{\tau} \cdot (M_{n-1} - M_n)$$

with τ as the mean transit time for any one compartment, and the last transit compartment serves as the response variable (with or without a baseline value). Assuming rapid binding relative to signaling events, RC can be replaced with the $E_{\max}$ equation to form a minimal model capable of mimicking PD onset delays without measurements of intermediary biomarkers (Mager and Jusko 2001b). Simulated profiles show a clear onset delay, a gradual rise to a peak response, and a linear decrease in pharmacological effects as the signal in the last compartment dissipates (Fig. 10). This platform model is flexible and can be used to describe an array of delayed response variables. There are numerous examples of using the last transit compartment as the response variable, including the parasympathomimetic activity of scopolamine and atropine (Perlstein et al. 2002), the myelosuppressive effect of many anticancer agents (Friberg et al. 2002), the anti-platelet effect of the ADC T-DM1 (Bender et al. 2012), and the C-reactive protein response to canakinumab, which was driven by predicted free concentrations of IL-1β (Ait-Oudhia et al. 2012). The sum of all the transit compartments has also been used to represent a distribution of cells in various stages of differentiation or cell death, such as the time-course of tumor volume following treatment with chemotherapeutic agents (Simeoni et al. 2004) and the platelet stimulatory effect of thrombopoietin (Harker et al. 2000). The last transit compartment has also been used as a surrogate signal for downstream PD effects, such as the elimination of tumor cells (Lobo and Balthasar 2002), and this model of signal distribution can be distinguished from the cell distribution model (Yang et al. 2010). As intermediary signaling or cell phenotypic biomarkers are measured, these simple transit compartment transduction models can be readily expanded into systems models that are describe later in this chapter and elsewhere in the book.

2.3 Advanced Pharmacodynamic Models

The basic PD models from Sect. 2.2 can be readily modified to incorporate drug- or system-specific complexities and/or combined to form small systems models of drug action. Such complexities include tolerance and/or rebound phenomena, opposing drug effects, receptor desensitization, receptor downregulation, non-constant biomarker baselines, physiological limits, disease progression, PD drug interactions, and many others. The approaches to integrate additional complexities vary greatly and can range from simple empirical functions altering some element of the basic PD models to coupling several basic PD models to account for feedback back and other control processes in a mechanistic manner.

2.3.1 Extended Indirect Response and Transduction Models

The turnover processes in the indirect response and transduction models are well suited to the incorporation of empirical and pharmacologically based functions of further complexities. For example, the basic indirect response models assume that the baseline of the response variable is constant in the absence of drug and defined by the ratio of the production and loss rate constants. However, there are many biomarkers that exhibit a circadian rhythm or some other irregular pattern (Ayyar and Sukumaran 2021). Endogenous cortisol is a steroid hormone produced by the adrenal glands, which shows a clear circadian rhythm, and its production is inhibited by the systemic exposure to exogenous corticosteroids. This response has been described using indirect response Model 1 with the zero-order production rate constant replaced with a circadian time-dependent function (Lew et al. 1993):

$$K_{in} = R_m + R_b \cdot \cos[(t - t_z) \cdot 2\pi/24]$$

with R_m as the mean input rate, R_b is the amplitude of the input rate, t_z is the peak time (baseline acrophase), and 2π /24 converts time into radians. Functions with additional harmonics can also be used, based on Fourier analysis, to better describe cortisol kinetics and improve model predictive performance (Chakraborty et al. 1999). This technique is not unique to cortisol and can be applied to other response variables with a circadian baseline profile, including those with circadian removal processes (Ayyar et al. 2019).

The addition of a precursor compartment is another highly useful modification to the basic indirect response models. Movin-Osswald and Hammarlund-Udenaes first

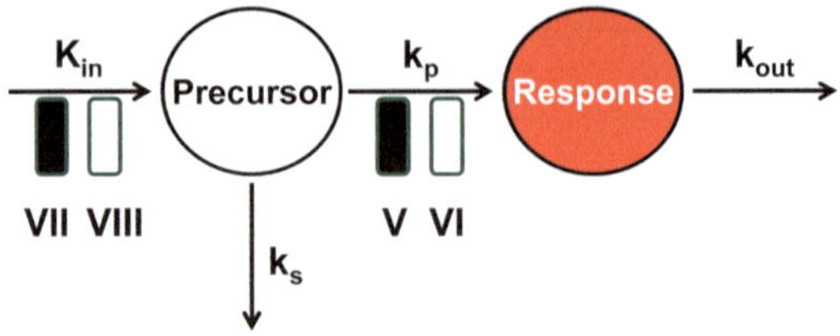

Fig. 11 Family of four precursor-dependent indirect response models. Models V and VI are for drugs that inhibit or stimulate the conversion of a precursor to the response variable, and models VII and VIII are defined by drug inhibition or stimulation of the production of a precursor state variable. Rectangles are as defined in Fig. 8

applied an integrated precursor indirect response model in which remoxipride concentrations stimulated the conversion of a precursor variable to prolactin (Movin-Osswald and Hammarlund-Udenaes 1995). A general model structure for a family of precursor-dependent indirect response models is shown in Fig. 11 and can be defined by two turnover equations for a precursor compartment (P) linked to the response variable:

$$\frac{dP}{dt} = K_{\text{in}} \cdot \{1 \pm H_1(C)\} - \left(k_s + k_p \cdot \{1 \pm H_2(C)\}\right) \cdot P$$

$$\frac{dR}{dt} = k_p \cdot \{1 \pm H_2(C)\} \cdot P - k_{\text{out}} \cdot R$$

with K_{in} redefined as the zero-order production of the precursor, k_p is the first-order production rate constant of the response, and k_s is an optional first-order removal rate constant of the precursor (the need for which can be tested using model-fitting criteria). Models 5 and 6 involve the inhibition or stimulation of precursor conversion (acting on k_p with $H_1(C) = 0$), and Models 7 and 8 can be used to describe drugs that inhibit or stimulate the production of a precursor. The model for remoxipride (Model 6) produces an increase in the production of the response variable (i.e., prolactin) but will simultaneously deplete the precursor pool. If a second dose of the drug is administered before the precursor pool has time to replenish, then the second dose will not elicit the same level of response, which is one form or cause of drug tolerance. Conversely, for cases when a drug inhibits the conversion of a precursor (Model 5), the response variable will decrease while the precursor pool increases. This model was applied by Magee et al. to describe the T-cell trafficking effects of prednisolone (Magee et al. 2001) and has the characteristic of a rebound phenomenon, in which the response variable increases beyond the original baseline value, as drug concentrations decrease well below of the IC_{50}, before eventually returning to the baseline condition. The mathematical properties of Models 5 and 6 have been reviewed (Sharma et al. 1998). Some examples of Model 7 include the inhibition of leukocytes by paclitaxel (Minami et al. 1998), susalimod inhibition of experimentally induced tumor necrosis factor-α (Gozzi et al. 1999), parathyroid hormone modulation by a negative allosteric modulator of the calcium-sensing receptor

(Abraham et al. 2011), and the inhibition of iNOS mRNA expression and nitric oxide by methylprednisolone during LPS-induced inflammation, which required two precursor compartments (Sukumaran et al. 2012). Model 8 is mathematically equivalent to two-compartment transduction model and has been used to describe the stimulation of a neopterin precursor (neopterin triphosphate) by interferon-β (Mager et al. 2003a).

Feedback mechanisms are another important component of many adaptive pharmacological systems, and both indirect response and transduction models have been modified to include feedback signals. For example, the transduction model of anticancer drug-induced myelosuppression by Friberg et al. contains a simple feedback expression: $(R(0)/R)^{\gamma}$, which is multiplied by the cellular proliferation in the first transit-compartment, such that as the response variable decreases, the ratio becomes greater than 1 stimulating cell production and thereby countering the anti-neutrophil effect of the drug, and produces a rebound effect as drug is washed out of the system (Friberg et al. 2002). In another example, Wakelkamp et al. utilized a moderator compartment to induce a tolerance signal to offset furosemide effects that were described with indirect response Model 2 (Wakelkamp et al. 1996). This concept was expanded by Gabrielsson and Peletier who described a family of nonlinear feedback models combining indirect response models with a modulator (M) (Gabrielsson and Peletier 2008):

$$\frac{dM}{dt} = k_{\text{tol}} \cdot (R - M)$$

with k_{tol} as a first-order rate constant controlling the production and loss of the moderator signal, which is included on either the production or loss of the response variable in opposition to the drug effect and can accommodate rebound and tolerance phenomena. A more general approach was introduced by Zhang and D'Argenio who coupled traditional indirect response models with methods from classical feedback control of engineered systems (Zhang and D'Argenio 2016). They defined a control input function, $u(t)$, as:

$$u(t) = G_p e(t) + G_i \int_0^t e(\tau) d\tau + G_d \frac{de(t)}{dt}$$

with the components referring to proportional (P), integral (I), and derivative (D) control terms, and the constants G_p, G_i, and G_d are their respective gains, which reflect their overall contribution to $u(t)$. Simulations were provided to show the signature profiles of select models, and the applied examples demonstrate superior model performance for various feedback systems. In addition to improving model-fitting criteria, these feedback models also have implications for the proper estimation and interpretation of EC_{50} values (e.g., a small change in the response variable might still reflect a potent drug if acting upon substantial feedback signals).

Turnover models can also be coupled to describe the modulating effect of specific system components. A classic example is the dynamic interplay of glucose-insulin regulation (Ackerman et al. 1965; Bolie 1960). One turnover equation each is specified for glucose and insulin, with glucose stimulating insulin production and insulin stimulating the removal (or tissue uptake) of plasma glucose. Such a platform can also be extended to evaluate drug effects, such as the terbutaline stimulation of glucose release (Lima et al. 2004) and the stimulation of glucose-dependent insulin release by exendin-4 and GLP-1 (Cao et al. 2012; Mager et al. 2004). Thus, indirect response models are versatile and only a small fraction of possibilities have been described in this section. Additional examples of modified indirect response models include (among others) cell-lifespan concepts (Krzyzanski et al. 1999), disease progression modeling (Post et al. 2005), physiological limits to response variables (Yao et al. 2006), coupling biophase distribution to an indirect response model (Cleton et al. 1999), replacing the $E_{\max}$ or Hill function with the operational model of agonism (Van der Graaf et al. 1999), use of a set-point temperature to a heat production/loss turnover model (Zuideveld et al. 2001), and the application of a distribution pool for the response variable (Krzyzanski and Jusko 2001).

2.3.2 Pharmacodynamic Drug Interactions

Pharmacodynamic drug-drug interactions (DDIs) can manifest when the pharmacological effect of one drug is affected by that of another. Numerous mechanisms can underly such reactions from competitive binding interactions at the same pharmacological target to uncompetitive and noncompetitive binding and other interconnected pathways. Such interactions may be desirable/designed to improve clinical outcomes or mitigate unexpected results or adverse drug effects. PD DDIs are typically classified as additive, synergistic, or antagonistic, albeit these terms are frequently misinterpreted. For example, additivity refers to the case when the effect of a drug combination can be represented as the sum of two independent drug effects, and synergy and antagonism are greater than or less than additive effects. Whereas synergy may be desirable, the combination of two drugs could be antagonistic or additive but still represent greater pharmacological effects than with either agent alone. Various mathematical modeling approaches have been used to characterize PD DDIs, which have been reviewed across the spectrum of in vitro, preclinical, and clinical studies (Fig. 12) (Jonker et al. 2005; Koch et al. 2016; Niu et al. 2019). The most accepted functions to define "no interaction" are Loewe additivity (Loewe and Muischnek 1926) and Bliss independence (Bliss 1939). Both approaches differ in their underlying assumptions and neither method has emerged as clearly superior to the other under various conditions.

In terms of mathematical modeling, Gaddum derived an expression for competitive pharmacodynamic interactions (Gaddum 1937):

$$E_{1+2} = \frac{E_{\max,1} \cdot \frac{C_1}{EC_{50,1}} + E_{\max,2} \cdot \frac{C_2}{EC_{50,2}}}{\frac{C_1}{EC_{50,1}} + \frac{C_2}{EC_{50,2}} + 1}$$

with the subscripts corresponding to Drug 1 or Drug 2, and parameters can be identified from experiments with single agents. Ariëns et al. derived an extended

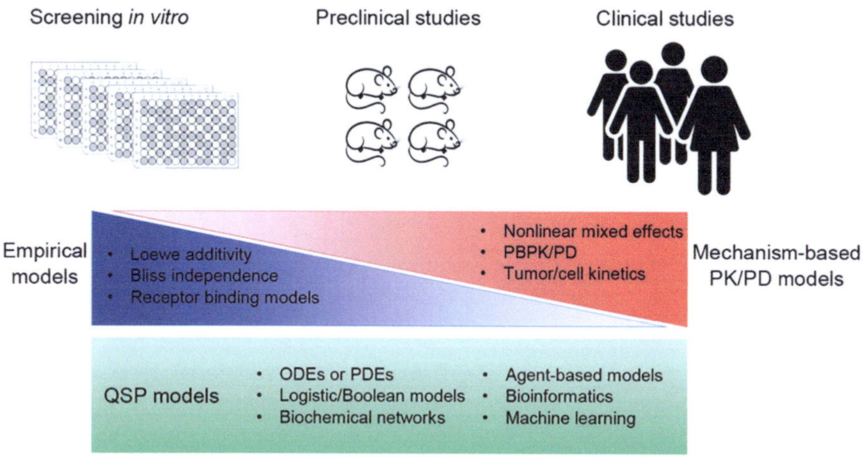

Fig. 12 Array of modeling approaches for analyzing pharmacodynamic drug-drug interactions. (Reprinted from Niu et al. (2019) with permission)

expression for noncompetitive interactions (Ariens et al. 1957). Should such equations not account for the observed response to the drug combination, an empirical interaction parameter (Ψ) can be introduced as an estimated multiplier on one of the drug EC_{50} values. Bliss independence or a non-competitive interaction can be re-expressed as (Chakraborty and Jusko 2002):

$$f_{u,1+2} = \left(1 - \frac{I_{\max,1} \cdot \left(\frac{C_1}{IC_{50,1}}\right)^{\gamma_1}}{\left(\frac{C_1}{IC_{50,1}}\right)^{\gamma_1} + 1}\right) \cdot \left(1 - \frac{I_{\max,2} \cdot \left(\frac{C_2}{IC_{50,2} \cdot \Psi}\right)^{\gamma_2}}{\left(\frac{C_2}{IC_{50,2} \cdot \Psi}\right)^{\gamma_2} + 1}\right)$$

with $f_{u,1+2}$ as the fraction of unaffected cells remaining after combination drug exposure. The estimated interaction parameter (Ψ) can then be used to categorize the nature of the PD DDI as Bliss synergism ($\Psi < 1$), additivity ($\Psi = 1$), or antagonism ($\Psi > 1$). This approach has been further extended to consider the effect of combining drugs in a *sequential* manner relative to *simultaneous* exposure and was used to determine that the sequential exposure of bortezomib for 24 h, followed by vironostat for an additional 24 h, produced synergistic cytotoxic effects (instead of simply additive during simultaneous exposure or the reverse sequence) against the U266 human multiple myeloma cell line (Nanavati and Mager 2017). Competitive and noncompetitive interaction equations have also been integrated into indirect response models (Earp et al. 2004), and interestingly, combinations that simultaneously stimulate production and inhibit loss or inhibit production and stimulate loss result in a natural form of synergy without the pre-specification of an empirical interaction parameter.

An aspirational goal of mechanistic modeling of PD DDIs should be the recapitulation of PD interaction data without the need for empirical interaction parameters.

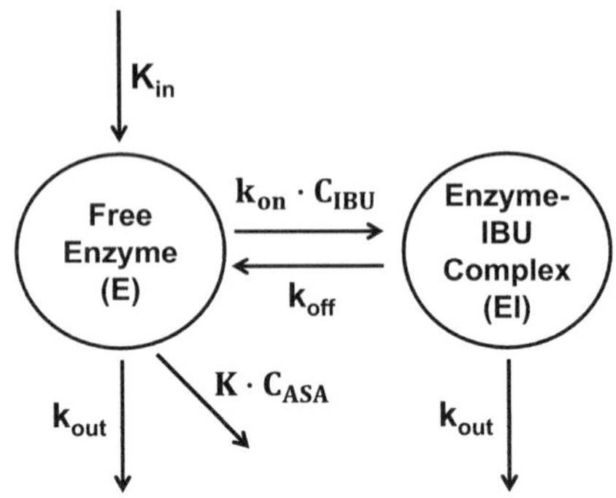

Fig. 13 Mechanism-based pharmacodynamic model of COX-1 inhibition by ibuprofen (IBU) and aspirin (ASA). Parameters for turnover, drug-target binding, and irreversible inactivation are defined in the text for the respective models. (Reproduced from Hong et al. (2008) with permission)

For example, Hong et al. combined a simple binding model with an irreversible inactivation turnover model to explain the blockade of aspirin-mediated inhibition of platelet aggregation by ibuprofen in healthy volunteers (Fig. 13) (Hong et al. 2008). The presumed mechanism was that ibuprofen binds to the COX-1 enzyme in a reversible manner, and owing to its longer half-life relative to aspirin, ibuprofen protects the enzyme from irreversible inactivation by systemic aspirin exposure. The final population model well-described the data without an interaction parameter, and model simulations provided support for only intermittent ibuprofen exposure to preserve the anti-platelet function of aspirin-based prophylaxis. Like the use of physiologically based PK models for PK DDIs, more holistic systems pharmacology models are particularly suited for explaining PD DDIs from first principles, and a good example is the QSP model by Kosinsky et al. to describe the synergy between radiation and anti-PD-L1 checkpoint inhibitor therapy and to optimize dosing regimens for this combination anti-cancer treatment (Kosinsky et al. 2018). Of particular interest is the renewed recognition that independent drug action appears to explain many chemotherapy combination regimens, especially given that it is impossible to assess synergy and other interaction mechanisms in individual patients in routine clinical trials (Palmer and Sorger 2017; Plana et al. 2022). More informative biomarkers are needed to individualize patient-specific QSP models (Hansen and Iyengar 2013; Iyengar et al. 2012), which could be used to identify optimal combination therapies for special subpopulations and individual patients.

2.3.3 Small Mechanistic PK/PD Systems Models

The fundamental and extended PD models in Sects. 2.2 and 2.3 can be combined to develop small systems models of drug action (Jusko 2013, 2016). Depending on the degree of interconnectedness, such models might be developed in a sequential manner or require simultaneous fitting and estimation or calibration of model parameters. The fifth generation model of corticosteroid gene-mediated effects (Ramakrishnan et al. 2002) is a classic example of an advanced PK/PD or small-scale systems model built by combining a collection of fundamental PK/PD sub-models (Fig. 14). On the whole, the model appears complex and connects methylprednisolone exposure (C) in rats after IV injection to the time-course of several molecular markers in rat liver including the glucocorticoid mRNA

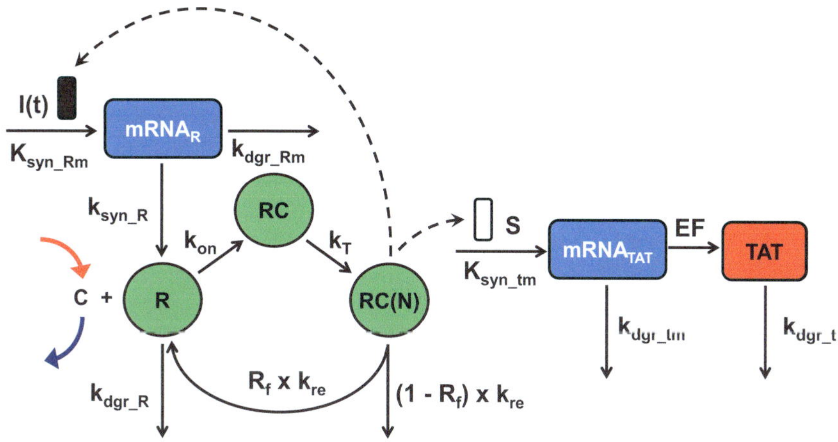

Fig. 14 Model diagram for the 5th generation model of corticosteroid gene-mediated pharmacodynamics. First-order rate constants are represented by k_i, zero-order rate constants are represented by K_i, I(t) is a time-dependent inhibition function, S is a linear stimulation coefficient, EF is a function that relates the expression of the messenger RNA of the target enzyme with enzyme activity, and all state variables are defined in the text. (Adapted from Ramakrishnan et al. (2002))

($mRNA_R$), free glucocorticoid receptor (R), the mRNA for the target of interest (i.e., tyrosine aminotransferase enzyme or $mRNA_{TAT}$), and TAT enzyme activity. However, the mathematical representation of each model element can be traced to PD models introduced in the previous section of this chapter. Methylprednisolone PK was defined by a simple biexponential equation, and the drug-receptor complex (RC) is formed with a *second-order rate constant of association* (k_{on}). A *transit compartment* approach is used to mimic the delay in the translocation of RC into the nucleus, and the receptor can be partially recycled (R_f) or undergo degradation ($1\text{-}R_f$). A *turnover model* defines the formation of free receptor driven by $mRNA_R$. The RC complex in the nucleus or RC(N) is the main driving force for subsequent drug action such as the inhibition of the production of $mRNA_R$, which follows *indirect response model* 1 and serves to capture a form of drug tolerance. Finally, RC (N) stimulates the production of $mRNA_{TAT}$ and subsequent translation into increased TAT activity in a manner consistent with a *precursor-dependent indirect response model*. Thus, the structure and features of basic and advanced PD models can be readily assembled into more detailed mechanistic models to recapitulate additional mediators of drug action and emergent system properties.

Another example of a highly useful small systems model is the integrated glucose-insulin-glucagon model that was developed to assess exposure-response relationships for a novel oral glucokinase activator (Schneck et al. 2013). Drug PK was well described using a simple one-compartment model, and the pharmacological system was represented by an extended version of a coupled glucose-insulin model (Choy et al. 2013; Silber et al. 2010). The final model characterizes the complex interplay of insulin, glucagon, and glucose, featuring glucose-dependent insulin production, insulin-dependent stimulation of glucose uptake (removal from circulation), the inhibitory effect of both glucose and insulin on glucagon production, glucagon stimulation of glucose production and the inhibition of glucagon production in the presence of elevated glucagon, and the effects of exogenous glucose and protein intake with meals. Sufficient mechanistic detail was included such that the contribution of both mechanisms of drug action (i.e., enhanced insulin release and inhibition of hepatic glucose production) could be tested and established through model simulations. The platform model should also be sufficient for designing and testing combination drug regimens for type 2 diabetes mellitus.

Small mechanistic systems models are also feasible for protein therapeutics. An integrated TMDD-C reactive protein (CRP) model was developed for five biological drugs (i.e., anakinra, canakinumab, infliximab, secukinumab, and tocilizumab) for rheumatoid arthritis (Nakada and Mager 2022). The PK of each drug was defined by minimal PBPK models featuring TMDD (Cao and Jusko 2014), and the systems model included the cross-talk among interacting targeted cytokines as well as the stimulating effects of endogenous IL-6 and IL-1β on CRP production represented by a 4-compartment *transduction model*. In addition to characterizing exposure-response data from extensive clinical studies for each drug, simulations of the final model served to identify interesting, unexpected covariate relationships, such as a potential role of IL-1β expression in altering the efficacy of anti-TNF drugs, and suggested that certain combinations of biological agents might be useful for the treatment of rheumatoid arthritis in patients refractory to single biologics. The three case studies described in this section highlight the use of pharmacodynamic first principles, such as PK driving functions, drug target occupancy, physiological turnover processes, and feedback adaptation, to study specific aspects of drug action. The transition toward quantitative systems pharmacology (QSP) is thus a natural extension to the breadth of available mathematical representations of pharmacological effects and homeostatic mechanisms under healthy and diseased conditions.

3 Quantitative Systems Pharmacology Modeling

Over the past decade, there has been a marked increase in the development and application of QSP models; however, many still struggle with a common definition of the field. Whereas empirical relationships and large-scale networks of pharmacological systems are on the opposite ends of the spectrum, the distinction among the terms advanced PK/PD, small systems models, and QSP is often debated. One approach is to simply consider the number of state variables and the identifiability of the model parameters. As reviewed in this chapter, classical PK/PD models often feature a minimal set of readily identifiable parameters for a limited range of variables (e.g., drug concentrations in one or two relevant tissue or fluid spaces and a small number of biomarkers or clinical outcomes). In Sect. 2.3.3, the number of state variables and interacting components for small systems models increases, and often several parameters will be fixed based on reported measured values or estimated values from previous modeling efforts. In contrast, the QSP models and methods described in the rest of this book are characterized by large numbers of state variables and challenges associated with non-unique parameter sets. Arguably, it might be best to focus more on what a model, regardless of size and complexity, was designed to achieve and how it accomplishes the task(s). A good working definition of QSP is, "an approach to translational medicine that combines computational and experimental methods to elucidate, validate, and apply new pharmacological concepts to the development and use of small molecule and biologic drugs" (Sorger et al. 2011). The goals and objectives of QSP are to, "create the knowledge needed to change complex cellular networks in a specified way with mono or combination therapy, alter the pathophysiology of disease so as to maximize therapeutic benefit and minimize toxicity, and implement a 'precision medicine' approach to improving the health of individual patients" (Sorger et al. 2011). Such an ambitious agenda cannot be addressed with traditional PK/PD or small systems models alone but rather with multiscale hybrid formalisms and platforms that are informed, developed, and applied by interdisciplinary teams. The idea is to use all these tools in a strategic manner to accelerate the approval of safe and effective therapies and to identify sources of inter-patient variability in a quantitative manner so as to inform treatment decisions for subpopulations and individual patients.

Similar to Teorell's vision for a physiological basis for drug disposition in the 1930s, *systems concepts* have been a part of biology for over 70 years, as evidenced by the work of pioneers like Ludwig von Bertalanffy (von Bertalanffy 1950). It again would take advancements in computation and analytical techniques before large-scale models of biological systems would become possible such as the seminal work of Yates and Urquhart who developed a complex systems model of adrenocortical regulation (Yates and Urquhart 1962). Qualitative systems models were also developed around that time, such as the finite state Boolean logic model of Gann (Gann et al. 1968), which is a modeling technique often attributed to Kauffman (Kauffman 1969). The field of systems pharmacology is thus rooted in systems physiology and bioengineering (Yates 1973). In the early 2000s, systems biology re-emerged as the advent of sequencing the human genome portended a new era of big data and the need for robust mathematical and computational formalisms in attempts to analyze and interpret large datasets of complex systems (Kitano 2002a, b). Also, there was further recognition that pathophysiological and pharmacological effects do not manifest from linear transduction pathways but rather through their influence on integrated biochemical networks (Bhalla and Iyengar 1999; Weng et al. 1999). In the decade that followed, QSP emerged as a new discipline, integrating systems biology and PK/PD concepts, with substantial promise and implications for the future of academic research into pharmacodynamics, drug development, regulatory sciences, and clinical therapeutics (Sorger et al. 2011; Vicini and van der Graaf 2013). Systems modeling is now having a significant impact on drug discovery, development, and pharmacotherapy (Ayyar and Jusko 2020; Fey et al. 2015; Knight-Schrijver et al. 2016; Milligan et al. 2013; Peterson and Riggs 2015; Putnins et al. 2022), and methods, applications, and challenges of QSP are well addressed in the chapters of this book.

4 Prospectus

This is an exciting time for QSP as new approaches are developed to realize its full potential. One interesting paradigm is the combination of multiple model types (i.e., hybrid models) to leverage their individual advantages and minimize their limitations (Aghamiri et al. 2022; Mentre et al. 2020). Different models can be applied in sequence or simultaneously, thereby allowing one model type to inform another. Systems level modeling (e.g., PBPK and QSP) coupled with biostatistics, machine learning, and other algorithms should help address drug attrition to poor efficacy and safety issues, accelerate the development process, and improve clinical outcomes through precision and individualized drug therapy.

Acknowledgements This chapter was supported, in part, by Grants R01GM57980 and R21CA234775, and the University at Buffalo Center for Protein Therapeutics and the Center of Excellence in Pharmacokinetics and Pharmacodynamics. The author is deeply grateful for the mentorship and friendship of Dr. William Jusko (Distinguished Professor, University at Buffalo, SUNY) who introduced him to the pharmacological modeling concepts described in this chapter. Many of the advancements in PK/PD and QSP can be traced to his pioneering research (Mager and Straubinger 2024). The author also appreciates the extraordinary patience, understanding, and support of the Editors of this book – Drs. Mary Spilker and Erica Bradshaw.

References

Abraham AK, Maurer TS, Kalgutkar AS, Gao X, Li M, Healy DR, Petersen DN, Griffith DA, Mager DE (2011) Pharmacodynamic model of parathyroid hormone modulation by a negative allosteric modulator of the calcium-sensing receptor. AAPS J 13:265–273. https://doi.org/10.1208/s12248-011-9266-9

Ackerman E, Gatewood LC, Rosevear JW, Molnar GD (1965) Model studies of blood-glucose regulation. Bull Math Biophys 27(Suppl):21–37. https://doi.org/10.1007/BF02477259

Aghamiri SS, Amin R, Helikar T (2022) Recent applications of quantitative systems pharmacology and machine learning models across diseases. J Pharmacokinet Pharmacodyn 49:19–37. https://doi.org/10.1007/s10928-021-09790-9

Ait-Oudhia S, Lowe PJ, Mager DE (2012) Bridging clinical outcomes of Canakinumab treatment in patients with rheumatoid arthritis with a population model of IL-1beta kinetics. CPT Pharmacometrics Syst Pharmacol 1:e5. https://doi.org/10.1038/psp.2012.6

An G (2020) Concept of pharmacologic target-mediated drug disposition in large-molecule and small-molecule compounds. J Clin Pharmacol 60:149–163. https://doi.org/10.1002/jcph.1545

Andersen SR, Lambrecht LJ, Swan SK, Cutler DL, Radwanski E, Affrime MB, Garaud JJ (1999) Disposition of recombinant human interleukin-10 in subjects with various degrees of renal function. J Clin Pharmacol 39:1015–1020. https://doi.org/10.1177/00912709922011773

Ariëns EJ (1954) Affinity and intrinsic activity in the theory of competitive inhibition. I. Problems and theory. Arch Int Pharmacodyn Ther 99:32–49

Ariëns EJ (1964) Molecular pharmacology: the mode of action of biologically active compounds. Academic, New York

Ariens EJ, Van Rossum JM, Simonis AM (1957) Affinity, intrinsic activity and drug interactions. Pharmacol Rev 9:218–236

Ayyar VS, Jusko WJ (2020) Transitioning from basic toward systems pharmacodynamic models: lessons from corticosteroids. Pharmacol Rev 72:414–438. https://doi.org/10.1124/pr.119.018101

Ayyar VS, Sukumaran S (2021) Circadian rhythms: influence on physiology, pharmacology, and therapeutic interventions. J Pharmacokinet Pharmacodyn 48:321–338. https://doi.org/10.1007/s10928-021-09751-2

Ayyar VS, Krzyzanski W, Jusko WJ (2019) Indirect pharmacodynamic models for responses with circadian removal. J Pharmacokinet Pharmacodyn 46:89–101. https://doi.org/10.1007/s10928-019-09620-z

Bajaj G, Suryawanshi S, Roy A, Gupta M (2019) Evaluation of covariate effects on pharmacokinetics of monoclonal antibodies in oncology. Br J Clin Pharmacol 85:2045–2058. https://doi.org/10.1111/bcp.13996

Bauer RJ, Dedrick RL, White ML, Murray MJ, Garovoy MR (1999) Population pharmacokinetics and pharmacodynamics of the anti-CD11a antibody hu1124 in human subjects with psoriasis. J Pharmacokinet Biopharm 27:397–420. https://doi.org/10.1023/a:1020917122093

Baxter LT, Zhu H, Mackensen DG, Jain RK (1994) Physiologically based pharmacokinetic model for specific and nonspecific monoclonal antibodies and fragments in normal tissues and human tumor xenografts in nude mice. Cancer Res 54:1517–1528

Beckett AH, Rowland M (1965) Urinary excretion of methylamphetamine in man. Nature 206: 1260–1261. https://doi.org/10.1038/2061260a0

Bender BC, Schaedeli-Stark F, Koch R, Joshi A, Chu YW, Rugo H, Krop IE, Girish S, Friberg LE, Gupta M (2012) A population pharmacokinetic/pharmacodynamic model of thrombocytopenia characterizing the effect of trastuzumab emtansine (T-DM1) on platelet counts in patients with HER2-positive metastatic breast cancer. Cancer Chemother Pharmacol 70:591–601. https://doi.org/10.1007/s00280-012-1934-7

Benet LZ, Galeazzi RL (1979) Noncompartmental determination of the steady-state volume of distribution. J Pharm Sci 68:1071–1074. https://doi.org/10.1002/jps.2600680845

Bhalla US, Iyengar R (1999) Emergent properties of networks of biological signaling pathways. Science 283:381–387

Bischoff KB, Dedrick RL (1968) Thiopental pharmacokinetics. J Pharm Sci 57:1346–1351. https://doi.org/10.1002/jps.2600570814

Bischoff KB, Dedrick RL, Zaharko DS, Longstreth JA (1971) Methotrexate pharmacokinetics. J Pharm Sci 60:1128–1133. https://doi.org/10.1002/jps.2600600803

Bliss CI (1939) The toxicity of poisons applied jointly 1. J Pharm Pharmacol 26:585–615

Bolie VW (1960) Coefficients of normal blood glucose regulation. J Appl Physiol 16:783–788

Bonate PL (2006) Pharmacokinetic-pharmacodynamic modeling and simulation. Springer, New York

Brown RP, Delp MD, Lindstedt SL, Rhomberg LR, Beliles RP (1997) Physiological parameter values for physiologically based pharmacokinetic models. Toxicol Ind Health 13:407–484. https://doi.org/10.1177/074823379701300401

Brunton LL, Knollmann BrC (2023) Goodman & Gilman's the pharmacological basis of therapeutics, 14th edn. McGraw Hill, New York, p 1 online resource

Cao Y, Jusko WJ (2012) Applications of minimal physiologically-based pharmacokinetic models. J Pharmacokinet Pharmacodyn 39:711–723. https://doi.org/10.1007/s10928-012-9280-2

Cao Y, Jusko WJ (2014) Incorporating target-mediated drug disposition in a minimal physiologically-based pharmacokinetic model for monoclonal antibodies. J Pharmacokinet Pharmacodyn 41:375–387. https://doi.org/10.1007/s10928-014-9372-2

Cao Y, Gao W, Jusko WJ (2012) Pharmacokinetic/pharmacodynamic modeling of GLP-1 in healthy rats. Pharm Res 29:1078–1086. https://doi.org/10.1007/s11095-011-0652-x

Cao Y, Balthasar JP, Jusko WJ (2013) Second-generation minimal physiologically-based pharmacokinetic model for monoclonal antibodies. J Pharmacokinet Pharmacodyn 40:597–607. https://doi.org/10.1007/s10928-013-9332-2

Chakraborty A, Jusko WJ (2002) Pharmacodynamic interaction of recombinant human interleukin-10 and prednisolone using in vitro whole blood lymphocyte proliferation. J Pharm Sci 91:1334–1342. https://doi.org/10.1002/jps.3000

Chakraborty A, Krzyzanski W, Jusko WJ (1999) Mathematical modeling of circadian cortisol concentrations using indirect response models: comparison of several methods. J Pharmacokinet Biopharm 27:23–43. https://doi.org/10.1023/a:1020678628317

Chakraborty A, Van LM, Skerjanec A, Floch D, Klein UR, Krammer G, Sunkara G, Howard D (2013) Pharmacokinetic and pharmacodynamic properties of canakinumab in patients with gouty arthritis. J Clin Pharmacol 53:1240–1251. https://doi.org/10.1002/jcph.162

Cheng H, Jusko WJ (1993) Pharmacokinetics of reversible metabolic systems. Biopharm Drug Dispos 14:721–766. https://doi.org/10.1002/bdd.2510140902

Chirmule N, Jawa V, Meibohm B (2012) Immunogenicity to therapeutic proteins: impact on PK/PD and efficacy. AAPS J 14:296–302. https://doi.org/10.1208/s12248-012-9340-y

Choy S, Henin E, van der Walt JS, Kjellsson MC, Karlsson MO (2013) Identification of the primary mechanism of action of an insulin secretagogue from meal test data in healthy volunteers based on an integrated glucose-insulin model. J Pharmacokinet Pharmacodyn 40:1–10. https://doi.org/10.1007/s10928-012-9281-1

Cirincione B, Edwards J, Mager DE (2017) Population pharmacokinetics of an extended-release formulation of exenatide following single- and multiple-dose administration. AAPS J 19:487–496. https://doi.org/10.1208/s12248-016-9975-1

Clark AJ (1926) The reaction between acetyl choline and muscle cells. J Physiol 61:530–546. https://doi.org/10.1113/jphysiol.1926.sp002314

Cleton A, de Greef HJ, Edelbroek PM, Voskuyl RA, Danhof M (1999) Application of a combined "effect compartment/indirect response model" to the central nervous system effects of tiagabine in the rat. J Pharmacokinet Biopharm 27:301–323. https://doi.org/10.1023/a:1020999114109

Clewell HJ 3rd, Andersen ME (1996) Use of physiologically based pharmacokinetic modeling to investigate individual versus population risk. Toxicology 111:315–329. https://doi.org/10.1016/0300-483x(96)03385-9

Covell DG, Barbet J, Holton OD, Black CD, Parker RJ, Weinstein JN (1986) Pharmacokinetics of monoclonal immunoglobulin G1, F(ab′)2, and Fab′ in mice. Cancer Res 46:3969–3978

Cox EH, Kerbusch T, Van der Graaf PH, Danhof M (1998) Pharmacokinetic-pharmacodynamic modeling of the electroencephalogram effect of synthetic opioids in the rat: correlation with the interaction at the mu-opioid receptor. J Pharmacol Exp Ther 284:1095–1103

Cummins CL, Jacobsen W, Benet LZ (2002) Unmasking the dynamic interplay between intestinal P-glycoprotein and CYP3A4. J Pharmacol Exp Ther 300:1036–1045. https://doi.org/10.1124/jpet.300.3.1036

Davies B, Morris T (1993) Physiological parameters in laboratory animals and humans. Pharm Res 10:1093–1095. https://doi.org/10.1023/a:1018943613122

Dayneka NL, Garg V, Jusko WJ (1993) Comparison of four basic models of indirect pharmacodynamic responses. J Pharmacokinet Biopharm 21:457–478. https://doi.org/10.1007/BF01061691

Diao L, Meibohm B (2013) Pharmacokinetics and pharmacokinetic-pharmacodynamic correlations of therapeutic peptides. Clin Pharmacokinet 52:855–868. https://doi.org/10.1007/s40262-013-0079-0

Dingman R, Balu-Iyer SV (2019) Immunogenicity of protein pharmaceuticals. J Pharm Sci 108:1637–1654. https://doi.org/10.1016/j.xphs.2018.12.014

Dua P, Hawkins E, van der Graaf PH (2015) A tutorial on target-mediated drug disposition (TMDD) models. CPT Pharmacometrics Syst Pharmacol 4:324–337. https://doi.org/10.1002/psp4.41

Dutta S, Matsumoto Y, Ebling WF (1996) Is it possible to estimate the parameters of the sigmoid Emax model with truncated data typical of clinical studies? J Pharm Sci 85:232–239. https://doi.org/10.1021/js950067y

Earp J, Krzyzanski W, Chakraborty A, Zamacona MK, Jusko WJ (2004) Assessment of drug interactions relevant to pharmacodynamic indirect response models. J Pharmacokinet Pharmacodyn 31:345–380. https://doi.org/10.1007/s10928-004-8319-4

Ebling WF, Szefler SJ, Jusko WJ (1985) Methylprednisolone disposition in rabbits. Analysis, prodrug conversion, reversible metabolism, and comparison with man. Drug Metab Dispos 13:296–304

Ette EI, Williams PJ (2007) Pharmacometrics: the science of quantitative pharmacology. Wiley, Hoboken

Eudy RJ, Riggs MM, Gastonguay MR (2015) A priori identifiability of target-mediated drug disposition models and approximations. AAPS J 17:1280–1284. https://doi.org/10.1208/s12248-015-9795-8

Fey D, Halasz M, Dreidax D, Kennedy SP, Hastings JF, Rauch N, Munoz AG, Pilkington R, Fischer M, Westermann F, Kolch W, Kholodenko BN, Croucher DR (2015) Signaling pathway models as biomarkers: patient-specific simulations of JNK activity predict the survival of neuroblastoma patients. Sci Signal 8:ra130. https://doi.org/10.1126/scisignal.aab0990

Fick A (1855) On liquid diffusion. London Edinburgh Philos Mag J Sci 10:31–39

Friberg LE, Henningsson A, Maas H, Nguyen L, Karlsson MO (2002) Model of chemotherapy-induced myelosuppression with parameter consistency across drugs. J Clin Oncol 20:4713–4721. https://doi.org/10.1200/JCO.2002.02.140

Gabrielsson J, Peletier LA (2008) A flexible nonlinear feedback system that captures diverse patterns of adaptation and rebound. AAPS J 10:70–83. https://doi.org/10.1208/s12248-008-9007-x

Gabrielsson J, Weiner D (2016) Pharmacokinetic and pharmacodynamic data analysis: concepts and applications, 5th edition, revised and expanded. edn. Apotekarsocieteten, Stockholm

Gaddum JH (1937) The quantitative effects of antagonistic drugs. J Physiol 89:7P–9P

Gann D, Schoeffler JD, Ostrander L (1968) A finite state model for the control of adrenal corticosteroid secretion. In: Mesarović MD (ed) Systems theory and biology. Springer-Verlag, New York, pp 185–200

Garg A, Balthasar JP (2007) Physiologically-based pharmacokinetic (PBPK) model to predict IgG tissue kinetics in wild-type and FcRn-knockout mice. J Pharmacokinet Pharmacodyn 34:687–709. https://doi.org/10.1007/s10928-007-9065-1

Gibaldi M, Perrier D (1982) Pharmacokinetics, 2nd edn. M. Dekker, New York

Gibbs JP, Doshi S, Kuchimanchi M, Grover A, Emery MG, Dodds MG, Gibbs MA, Somaratne R, Wasserman SM, Blom D (2017) Impact of target-mediated elimination on the dose and regimen of Evolocumab, a human monoclonal antibody against proprotein convertase Subtilisin/Kexin Type 9 (PCSK9). J Clin Pharmacol 57:616–626. https://doi.org/10.1002/jcph.840

Gibiansky L, Gibiansky E, Kakkar T, Ma P (2008) Approximations of the target-mediated drug disposition model and identifiability of model parameters. J Pharmacokinet Pharmacodyn 35: 573–591. https://doi.org/10.1007/s10928-008-9102-8

Gillette JR (1971) Factors affecting drug metabolism. Ann N Y Acad Sci 179:43–66. https://doi.org/10.1111/j.1749-6632.1971.tb46890.x

Gisleskog PO, Hermann D, Hammarlund-Udenaes M, Karlsson MO (1998) A model for the turnover of dihydrotestosterone in the presence of the irreversible 5 alpha-reductase inhibitors GI198745 and finasteride. Clin Pharmacol Ther 64:636–647. https://doi.org/10.1016/S0009-9236(98)90054-6

Gozzi P, Pahlman I, Palmer L, Gronberg A, Persson S (1999) Pharmacokinetic-pharmacodynamic modeling of the immunomodulating agent susalimod and experimentally induced tumor necrosis factor-alpha levels in the mouse. J Pharmacol Exp Ther 291:199–203

Graham H, Walker M, Jones O, Yates J, Galetin A, Aarons L (2012) Comparison of in-vivo and in-silico methods used for prediction of tissue: plasma partition coefficients in rat. J Pharm Pharmacol 64:383–396. https://doi.org/10.1111/j.2042-7158.2011.01429.x

Grimstein M, Yang Y, Zhang X, Grillo J, Huang SM, Zineh I, Wang Y (2019) Physiologically based pharmacokinetic modeling in regulatory science: An update from the U.S. Food and Drug Administration's Office of Clinical Pharmacology. J Pharm Sci 108:21–25. https://doi.org/10.1016/j.xphs.2018.10.033

Hansen J, Iyengar R (2013) Computation as the mechanistic bridge between precision medicine and systems therapeutics. Clin Pharmacol Ther 93:117–128. https://doi.org/10.1038/clpt.2012.199

Harker LA, Roskos LK, Marzec UM, Carter RA, Cherry JK, Sundell B, Cheung EN, Terry D, Sheridan W (2000) Effects of megakaryocyte growth and development factor on platelet production, platelet life span, and platelet function in healthy human volunteers. Blood 95: 2514–2522

Hill AV (1910) The possible effects of the aggregation of the molecules of haemoglobin on its dissociation curves. J Physiol 40:iv–vii

Hinderling PH, Schmidlin O, Seydel JK (1984) Quantitative relationships between structure and pharmacokinetics of beta-adrenoceptor blocking agents in man. J Pharmacokinet Biopharm 12: 263–287. https://doi.org/10.1007/BF01061721

Hong Y, Gengo FM, Rainka MM, Bates VE, Mager DE (2008) Population pharmacodynamic modelling of aspirin- and ibuprofen-induced inhibition of platelet aggregation in healthy subjects. Clin Pharmacokinet 47:129–137. https://doi.org/10.2165/00003088-200847020-00006

Houston JB (1994) Utility of in vitro drug metabolism data in predicting in vivo metabolic clearance. Biochem Pharmacol 47:1469–1479. https://doi.org/10.1016/0006-2952(94)90520-7

Huang SM, Abernethy DR, Wang Y, Zhao P, Zineh I (2013) The utility of modeling and simulation in drug development and regulatory review. J Pharm Sci 102:2912–2923. https://doi.org/10.1002/jps.23570

Iyengar R, Zhao S, Chung SW, Mager DE, Gallo JM (2012) Merging systems biology with pharmacodynamics. Sci Transl Med 4:126ps7. https://doi.org/10.1126/scitranslmed.3003563

Jacquez JA, Bellman R, Kalaba R (1960) Some mathematical aspects of chemotherapy-II: the distribution of a drug in the body. Bull Math Biophys 22:309–322

Jamei M (2016) Recent advances in development and application of physiologically-based pharmacokinetic (PBPK) models: a transition from academic curiosity to regulatory acceptance. Curr Pharmacol Rep 2:161–169. https://doi.org/10.1007/s40495-016-0059-9

Jin F, Krzyzanski W (2004) Pharmacokinetic model of target-mediated disposition of thrombopoietin. AAPS PharmSci 6:E9. https://doi.org/10.1208/ps060109

Jonker DM, Visser SA, van der Graaf PH, Voskuyl RA, Danhof M (2005) Towards a mechanism-based analysis of pharmacodynamic drug-drug interactions in vivo. Pharmacol Ther 106:1–18. https://doi.org/10.1016/j.pharmthera.2004.10.014

Jonsson EN, Macintyre F, James I, Krams M, Marshall S (2005) Bridging the pharmacokinetics and pharmacodynamics of UK-279,276 across healthy volunteers and stroke patients using a mechanistically based model for target-mediated disposition. Pharm Res 22:1236–1246. https://doi.org/10.1007/s11095-005-5264-x

Joshi A, Bauer R, Kuebler P, White M, Leddy C, Compton P, Garovoy M, Kwon P, Walicke P, Dedrick R (2006) An overview of the pharmacokinetics and pharmacodynamics of Efalizumab: a monoclonal antibody approved for use in psoriasis. J Clin Pharmacol 46:10–20. https://doi.org/10.1177/0091270005283282

Jusko WJ (1971) Pharmacodynamics of chemotherapeutic effects: dose-time-response relationships for phase-nonspecific agents. J Pharm Sci 60:892–895. https://doi.org/10.1002/jps.2600600618

Jusko WJ (1973) A pharmacodynamic model for cell-cycle-specific chemotherapeutic agents. J Pharmacokinet Biopharm 1:175–200

Jusko WJ (1989) Pharmacokinetics of capacity-limited systems. J Clin Pharmacol 29:488–493. https://doi.org/10.1002/j.1552-4604.1989.tb03369.x

Jusko WJ (1995) Guidelines for collection and analysis of pharmacokinetic data. In: Burton ME, Shaw LM, Schentag JJ, Evans WE (eds) Applied pharmacokinetics & pharmacodynamics: principles of therapeutic drug monitoring, 4th edn. Lippincott Williams & Wilkins, Baltimore

Jusko WJ (2013) Moving from basic toward systems pharmacodynamic models. J Pharm Sci 102: 2930–2940. https://doi.org/10.1002/jps.23590

Jusko WJ (2016) Foundations of pharmacodynamic systems analysis. In: Mager DE, Kimko HC (eds) Systems pharmacology and pharmacodynamics. Springer Nature, Cham, pp 161–176

Jusko WJ, Ko HC (1994) Physiologic indirect response models characterize diverse types of pharmacodynamic effects. Clin Pharmacol Ther 56:406–419. https://doi.org/10.1038/clpt.1994.155

Kagan L (2014) Pharmacokinetic modeling of the subcutaneous absorption of therapeutic proteins. Drug Metab Dispos 42:1890–1905. https://doi.org/10.1124/dmd.114.059121

Kagan L, Mager DE (2013) Mechanisms of subcutaneous absorption of rituximab in rats. Drug Metab Dispos 41:248–255. https://doi.org/10.1124/dmd.112.048496

Katashima M, Yamamoto K, Tokuma Y, Hata T, Sawada Y, Iga T (1998) Comparative pharmacokinetic/pharmacodynamic analysis of proton pump inhibitors omeprazole, lansoprazole and pantoprazole, in humans. Eur J Drug Metab Pharmacokinet 23:19–26. https://doi.org/10.1007/BF03189822

Kauffman SA (1969) Metabolic stability and epigenesis in randomly constructed genetic nets. J Theor Biol 22:437–467

Kenakin T (1997) Pharmacologic analysis of drug-receptor interactions. Lippincott-Raven, Philadelphia

Kitano H (2002a) Computational systems biology. Nature 420:206–210. https://doi.org/10.1038/nature01254

Kitano H (2002b) Systems biology: a brief overview. Science 295:1662–1664. https://doi.org/10.1126/science.1069492

Knight-Schrijver VR, Chelliah V, Cucurull-Sanchez L, Le Novere N (2016) The promises of quantitative systems pharmacology modelling for drug development. Comput Struct Biotechnol J 14:363–370. https://doi.org/10.1016/j.csbj.2016.09.002

Koch G, Schropp J, Jusko WJ (2016) Assessment of non-linear combination effect terms for drug-drug interactions. J Pharmacokinet Pharmacodyn 43:461–479. https://doi.org/10.1007/s10928-016-9490-0

Kosinsky Y, Dovedi SJ, Peskov K, Voronova V, Chu L, Tomkinson H, Al-Huniti N, Stanski DR, Helmlinger G (2018) Radiation and PD-(L)1 treatment combinations: immune response and dose optimization via a predictive systems model. J Immunother Cancer 6:17. https://doi.org/10.1186/s40425-018-0327-9

Krzyzanski W, Jusko WJ (1998) Mathematical formalism and characteristics of four basic models of indirect pharmacodynamic responses for drug infusions. J Pharmacokinet Biopharm 26:385–408. https://doi.org/10.1023/a:1021060000789

Krzyzanski W, Jusko WJ (2001) Indirect pharmacodynamic models for responses with multicompartmental distribution or polyexponential disposition. J Pharmacokinet Pharmacodyn 28:57–78. https://doi.org/10.1023/a:1011517718990

Krzyzanski W, Ramakrishnan R, Jusko WJ (1999) Basic pharmacodynamic models for agents that alter production of natural cells. J Pharmacokinet Biopharm 27:467–489. https://doi.org/10.1023/a:1023249813106

Krzyzanski W, Harrold JM, Wu LS, Perez-Ruixo JJ (2016) A cell-level model of pharmacodynamics-mediated drug disposition. J Pharmacokinet Pharmacodyn 43:513–527. https://doi.org/10.1007/s10928-016-9491-z

Leader B, Baca QJ, Golan DE (2008) Protein therapeutics: a summary and pharmacological classification. Nat Rev Drug Discov 7:21–39. https://doi.org/10.1038/nrd2399

Levy G (1964) Relationship between elimination rate of drugs and rate of decline of their pharmacologic effects. J Pharm Sci 53:342–343. https://doi.org/10.1002/jps.2600530325

Levy G (1994a) Mechanism-based pharmacodynamic modeling. Clin Pharmacol Ther 56:356–358. https://doi.org/10.1038/clpt.1994.149

Levy G (1994b) Pharmacologic target-mediated drug disposition. Clin Pharmacol Ther 56:248–252. https://doi.org/10.1038/clpt.1994.134

Lew KH, Ludwig EA, Milad MA, Donovan K, Middleton E Jr, Ferry JJ, Jusko WJ (1993) Gender-based effects on methylprednisolone pharmacokinetics and pharmacodynamics. Clin Pharmacol Ther 54:402–414. https://doi.org/10.1038/clpt.1993.167

Li H, Mager DE, Sandmaier BM, Storer BE, Boeckh MJ, Bemer MJ, Phillips BR, Risler LJ, McCune JS (2014) Pharmacokinetic and pharmacodynamic analysis of inosine monophosphate dehydrogenase activity in hematopoietic cell transplantation recipients treated with mycophenolate mofetil. Biol Blood Marrow Transplant 20:1121–1129. https://doi.org/10.1016/j.bbmt.2014.03.032

Lima JJ, Matsushima N, Kissoon N, Wang J, Sylvester JE, Jusko WJ (2004) Modeling the metabolic effects of terbutaline in beta2-adrenergic receptor diplotypes. Clin Pharmacol Ther 76:27–37. https://doi.org/10.1016/j.clpt.2004.03.006

Lin JH, Sugiyama Y, Awazu S, Hanano M (1982) In vitro and in vivo evaluation of the tissue-to-blood partition coefficient for physiological pharmacokinetic models. J Pharmacokinet Biopharm 10:637–647. https://doi.org/10.1007/BF01062545

Linton AL, Luke RG, Briggs JD (1967) Methods of forced diuresis and its application in barbiturate poisoning. Lancet 2:377–379. https://doi.org/10.1016/s0140-6736(67)92003-x

Liu S, Li Y, Li Z, Wu S, Harrold JM, Shah DK (2024) Translational two-pore PBPK model to characterize whole-body disposition of different-size endogenous and exogenous proteins. J Pharmacokinet Pharmacodyn 51:449. https://doi.org/10.1007/s10928-024-09922-x

Lobo ED, Balthasar JP (2002) Pharmacodynamic modeling of chemotherapeutic effects: application of a transit compartment model to characterize methotrexate effects in vitro. AAPS PharmSci 4:E42. https://doi.org/10.1208/ps040442

Loewe S, Muischnek H (1926) Effect of combinations: mathematical basis of problem. Arch Exp Pathol Pharmacol 114:313–326

Magee MH, Blum RA, Lates CD, Jusko WJ (2001) Prednisolone pharmacokinetics and pharmacodynamics in relation to sex and race. J Clin Pharmacol 41:1180–1194. https://doi.org/10.1177/00912700122012733

Mager DE (2006) Target-mediated drug disposition and dynamics. Biochem Pharmacol 72:1–10. https://doi.org/10.1016/j.bcp.2005.12.041

Mager DE, Jusko WJ (2001a) General pharmacokinetic model for drugs exhibiting target-mediated drug disposition. J Pharmacokinet Pharmacodyn 28:507–532. https://doi.org/10.1023/a:1014414520282

Mager DE, Jusko WJ (2001b) Pharmacodynamic modeling of time-dependent transduction systems. Clin Pharmacol Ther 70:210–216. https://doi.org/10.1067/mcp.2001.118244

Mager DE, Krzyzanski W (2005) Quasi-equilibrium pharmacokinetic model for drugs exhibiting target-mediated drug disposition. Pharm Res 22:1589–1596. https://doi.org/10.1007/s11095-005-6650-0

Mager DE, Straubinger RM (2024) Contributions of William Jusko to development of pharmacokinetic and pharmacodynamic models and methods. J Pharm Sci 113:2–10. https://doi.org/10.1016/j.xphs.2023.09.019

Mager DE, Neuteboom B, Efthymiopoulos C, Munafo A, Jusko WJ (2003a) Receptor-mediated pharmacokinetics and pharmacodynamics of interferon-beta1a in monkeys. J Pharmacol Exp Ther 306:262–270. https://doi.org/10.1124/jpet.103.049502

Mager DE, Wyska E, Jusko WJ (2003b) Diversity of mechanism-based pharmacodynamic models. Drug Metab Dispos 31:510–518. https://doi.org/10.1124/dmd.31.5.510

Mager DE, Abernethy DR, Egan JM, Elahi D (2004) Exendin-4 pharmacodynamics: insights from the hyperglycemic clamp technique. J Pharmacol Exp Ther 311:830–835. https://doi.org/10.1124/jpet.104.069765

Mahmood I, Green MD (2007) Drug interaction studies of therapeutic proteins or monoclonal antibodies. J Clin Pharmacol 47:1540–1554. https://doi.org/10.1177/0091270007308616

Martinez MN, Amidon GL (2002) A mechanistic approach to understanding the factors affecting drug absorption: a review of fundamentals. J Clin Pharmacol 42:620–643. https://doi.org/10.1177/00970002042006005

McLennan DN, Porter CJ, Charman SA (2005) Subcutaneous drug delivery and the role of the lymphatics. Drug Discov Today Technol 2:89–96. https://doi.org/10.1016/j.ddtec.2005.05.006

Meibohm B, Zhou H (2012) Characterizing the impact of renal impairment on the clinical pharmacology of biologics. J Clin Pharmacol 52:54S–62S. https://doi.org/10.1177/0091270011413894

Mentre F, Friberg LE, Duffull S, French J, Lauffenburger DA, Li L, Mager DE, Sinha V, Sobie E, Zhao P (2020) Pharmacometrics and systems pharmacology 2030. Clin Pharmacol Ther 107: 76–78. https://doi.org/10.1002/cpt.1683

Milligan PA, Brown MJ, Marchant B, Martin SW, van der Graaf PH, Benson N, Nucci G, Nichols DJ, Boyd RA, Mandema JW, Krishnaswami S, Zwillich S, Gruben D, Anziano RJ, Stock TC, Lalonde RL (2013) Model-based drug development: a rational approach to efficiently accelerate drug development. Clin Pharmacol Ther 93:502–514. https://doi.org/10.1038/clpt.2013.54

Minami H, Sasaki Y, Saijo N, Ohtsu T, Fujii H, Igarashi T, Itoh K (1998) Indirect-response model for the time course of leukopenia with anticancer drugs. Clin Pharmacol Ther 64:511–521. https://doi.org/10.1016/S0009-9236(98)90134-5

Minematsu T, Ohtani H, Yamada Y, Sawada Y, Sato H, Iga T (2001) Quantitative relationship between myocardial concentration of tacrolimus and QT prolongation in Guinea pigs: pharmacokinetic/pharmacodynamic model incorporating a site of adverse effect. J Pharmacokinet Pharmacodyn 28:533–554. https://doi.org/10.1023/a:1014460404352

Movin-Osswald G, Hammarlund-Udenaes M (1995) Prolactin release after remoxipride by an integrated pharmacokinetic-pharmacodynamic model with intra- and interindividual aspects. J Pharmacol Exp Ther 274:921–927

Nagashima R, O'Reilly RA, Levy G (1969) Kinetics of pharmacologic effects in man: the anticoagulant action of warfarin. Clin Pharmacol Ther 10:22–35. https://doi.org/10.1002/cpt196910122

Nakada T, Mager DE (2022) Systems model identifies baseline cytokine concentrations as potential predictors of rheumatoid arthritis inflammatory response to biologics. Br J Pharmacol 179: 4063–4077. https://doi.org/10.1111/bph.15845

Nanavati C, Mager DE (2017) Sequential exposure of Bortezomib and Vorinostat is synergistic in multiple myeloma cells. Pharm Res 34:668–679. https://doi.org/10.1007/s11095-017-2095-5

Nestorov I (2003) Whole body pharmacokinetic models. Clin Pharmacokinet 42:883–908. https://doi.org/10.2165/00003088-200342100-00002

Nestorov IA, Aarons LJ, Arundel PA, Rowland M (1998) Lumping of whole-body physiologically based pharmacokinetic models. J Pharmacokinet Biopharm 26:21–46. https://doi.org/10.1023/a:1023272707390

Nieforth KA, Nadeau R, Patel IH, Mould D (1996) Use of an indirect pharmacodynamic stimulation model of MX protein induction to compare in vivo activity of interferon alfa-2a and a polyethylene glycol-modified derivative in healthy subjects. Clin Pharmacol Ther 59:636–646. https://doi.org/10.1016/S0009-9236(96)90003-X

Niu J, Straubinger RM, Mager DE (2019) Pharmacodynamic drug-drug interactions. Clin Pharmacol Ther 105:1395–1406. https://doi.org/10.1002/cpt.1434

Palmer AC, Sorger PK (2017) Combination cancer therapy can confer benefit via patient-to-patient variability without drug additivity or synergy. Cell 171:1678–1691 e13. https://doi.org/10.1016/j.cell.2017.11.009

Pang KS, Rowland M (1977) Hepatic clearance of drugs. I. Theoretical considerations of a "well-stirred" model and a "parallel tube" model. Influence of hepatic blood flow, plasma and blood cell binding, and the hepatocellular enzymatic activity on hepatic drug clearance. J Pharmacokinet Biopharm 5:625–653. https://doi.org/10.1007/BF01059688

Pang KS, Han YR, Noh K, Lee PI, Rowland M (2019) Hepatic clearance concepts and misconceptions: why the well-stirred model is still used even though it is not physiologic reality? Biochem Pharmacol 169:113596. https://doi.org/10.1016/j.bcp.2019.07.025

Peletier LA, Gabrielsson J (2012) Dynamics of target-mediated drug disposition: characteristic profiles and parameter identification. J Pharmacokinet Pharmacodyn 39:429–451. https://doi.org/10.1007/s10928-012-9260-6

Perlstein I, Stepensky D, Krzyzanski W, Hoffman A (2002) A signal transduction pharmacodynamic model of the kinetics of the parasympathomimetic activity of low-dose scopolamine and atropine in rats. J Pharm Sci 91:2500–2510. https://doi.org/10.1002/jps.10243
Peterson MC, Riggs MM (2015) FDA advisory meeting clinical pharmacology review utilizes a quantitative systems pharmacology (QSP) model: a watershed moment? CPT Pharmacometrics Syst Pharmacol 4:e00020. https://doi.org/10.1002/psp4.20
Plana D, Palmer AC, Sorger PK (2022) Independent drug action in combination therapy: implications for precision oncology. Cancer Discov 12:606–624. https://doi.org/10.1158/2159-8290.CD-21-0212
Post TM, Freijer JI, DeJongh J, Danhof M (2005) Disease system analysis: basic disease progression models in degenerative disease. Pharm Res 22:1038–1049. https://doi.org/10.1007/s11095-005-5641-5
Putnins M, Campagne O, Mager DE, Androulakis IP (2022) From data to QSP models: a pipeline for using Boolean networks for hypothesis inference and dynamic model building. J Pharmacokinet Pharmacodyn 49:101–115. https://doi.org/10.1007/s10928-021-09797-2
Radwanski E, Chakraborty A, Van Wart S, Huhn RD, Cutler DL, Affrime MB, Jusko WJ (1998) Pharmacokinetics and leukocyte responses of recombinant human interleukin-10. Pharm Res 15:1895–1901. https://doi.org/10.1023/a:1011918425629
Ramakrishnan R, DuBois DC, Almon RR, Pyszczynski NA, Jusko WJ (2002) Fifth-generation model for corticosteroid pharmacodynamics: application to steady-state receptor down-regulation and enzyme induction patterns during seven-day continuous infusion of methylprednisolone in rats. J Pharmacokinet Pharmacodyn 29:1–24. https://doi.org/10.1023/a:1015765201129
Rane A, Wilkinson GR, Shand DG (1977) Prediction of hepatic extraction ratio from in vitro measurement of intrinsic clearance. J Pharmacol Exp Ther 200:420–424
Renkin EM (1955) Effects of blood flow on diffusion kinetics in isolated, perfused hindlegs of cats; a double circulation hypothesis. Am J Phys 183:125–136. https://doi.org/10.1152/ajplegacy.1955.183.1.125
Richter WF, Jacobsen B (2014) Subcutaneous absorption of biotherapeutics: knowns and unknowns. Drug Metab Dispos 42:1881–1889. https://doi.org/10.1124/dmd.114.059238
Richter WF, Bhansali SG, Morris ME (2012) Mechanistic determinants of biotherapeutics absorption following SC administration. AAPS J 14:559–570. https://doi.org/10.1208/s12248-012-9367-0
Rippe B, Haraldsson B (1994) Transport of macromolecules across microvascular walls: the two-pore theory. Physiol Rev 74:163–219. https://doi.org/10.1152/physrev.1994.74.1.163
Roberts MS, Rowland M (1986) A dispersion model of hepatic elimination: 1. Formulation of the model and bolus considerations. J Pharmacokinet Biopharm 14:227–260. https://doi.org/10.1007/BF01106706
Rostami-Hodjegan A, Toon S (2020) Physiologically based pharmacokinetics as a component of model-informed drug development: where we were, where we are, and where we are heading. J Clin Pharmacol 60(Suppl 1):S12–S16. https://doi.org/10.1002/jcph.1654
Rowland M (1972) Influence of route of administration on drug availability. J Pharm Sci 61:70–74. https://doi.org/10.1002/jps.2600610111
Rowland M, Tozer TN (2010) Clinical pharmacokinetics and pharmacodynamics: concepts and applications, 4th edn. Lippincot Williams & Wilkins, Baltimore
Rowland M, Benet LZ, Graham GG (1973) Clearance concepts in pharmacokinetics. J Pharmacokinet Biopharm 1:123–136. https://doi.org/10.1007/BF01059626
Rowland M, Peck C, Tucker G (2011) Physiologically-based pharmacokinetics in drug development and regulatory science. Annu Rev Pharmacol Toxicol 51:45–73. https://doi.org/10.1146/annurev-pharmtox-010510-100540
Rowland M, Lesko LJ, Rostami-Hodjegan A (2015) Physiologically based pharmacokinetics is impacting drug development and regulatory decision making. CPT Pharmacometrics Syst Pharmacol 4:313–315. https://doi.org/10.1002/psp4.52

Sato H, Sugiyama Y, Sawada Y, Iga T, Hanano M (1987) Physiologically based pharmacokinetics of radioiodinated human beta-endorphin in rats. An application of the capillary membrane-limited model. Drug Metab Dispos 15:540–550

Savic RM, Jonker DM, Kerbusch T, Karlsson MO (2007) Implementation of a transit compartment model for describing drug absorption in pharmacokinetic studies. J Pharmacokinet Pharmacodyn 34:711–726. https://doi.org/10.1007/s10928-007-9066-0

Schneck KB, Zhang X, Bauer R, Karlsson MO, Sinha VP (2013) Assessment of glycemic response to an oral glucokinase activator in a proof of concept study: application of a semi-mechanistic, integrated glucose-insulin-glucagon model. J Pharmacokinet Pharmacodyn 40:67–80. https://doi.org/10.1007/s10928-012-9287-8

Schropp J, Khot A, Shah DK, Koch G (2019) Target-mediated drug disposition model for bispecific antibodies: properties, approximation, and optimal dosing strategy. CPT Pharmacometrics Syst Pharmacol 8:177–187. https://doi.org/10.1002/psp4.12369

Shah DK (2015) Pharmacokinetic and pharmacodynamic considerations for the next generation protein therapeutics. J Pharmacokinet Pharmacodyn 42:553–571. https://doi.org/10.1007/s10928-015-9447-8

Shah DK, Betts AM (2012) Towards a platform PBPK model to characterize the plasma and tissue disposition of monoclonal antibodies in preclinical species and human. J Pharmacokinet Pharmacodyn 39:67–86. https://doi.org/10.1007/s10928-011-9232-2

Shah DK, Betts AM (2013) Antibody biodistribution coefficients: inferring tissue concentrations of monoclonal antibodies based on the plasma concentrations in several preclinical species and human. MAbs 5:297–305. https://doi.org/10.4161/mabs.23684

Shah DK, Li Z, Mager DE (2022) Targeted drug delivery systems and biotechnology products. In: Ducharme MP, Shargel L (eds) Shargel and Yu's applied biopharmaceutics and pharmacokinetics, 8th edn. McGraw Hill, New York, pp 351–376

Sharma A, Jusko WJ (1998) Characteristics of indirect pharmacodynamic models and applications to clinical drug responses. Br J Clin Pharmacol 45:229–239. https://doi.org/10.1046/j.1365-2125.1998.00676.x

Sharma A, Ebling WF, Jusko WJ (1998) Precursor-dependent indirect pharmacodynamic response model for tolerance and rebound phenomena. J Pharm Sci 87:1577–1584. https://doi.org/10.1021/js980171q

Sheiner LB, Stanski DR, Vozeh S, Miller RD, Ham J (1979) Simultaneous modeling of pharmacokinetics and pharmacodynamics: application to d-tubocurarine. Clin Pharmacol Ther 25:358–371. https://doi.org/10.1002/cpt1979253358

Silber HE, Jauslin PM, Frey N, Karlsson MO (2010) An integrated model for the glucose-insulin system. Basic Clin Pharmacol Toxicol 106:189–194. https://doi.org/10.1111/j.1742-7843.2009.00510.x

Simeoni M, Magni P, Cammia C, De Nicolao G, Croci V, Pesenti E, Germani M, Poggesi I, Rocchetti M (2004) Predictive pharmacokinetic-pharmacodynamic modeling of tumor growth kinetics in xenograft models after administration of anticancer agents. Cancer Res 64:1094–1101. https://doi.org/10.1158/0008-5472.can-03-2524

Sorger PK, Allerheiligen SRB, Abernethy DR, Altman RB, Brouwer KLR, Califano A, D'Argenio DZ, Iyenger R, Jusko WJ, Lalonde R, Lauffenburger DA, Shoichet B, Stevens JL, Subramaniam S, Van der Graaf P, Vicini P (2011) Quantitative and systems pharmacology in the postgenomic era: new approaches to discovering and understanding therapeutic drugs and mechanisms. An NIH White Paper from the QSP Workshop Group. http://www.nigms.nih.gov/Training/Documents/SystemsPharmaWPSorger2011.pdf

Staniforth DH, Jackson D, Clarke HL, Horton R (1983) Amoxycillin/clavulanic acid: the effect of probenecid. J Antimicrob Chemother 12:273–275. https://doi.org/10.1093/jac/12.3.273

Stephenson RP (1956) A modification of receptor theory. Br J Pharmacol Chemother 11:379–393. https://doi.org/10.1111/j.1476-5381.1956.tb00006.x

Sugiyama Y, Aoki Y (2023) A 20-year research overview: quantitative prediction of hepatic clearance using the in vitro-in vivo extrapolation approach based on physiologically based pharmacokinetic modeling and extended clearance concept. Drug Metab Dispos 51:1067–1076. https://doi.org/10.1124/dmd.123.001344

Sukumaran S, Lepist EI, DuBois DC, Almon RR, Jusko WJ (2012) Pharmacokinetic/pharmacodynamic modeling of methylprednisolone effects on iNOS mRNA expression and nitric oxide during LPS-induced inflammation in rats. Pharm Res 29:2060–2069. https://doi.org/10.1007/s11095-012-0733-5

Supersaxo A, Hein WR, Steffen H (1990) Effect of molecular weight on the lymphatic absorption of water-soluble compounds following subcutaneous administration. Pharm Res 7:167–169. https://doi.org/10.1023/a:1015880819328

Teorell T (1937) Kinetics of distribution of substances administered to the body I. The extravascular modes of administration. Arch Int Pharmacodyn 57:205–225

Tucker GT (1981) Measurement of the renal clearance of drugs. Br J Clin Pharmacol 12:761–770. https://doi.org/10.1111/j.1365-2125.1981.tb01304.x

Van Der Graaf PH, Van Schaick EA, Math-ot RA, Ijzerman AP, Danhof M (1997) Mechanism-based pharmacokinetic-pharmacodynamic modeling of the effects of N6-cyclopentyladenosine analogs on heart rate in rat: estimation of in vivo operational affinity and efficacy at adenosine A1 receptors. J Pharmacol Exp Ther 283:809–816

Van der Graaf PH, Van Schaick EA, Visser SA, De Greef HJ, Ijzerman AP, Danhof M (1999) Mechanism-based pharmacokinetic-pharmacodynamic modeling of antilipolytic effects of adenosine A(1) receptor agonists in rats: prediction of tissue-dependent efficacy in vivo. J Pharmacol Exp Ther 290:702–709

Vicini P, van der Graaf PH (2013) Systems pharmacology for drug discovery and development: paradigm shift or flash in the pan? Clin Pharmacol Ther 93:379–381. https://doi.org/10.1038/clpt.2013.40

Visser SA, Wolters FL, Gubbens-Stibbe JM, Tukker E, Van Der Graaf PH, Peletier LA, Danhof M (2003) Mechanism-based pharmacokinetic/pharmacodynamic modeling of the electroencephalogram effects of GABAA receptor modulators: in vitro-in vivo correlations. J Pharmacol Exp Ther 304:88–101. https://doi.org/10.1124/jpet.102.042341

von Bertalanffy L (1950) An outline of general system theory. Br J Phil Sci 1:134–165

Wagner JG (1968) Kinetics of pharmacologic response. I. Proposed relationships between response and drug concentration in the intact animal and man. J Theor Biol 20:173–201. https://doi.org/10.1016/0022-5193(68)90188-4

Wagner JG (1971) Biopharmaceutics and relevant pharmacokinetics, 1st edn. Drug Intelligence Publications, Hamilton

Wagner JG, Northam JI, Alway CD, Carpenter OS (1965) Blood levels of drug at the equilibrium state after multiple dosing. Nature 207:1301–1302. https://doi.org/10.1038/2071301a0

Wakelkamp M, Alvan G, Gabrielsson J, Paintaud G (1996) Pharmacodynamic modeling of furosemide tolerance after multiple intravenous administration. Clin Pharmacol Ther 60:75–88. https://doi.org/10.1016/S0009-9236(96)90170-8

Wang W, Wang EQ, Balthasar JP (2008) Monoclonal antibody pharmacokinetics and pharmacodynamics. Clin Pharmacol Ther 84:548–558. https://doi.org/10.1038/clpt.2008.170

Weng G, Bhalla US, Iyengar R (1999) Complexity in biological signaling systems. Science 284:92–96

Woo S, Krzyzanski W, Jusko WJ (2006) Pharmacokinetic and pharmacodynamic modeling of recombinant human erythropoietin after intravenous and subcutaneous administration in rats. J Pharmacol Exp Ther 319:1297–1306. https://doi.org/10.1124/jpet.106.111377

Xu L, Eiseman JL, Egorin MJ, D'Argenio DZ (2003) Physiologically-based pharmacokinetics and molecular pharmacodynamics of 17-(allylamino)-17-demethoxygeldanamycin and its active metabolite in tumor-bearing mice. J Pharmacokinet Pharmacodyn 30:185–219. https://doi.org/10.1023/a:1025542026488

Yamamoto K, Abe M, Katashima Y, Sawada Y, Iga T (1996) Pharmacodynamic analysis of antiplatelet effect of aspirin in the literature – modeling based on inhibition of cyclooxygenase in the platelet and the vessel wall endothelium. Jpn J Hosp Pharm 22:133–141

Yang J, Mager DE, Straubinger RM (2010) Comparison of two pharmacodynamic transduction models for the analysis of tumor therapeutic responses in model systems. AAPS J 12:1–10. https://doi.org/10.1208/s12248-009-9155-7

Yao Z, Krzyzanski W, Jusko WJ (2006) Assessment of basic indirect pharmacodynamic response models with physiological limits. J Pharmacokinet Pharmacodyn 33:167–193. https://doi.org/10.1007/s10928-006-9003-7

Yates FE (1973) Systems biology as a concept. In: Brown JHU, Gann DS (eds) Engineering principles in physiology. Academic, New York, pp 141–184

Yates FE, Urquhart J (1962) Control of plasma concentrations of adrenocortical hormones. Physiol Rev 42:359–433

Yoshida K, Budha N, Jin JY (2017) Impact of physiologically based pharmacokinetic models on regulatory reviews and product labels: frequent utilization in the field of oncology. Clin Pharmacol Ther 101:597–602. https://doi.org/10.1002/cpt.622

Zhang Y, D'Argenio DZ (2016) Feedback control indirect response models. J Pharmacokinet Pharmacodyn 43:343–358

Zhi JG, Nightingale CH, Quintiliani R (1988) Microbial pharmacodynamics of piperacillin in neutropenic mice of systematic infection due to Pseudomonas aeruginosa. J Pharmacokinet Biopharm 16:355–375. https://doi.org/10.1007/BF01062551

Zuideveld KP, Maas HJ, Treijtel N, Hulshof J, van der Graaf PH, Peletier LA, Danhof M (2001) A set-point model with oscillatory behavior predicts the time course of 8-OH-DPAT-induced hypothermia. Am J Physiol Regul Integr Comp Physiol 281:R2059–R2071. https://doi.org/10.1152/ajpregu.2001.281.6.R2059

A Framework for Quantitative Systems Pharmacology Model Execution

Victor Sokolov, Kirill Peskov, and Gabriel Helmlinger

Contents

V. Sokolov (✉)
M&S Decisions FZ LLC, Dubai, UAE

Marchuk Institute of Numerical Mathematics of Russian Academy of Sciences, Moscow, Russia
e-mail: victor.sokolov@msdecisions.tech

K. Peskov
M&S Decisions FZ LLC, Dubai, UAE

Marchuk Institute of Numerical Mathematics of Russian Academy of Sciences, Moscow, Russia

Research Center of Model-Informed Drug Development, Sechenov First Moscow State Medical University, Moscow, Russia

G. Helmlinger
Dyne Therapeutics, Lexington, MA, USA

E. L. Bradshaw, M. E. Spilker (eds.), *Quantitative Systems Pharmacology*,
Handbook of Experimental Pharmacology 289, https://doi.org/10.1007/164_2024_738

Abstract

A mathematical model can be defined as a theoretical approximation of an observed pattern. The specific form of the model and the associated mathematical methods are typically dictated by the question(s) to be addressed by the model and the underlying data. In the context of research and development of new medicines, these questions often focus on the dose-exposure-response relationship.

The general workflow for model development and application can be delineated in three major elements: defining the model, qualifying the model, and performing simulations. These elements may vary significantly depending on modeling objectives. Quantitative systems pharmacology (QSP) models address the formidable challenge of quantitatively and mechanistically characterizing human and animal biology, pathophysiology, and therapeutic intervention.

QSP model development, by necessity, relies heavily on preexisting knowledge, requires a comprehensive understanding of current physiological concepts, and often makes use of heterogeneous and aggregated datasets from multiple sources. This reliance on diverse datasets presents an upfront challenge: the determination of an optimal model structure while balancing model complexity and uncertainty. Additionally, QSP model calibration is arduous due to data scarcity (particularly at the human subject level), which necessitates the use of a variety of parameter estimation approaches and sensitivity analyses, earlier in the modeling workflow as compared to, for example, population modeling. Finally, the interpretation of model-based predictions must be thoughtfully aligned with the data and the mathematical methods applied during model development.

The purpose of this chapter is to provide readers with a high-level yet comprehensive overview of a QSP modeling workflow, with an emphasis on the various challenges encountered in this process. The workflow is centered around the construction of ordinary differential equation models and may be extended beyond this framework. It includes the fundamentals of systematic literature reviews, the selection of appropriate structural model equations, the analysis of system behavior, model qualification, and the application of various types of model-based simulations. The chapter concludes with details on existing software options suitable for implementing the described methodologies.

This workflow may serve as a valuable resource to both newcomers and experienced QSP modelers, offering an introduction to the field as well as operating procedures and references for routine analyses.

Keywords

Model development · Model validation · Model-based simulations · Modeling workflow · Quantitative systems pharmacology

Abbreviations

ABM	Agent-based modeling
BFGS	Broyden-Fletcher-Goldfarb-Shanno
BSV	Between-subject variability
CI	Confidence interval
COU	Context of use
DDE	Delay differential equation
eFAST	Extended Fourier amplitude sensitivity test
ELS	Extended least-squares
FAST	Fourier amplitude sensitivity test
FIM	Fisher information matrix
FO	First order
FOCE	First-order conditional estimation
LSI	Local sensitivity index
MIDD	Model-informed drug development
NLME	Non-linear mixed effects modeling
NPDE	Normalized prediction distribution error
ODE	Ordinary differential equation
OF	Objective function
OLS	Ordinary least-squares
PBPK	Physiologically based pharmacokinetic modeling
PD	Pharmacodynamics
PDE	Partial differential equation
PI	Prediction interval
PK	Pharmacokinetics
PRCC	Partial rank correlation coefficients
PRISMA	Preferred reporting items for systematic reviews and meta-analyses
QSP	Quantitative systems pharmacology
RSE	Relative standard error
SAEM	Stochastic approximation expectation maximization
SBML	Systems biology markup language
SD	Standard deviation
SDE	Stochastic differential equation
SE	Standard error
VPC	Visual predictive check
VPP	Virtual patient population
WLS	Weighted least-squares
-2LL	-2 log-likelihood

1 Introduction

A mathematical model is an integrative construct of select aspects of reality, or rather our understanding of these, which can be used to evaluate numerous "what-if" scenarios for a subject of interest. In other words, a model is an approximation of reality, with its selected components being determined by the scenarios we want to investigate, the data at hand, and limits in our knowledge. In model-informed drug development (MIDD), dose-to-response relationships may be considered as one, if not the primary subject of interest (Sheiner 1997; Peck et al. 1994). The most straightforward approach to modeling such relationships is to establish a link between drug dose and a measure of response with an empirical model (e.g., a linear, Emax, logistic, or other model; see Chapter 2: Breadth of Pharmacology Modeling: Fundamentals of Pharmacokinetics, Pharmacodynamics and Mechanistic Modeling for examples) (Fahrmeir et al. 2021). While such models are relatively easy to understand and develop, the scope of their extrapolation is inevitably constrained by their simplicity. In this regard, they may be able to interpolate around the dataset they were derived from, but may not be able to extrapolate well beyond the dataset and thus may be unable to address questions such as the following: What if the dose were administered at a different time? Why do some patients respond better to a treatment than others? Is there a sufficient amount of drug at the site of action? What if the drug of interest was given in combination with another drug?

To address such increasingly elaborate questions, models need to be augmented with additional descriptors and mechanistic insights to capture biophysical and biochemical elements that shape the relationships between the drug dose, drug exposure, target engagement, biomarker response, and disease endpoint response, which may encompass temporal and spatial components, signaling pathways, molecular and cellular interactions, stochastic vs. deterministic behaviors, etc. Mathematical tools are leveraged to enable development of mechanistically oriented, systems-level models such as quantitative systems pharmacology (QSP) models. QSP modeling is particularly suited to answer drug development questions when the questions can be addressed by probing, mechanistically and/or quantitatively, deeper into the physiologic and pharmacologic system that underlies the over-arching dose-response relationship. There is an array of mathematical methods that can be employed in QSP modeling. For example, spatial components may be introduced using partial differential equations (PDEs), as in computational fluid dynamic models based on Navier-Stokes equations or in on- and off-lattice agent-based modeling (ABM), or via mass-balance using a set of ordinary differential equations (ODEs) (Wendt et al. 2008; Moise and Friedman 2021; Helbing 2012; Jones and Sleeman 2003). Time may be represented as an implicit or explicit component of a mathematical system (e.g., delay differential equations (DDEs)) (Rihan 2021). Boolean descriptors are often used to capture complex networks of interactions among various biological entities, and probabilistic approaches may be included to account for the stochastic behavior within a system (e.g., stochastic differential equations (SDEs)) (Wang et al. 2012; Allen 2011). Some modeling researchers

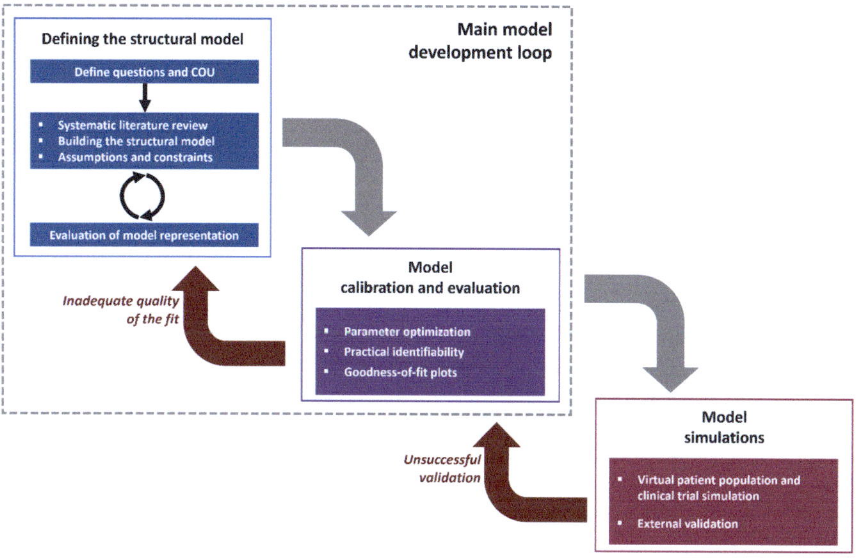

Fig. 1 Schematic representation of the QSP modeling workflow. (Main model development loop consists of defining the relevant questions, context of use (COU), systematic literature review, and compilation of a structural model which is then calibrated against the observed data. If the quality of the data fit is adequate based on multiple criteria outlined in the chapter, a model can be considered suitable for simulations. Further predictive ability of a model is tested by external validation. If a model fails to describe the data at any stage, revisions of the model structure and assumptions are warranted. Sensitivity analysis can be performed at any stage of model development (see Sect. 4.2))

have invoked a combinations of techniques such as hybrid ABM/ODE or ABM/PDE models (Gong et al. 2021; Marzban et al. 2021).

The wealth and diversity of mathematical tools and associated model constructs combined with challenging timelines for model development and practical uses in the industry can make QSP model assessment, calibration, and acceptance across stakeholders difficult without a systematic, best-practice modeling workflow. Thus, an "umbrella" workflow that provides a framework for the development, evaluation, and validation of QSP modeling and simulation is needed. The workflow proposed in this chapter (Fig. 1) is inspired by multiple previous and current endeavors in the field, as well as model verification and validation standards developed by the medical devices industry (Musuamba et al. 2021; Cucurull-Sanchez et al. 2019; Gadkar et al. 2016). It aims to (1) streamline model assessments, (2) define and document model assumptions, reproducibility, and limitations, and (3) aid in communication of model-based simulations within a context of use (COU) (Musuamba et al. 2021; Cucurull-Sanchez et al. 2019; Helmlinger et al. 2019; The American Society of Mechanical Engineers (ASME) n.d.). In this chapter, we opted to consider QSP modeling in its most common form as a set of ODEs (Nijsen et al. 2018) and, therefore, many of the technical methods are built upon established ODE modeling best practices. QSP-specific considerations will be called out where needed. Most of the workflow concepts and related implementation techniques described below are applicable to all QSP models, regardless of the underlying mathematical tools selected (and potential combination thereof). Furthermore, the proposed modeling workflow should be considered as a framework which aims at providing a comprehensive array of tools for QSP model development clustered by core functionalities and with options (depending on the problem at hand, the data, time and resource constraints, unknowns), rather than a rigid set of linear instructions to be followed stepwise.

The implementation and execution of the modeling workflow consist of three sub-modules which map to the next three sections in this chapter (Fig. 1). The first sub-module is dedicated to the principles underlying the choice of a model structure, both formally and heuristically, as well as the methods used to organize the search for and collection of the associated data. Parameter estimation and model calibration (i.e., in engineering terms, solving the inverse problem) are the main subjects of the second sub-module. The third sub-module provides a taxonomy of model simulations and associated inference, sensitivity analyses, and virtual populations. Finally, at the end of the chapter, a brief overview of available software tools for QSP modeling is provided.

2 Defining the Structural Model

Mathematical modeling of the biology, physiology, and pharmacology of living organisms poses two major challenges. Firstly, physiology – and human physiology in particular – is complex and a mathematical model may only reflect a part of the full system. Secondly, by modeling a system, a modeler needs to make assumptions of causality, to link elements to one another. Both endeavors require an in depth understanding of the current knowledge in the relevant field via qualitative information and quantitative data.

The first steps in the model building process are to clearly define the relevant questions and the context of use (COU). COU is the full, clear, and concise description of the way the methodology is to be used and the related purpose of the use. Two COU examples are the "optimization of the treatment regimen for first-in-human studies using a QSP model (developed based on in vitro, ex vivo, and/or in vivo animal data)" or "use of a QSP model to establish the mechanism behind unexpected treatment effect observed in post-marketing surveillance" (Bonate 2014). As such, COU provides an essential framework for any subsequent evaluation of the "usefulness" of a model and, together with relevant questions, defines the scope the modeling project.

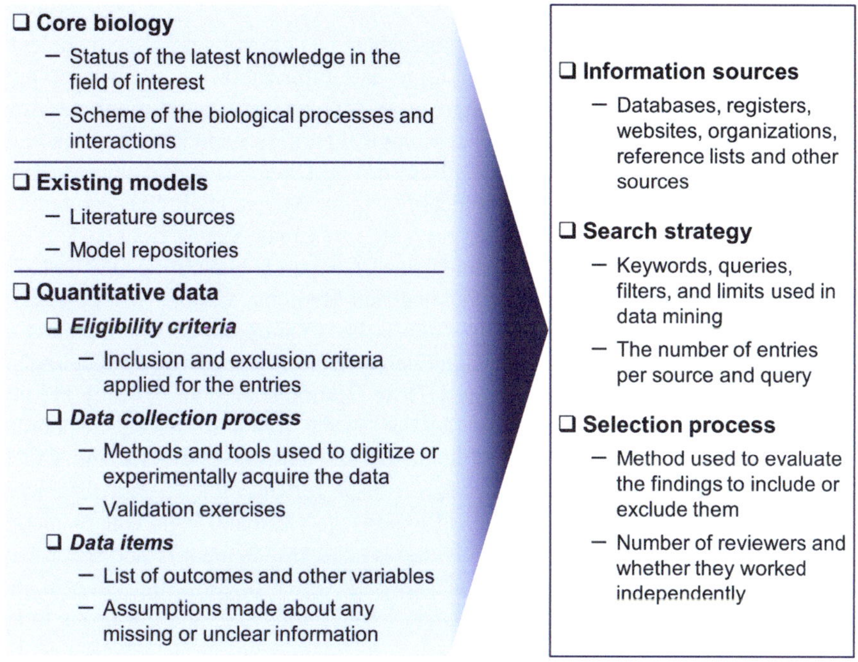

Fig. 2 Proposed QSP data mining process

2.1 Systematic Literature Review as a Basis for Model Development

A systematic review of the scientific literature is an essential part of QSP model development. Furthermore, transparent documentation of the model, including details on expert knowledge, experimental conditions, and data used for model development, has been identified by regulatory authorities as a major factor for robust model assessment (Bai et al. 2021). In this aspect, QSP resembles another quantitative modeling methodology, namely meta-analyses (Chan et al. 2022). Performing inferences based on meta-analyses is considered to be the pinnacle of evidence-based medicine (Murad et al. 2016) and, not surprisingly, such analyses are expected to follow the high-standard Preferred Reporting Items for Systematic Reviews and Meta-Analyses (PRISMA) guideline. Initially published in 2009, the PRISMA statement has been updated in 2020 and covers 27 items necessary to perform transparent and robust literature reviews and corresponding meta-analyses (Page et al. 2021). Inspired by this initiative, we propose to adapt this high standard for implementing QSP modeling workflows.

The QSP data mining process can be divided into three parts: core biology, existing models, and quantitative data (Fig. 2) with additional comments on documentation of the structural model and data mining process.

Core Biology

Since a QSP model captures a slice of knowledge in a field of interest, the modeler is expected to derive comprehensive and up-to-date information on relevant biology from scientifically reliable sources. Usually, this includes reading through multiple papers and textbooks to capture relevant biological processes and interactions within a connectivity diagram (e.g., using MS PowerPoint, Diagrams software https://app.diagrams.net/, Systems Biology Graphical Notation (Novère et al. 2009) compliant CellDesigner https://www.celldesigner.org/, etc.), as can be seen in Chapter 1: What Is QSP and Why Does It Exist?: A Brief History, Chapter 5: Integrating QSP and ML to Facilitate Drug Development and Personalized Medicine, Chapter 7: Application of Quantitative Systems Pharmacology Approaches to Support Pediatric Labeling in Rare Diseases, and Chapter 8: Quantitative Systems Pharmacology Modeling in Immuno-Oncology: Hypothesis Testing, Dose Optimization, and Efficacy Prediction. Initially the scheme may be as detailed as possible and represent elaborate pathway reconstructions within the scope of the relevant questions and COU. Furthermore, such a scheme can become a basis for discussions with the broad team of specialists, including biological experts, to facilitate alignment of model structure and relevant components. This step is necessary to identify and accurately capture the state-of-the-art information and data which underlie the basis of the model structure and may support necessary assumptions and simplifications to be assessed or verified by a potential reviewer at a later stage of model development.

Existing Models

Another feature of QSP models relates to their constant evolution and, hence, their "transportability, potential for improvements, hereditary" character. Today, multiple systems-level QSP models exist for a range of diseases and therapeutic areas. Some of these have been developed over a decade or more, while improving and being augmented and adapted for specific COU (see Chapter 6: Quantitative Systems Pharmacology Development and Application in Neuroscience, Chapter 7: Application of Quantitative Systems Pharmacology Approaches to Support Pediatric Labeling in Rare Diseases, Chapter 8: Quantitative Systems Pharmacology Modeling in Immuno-Oncology: Hypothesis Testing, Dose Optimization, and Efficacy Prediction, Chapter 9: Application of Mechanistic Mathematical Modeling to Toxicology: Quantitative Systems Toxicology (QST) and Chapter 10: The Role of Cross-Institutional and Interdisciplinary Collaboration in Defining and Executing a QSP Strategy for examples). Thus, prior to initiating the tedious and time-consuming process of de novo QSP model development, it is recommended to perform a systematic search for existing mathematical models in diseases and therapeutic areas of interest, across various sources (e.g., PubMed, DDMoRe http://repository.ddmore.eu/, bioRxiv https://www.biorxiv.org/search, BioModels https://www.ebi.ac.uk/biomodels/). Such existing models, combined with expert knowledge gleaned from the data mining process described above, will enable the modeler to evaluate and potentially use these models as a basis for further model development or for direct adoption (with minimal adjustments) toward a particular project with a similar COU.

Quantitative Data

Finally, once a sufficient understanding of the physiological system at hand and of preexisting modeling solutions has been achieved, quantitative data is needed to appropriately inform the development of the new or augmented QSP model. These data usually emanate from multiple sources and may include privately held data (e.g., by a pharmaceutical company) as well as publicly available data. A QSP model is well suited to absorb different types and levels of information, from qualitative facts and observations to aggregated- and subject-level data from various studies and sources. A bulk of these data are longitudinal measurements of pharmacokinetics (PK) and pharmacodynamic (PD) markers, from various preclinical and clinical studies, often with multiple arms or cohorts treated with placebo, standard of care, or novel therapeutics. In addition to time series data, QSP model development usually requires a variety of information associated with the state of the system or modeled population, e.g., baseline measurements of biological entities corresponding to the dependent variables within the model and their distributions, as well as experimentally quantifiable kinetics, such as half-lives or degradation rates of cells and molecules and binding constants, typically measurable via in vitro assays and possibly in vivo studies. This topic is addressed further in Chapter 4: Role of Data in Development and Application of Quantitative Systems Pharmacology Models, with examples of data types used in QSP models in Chapter 6: Quantitative Systems Pharmacology Development and Application in Neuroscience, Chapter 7: Application of Quantitative Systems Pharmacology Approaches to Support Pediatric Labeling in Rare Diseases, Chapter 8: Quantitative Systems Pharmacology Modeling in Immuno-Oncology: Hypothesis Testing, Dose Optimization, and Efficacy Prediction and Chapter 9: Application of Mechanistic Mathematical Modeling to Toxicology: Quantitative Systems Toxicology (QST).

Quantitative data should be annotated with eligibility criteria, a data collection process, and data items: (i) eligibility criteria are the inclusion and exclusion criteria applied for data entry selection; (ii) the data collection process uses methods and tools to digitize the data (e.g., PlotDigitizer https://plotdigitizer.com/), followed by a validation exercise (e.g., visual comparison of the original vs. digitized plots); (iii) data items define which outcomes and other variables (e.g., study demographics and intervention characteristics) have been collected and which assumptions have been made to address factors such as missing datapoints or ambiguous information. Data may thus be captured into a database such as an MS Excel file structured with three layers of information (study level, cohorts, outcomes), as depicted in Fig. 3. It is also useful to store relevant entries captured through the data mining process in bibliographic software (e.g., Zotero https://www.zotero.org/, Mendeley https://www.mendeley.com/).

The importance of a thorough data mining process should not be underestimated, both timewise and in terms of its future impact on model quality, applicability, and reproducibility. For a dedicated team of two modeling scientists with prior experience in the therapeutic area, the average time commitment for the data mining required in a particular project typically takes 2–3 months. Systematic reporting

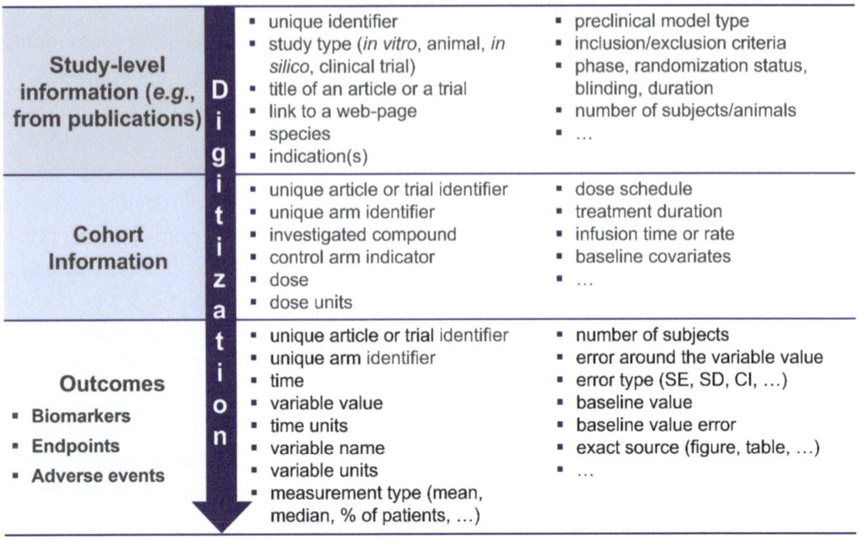

Fig. 3 Overall structure of the database

following a literature search procedure has not yet reached the status of common practices in QSP modeling, and choices related to the database structure as well as the comprehensiveness of the associated documentation often depend upon project needs and timelines. It should be remembered, however, that while onerous, this initial stage in the implementation of the QSP modeling workflow sets the foundation for all subsequent stages and is the primary determinant in defining the QSP model structure.

Documentation

Careful documentation should occur at each step of the model building and data mining process. As mentioned earlier, model building documentation should include assumptions related to which biological features were and were not included in the model and a rationale for the decisions with a similar commentary on the assessment of existing models. The data mining process documentation should include information sources, search strategy, and selection processes. Information sources capture the databases, registers, websites, organizations, reference lists, and other sources used to collect the information for model development (e.g., PubMed, Google Scholar, Web of Science, ClinicalTrials.gov, and Trialtrove), as well as the date of the last search within each source. The search strategy explicitly states keywords, queries, and filters used in data mining, as well as the number of entries *per* source and query. The selection process outlines the method(s) used to evaluate the findings to include or exclude select references or datasets, as well as the number of reviewers who will have worked independently to perform such selections. As the model progresses in its development and calibration, documentation of parameter value ranges, units, and any necessary unit conversions is critical as well as documenting the testing strategy and performance against calibration data while capturing subsequent model adjustments as needed.

2.2 Building the Structural Model: Use of Mathematics to Depict Biology

At its core, a QSP model can be defined as a mathematical system that integrates pharmacology and biology to predict response(s) at molecular, cellular, physiological, and/or outcomes levels (Bradshaw et al. 2019; Helmlinger et al. 2017). The structure of such a model is, in itself, a hypothesis, and it is challenging to specify typical forms of QSP models, in contrast to well-established PK and PK/PD models (one compartment, two compartment, direct response, indirect response; see Chapter 2: Breadth of Pharmacology Modeling: Fundamentals of Pharmacokinetics, Pharmacodynamics and Mechanistic Modeling for more details). Nevertheless, it is possible to define a few common principles and building blocks used to translate biological processes into sets of mathematical equations.

First, a system needs to include temporal and spatial components. Both can be taken into account by using ODEs. Based on the law of mass action, ODEs can be arranged into a multi-compartmental system mimicking the physiological distribution of biomarkers and the exogenous compounds or drugs as well as mechanistic features such as receptor density and protein turnover. Initial conditions of the state variables and the right-hand sides of the equations need then to be defined.

Below are a few mathematical representations of biological processes that are frequently encountered. In the simplest case, the right-hand side can be represented by a zero- or first-order rate or flux, as in the following turnover equation (Dayneka et al. 1993):

$$\frac{dX}{dt} = k_{\text{in}} - k_{\text{out}} * X(t), \tag{1}$$

where k_{in} and k_{out} are zero-order and first-order rate constants; t represents time; and $X(t)$ is a state variable. Biologically, this mathematical construct can be used to represent the temporal turnover of a protein ($X(t)$), where k_{out} is the first-order degradation rate of the protein and k_{in} is the zero-order production rate of the protein.

However, many molecule-to-molecule transformations and modifications are the product of enzyme machinery (transferases, lyases, isomerases) and follow the principles of enzyme kinetics (Koolman and Röhm 2013). The most basic representation of these principles can be captured using the Michaelis-Menten equation, derived for the catalytic reaction of one substrate and one product (Cornish-Bowden 2012):

$$\begin{aligned} \frac{dX}{dt} &= \frac{V_{\max} * S(t)}{k_m + S(t)}, \\ V_{\max} &= k_{\text{cat}} * E_0, \end{aligned} \tag{2}$$

where $V_{\max}$ is the maximum rate of the reaction, k_{cat} is the catalytic constant, Km is the Michaelis constant, t represents time, E_0 is the concentration of an enzyme, $S(t)$ is the concentration of a substrate, and $X(t)$ is a product of the reaction.

The steepness of the relationship between the product and the substrate can be captured by incorporating an additional degree of freedom into Eq. 2, which then becomes a Hill-type equation (Goutelle et al. 2008):

$$\frac{dX}{dt} = \frac{V_{\max} * S(t)^{Nh}}{Km^{Nh} + S(t)^{Nh}}, \tag{3}$$

where Nh is the Hill coefficient.

Equations 2 or 3 may be modified further to account for the presence of a drug agent with an associated mechanism of action, e.g., competitive inhibition (Eq. 4), non-competitive inhibition (Eq. 5), uncompetitive inhibition (Eq. 6), and others (Cornish-Bowden 2012; Segel 2001; Cleland 1963):

$$\frac{dX}{dt} = \frac{V_{\max} * S(t)}{K_m * \left(1 + \frac{I}{K_i}\right) + S(t)}, \tag{4}$$

$$\frac{dX}{dt} = \frac{V_{\max} * S(t)/\left(1 + \frac{I}{K_i}\right)}{K_m + S(t)}, \tag{5}$$

$$\frac{dX}{dt} = \frac{V_{\max} * S(t)}{K_m + S(t) * \left(1 + \frac{I}{K_i}\right)}, \tag{6}$$

where I is the concentration of an inhibitor drug and K_i is the inhibition constant.

The same principles of enzyme kinetics used to derive the Michaelis-Menten equation (by using a quasi steady-state approximation) are applicable to more complex systems of substrates, products, activators, or inhibitors, which are often encountered when modeling metabolic pathways, signaling cascades, and other networks with multiple feedback (Lo-Thong et al. 2020; Klipp and Liebermeister 2006; Schauer and Heinrich 1983). For example, to characterize the effect of two modulators (e.g., drugs), the following expression can be used:

$$\frac{dX}{dt} = \frac{E_{\max} * \left(\frac{M_1(t)}{EC50_1} + \frac{M_2(t)}{EC50_2} + \alpha * \frac{M_1(t) * M_2(t)}{EC50_1 * EC50_2}\right)}{1 + \left(\frac{M_1(t)}{EC50_1} + \frac{M_2(t)}{EC50_2} + \alpha * \frac{M_1(t) * M_2(t)}{EC50_1 * EC50_2}\right)}, \tag{7}$$

where $M_1(t)$ and $M_2(t)$ represent the change over time in the amount or concentration of modulators 1 and 2; $V_{\max}$, $EC50_1$, $EC50_2$, and α are parameters representing the maximal effect, half-maximal effect concentration for modulators 1 and 2, and a term defining the degree of synergy between the two modulators' actions, respectively.

Michaelis-Menten and Hill equations are two of many functional forms used to capture saturable rates. For example, the Verhulst equation may serve as an alternative and is usually applied to describe the growth of a cell population (Peleg et al. 2007):

$$\frac{dX}{dt} = q * \left(1 - \frac{X(t)}{K}\right) * X(t), \tag{8}$$

where q and K are parameters describing the growth rate and carrying capacity and $X(t)$ is the number of cells. In oncology applications, this model construct is also referred to as logistic growth and represents a proportional growth of cells that slows down at high cellular/tumor volumes.

Receptor-ligand interactions are also frequently encountered in biology and may be represented in mechanistic QSP models as follows (Finlay et al. 2020):

$$\begin{cases} \frac{dL}{dt} = k_{\text{in},L} - k_{\text{out},L} * L(t) - k_{\text{on}} * L(t) * R(t) + k_{\text{off}} * \text{Com}(t) \\ \frac{dR}{dt} = k_{\text{in},R} - k_{\text{out},R} * R(t) - k_{\text{on}} * L(t) * R(t) + k_{\text{off}} * \text{Com}(t) \\ \frac{d\text{Com}}{dt} = k_{\text{on}} * L(t) * R(t) - k_{\text{off}} * \text{Com}(t) \end{cases}, \tag{9}$$

where $k_{\text{in},\ L}$, $k_{\text{in},\ R}$, $k_{\text{out},\ L}$, $k_{\text{out},\ R}$ are rate constants describing the production and elimination of the ligand (L) and receptor (R), k_{on} is the association constant, k_{off} is the dissociation constant, and $L(t)$, $R(t)$, and Com(t) are the amounts of ligand, receptor, and ligand-receptor complex, respectively. Associated target engagement models have been extensively studied in multiple research papers (Mager and Jusko 2001; Mager and Krzyzanski 2005; Gibiansky et al. 2008).

Depending on the patterns in the data and other information relevant to a particular dataset, various equation forms describing functional relationships may be tested in search for an optimal model (see Sect. 3); also, the same equation structure may be used to describe different cellular and physiological processes, such as active transport and synergy. One rule of thumb here is to avoid expanding unnecessarily the system when the objectives of the project and the model simulations to address the corresponding questions do not require it. To cite George E.P. Box, "*All models are wrong, but some are useful*" (Box 1979), even the seemingly straightforward Michaelis-Menten equation (Eq. 2) is based on multiple assumptions such as $E_0 \gg S(t)$ and fast binding kinetics relative to the catalytic reaction. The choice of a model structure as well as mathematical simplifications based on assumptions and data availability should be grounded in biological context for any model-based analysis.

A major aspiration of QSP models is their generalizability for further model development and reuse to address a broader range of relevant questions and a different or augmented COU – a feature of QSP models which also goes hand-in-hand with their mechanistic foundation. Each mechanistic feature added to a model structure brings additional assumptions. The utility of QSP models thus lies in their careful balance between quantitative features vs. (the number of) mechanistic features, i.e., predictability vs. generalizability, to ensure models remain fit-for-purpose. As such, the inclusion of as many mechanistic components and processes as possible, when shaping the model structure, does not necessarily result in a "better" model. Another drawback in featuring excessive complexity during model development is purely practical, including the time required to properly develop and qualify a model and the efforts necessary to properly analyze model simulation results for various scenarios of interest, i.e., model tractability. Consequently, a QSP model always describes a limited set of processes, with a certain degree of scale and modularity. Some QSP models have been successfully generalized through select fit-for-purpose adjustments and extensions while building upon the original core model, as shown in several therapeutic areas (Jafarnejad et al. 2019; Ma et al. 2020; Milberg et al. 2019; Shah et al. 2021; Sokolov et al. 2020, 2023; Yakovleva et al. 2019).

Depending on the scale, each model may itself represent a module within a larger model. Furthermore, by structurally connecting different modules within a larger model, a broader context of use may be achieved. This brings forth the modular principle of QSP models, where modules are linked according to inter-dependent physiological, biological, pharmacological processes. The scope of each module can be driven by the distribution of processes (e.g., different organs), scale (e.g., cellular vs. molecular or even cross-species), or types of processes (e.g., B-cell immune responses vs anti-proliferative effects). This modular principle provides multiple advantages in model calibration (Sect. 3) and in the overall handling of the fully integrated model. *In fine*, the credibility of the overarching QSP model will be predicated upon both the level and quality of individual modules, the quality of module inter-dependencies to formulate the full model, and the integrated model performance (Musuamba et al. 2021).

2.3 Evaluating QSP Model Representation of System-Level Behavior

Once a set of equations has been laid out, analysis of the corresponding steady states can be used to understand potential behaviors of the system. A steady state is defined as a condition where the state variables are in balance and therefore the rate of change of the variable is zero ($dX_k/dt = 0$, for all state variables $X_k(t)$, $k = 1, \ldots, K$). While this can be achieved using phase plane analyses as illustrated in the Appendix, it can also be assessed via sensitivity analysis, which is covered in detail in Sect. 4.2. Sensitivity analysis, as a technique used in the evaluation of forward simulations, may be used in QSP modeling as soon as equations are coded in a model and can generate multiple trajectories of model state variables via Monte Carlo sampling of the parameter space, which may be performed using non-informative or informative parameter priors. Such an exercise may provide important model characterization, as it can simulate time courses associated with steady-state responses and allows for the evaluation of a range of theoretical outcomes with clinically (or preclinically) relevant scenarios. Sensitivity analysis at this stage can aid in the interpretation of how the biology is implemented in the model by informing on how select parameterizations alter model behaviors.

From a physiological perspective, a set of ODEs at steady state can be interpreted as a homeostatic baseline state of the system. A steady state is often desirable, prior to considering model simulations exploitable, except when time-dependent functions must be included (for example, to capture disease progression modeling or cyclical perturbations such as food consumption). The simplest option for reaching a steady state is to implement a "burn-in" period into all simulations, which gives the system time to balance itself before introducing any perturbations, such as a drug dose. A more elegant solution would be to assume the following and then define the initial conditions of the system:

$$X_k(0) = X_k(SS), \tag{10}$$

where $X_k(0)$ is the initial value of a state variable and $X_k(SS)$ is the state variable value at steady state, $k = 1, \ldots, K$.

By using the definition of a steady state ($dX_k/dt = 0$), initial conditions can be derived from the parameters and vice versa, avoiding the burn-in. As a simple example, take Eq. 1, where the response at baseline ($X(0)$) can be defined as $\frac{k_{\text{in}}}{k_{\text{out}}}$, or k_{in} can be set equal to $k_{\text{out}} * X(0) = k_{\text{out}} * X(SS)$. The process becomes more challenging once non-linear functions, more variables, and additional feedback mechanisms are implemented in the model.

Another important assessment of the QSP model is structural identifiability. Since all parameters in the model need to be quantified, either by fixing them using prior knowledge or by estimating their values against observed data (see Sect. 3), a structural (a priori) identifiability analysis can be performed. The general purpose of a structural identifiability analysis is to understand which parameters in the model can be, in theory, unambiguously estimated, provided that the input and output data are noise-free and as exhaustive as possible (Walter and Pronzato 1997). This approach is described in the Appendix; however, it can be difficult to achieve in practice, and a global sensitivity analysis is often the more pragmatic approach. A global sensitivity analysis (see Sect. 4.2) may substitute for a structural identifiability analysis given that it informs on which elements of the model can be altered or reduced when model outcomes are not sensitive to certain model parameters. However, such inference strongly depends on the range of parameters selected for simulations and therefore should be based on strong priors for the selected range (Hsieh et al. 2018).

In conclusion, the first major step in executing the QSP modeling workflow consists of a systematic literature review and construction of a model structure. The literature review should follow data mining guidelines to ensure transparency, reproducibility, and completeness, while the initial model structure represents a parsimonious quantitative mathematical approximation of the existing knowledge and data relevant to a system of interest (e.g., a disease indication with select pathophysiological processes) and its perturbations (e.g., using pharmacological intervention). However, most models require estimating parameters against observed data, which is the focus of the next section.

3 Model Calibration and Evaluation

One of the conceptual advantages of a mechanistic model, as compared to an empirical one, is the transition from theoretical macro-parameters that reflect larger-scale, aggregate factors that describe the overall behavior or trends of a system and are often closely tied to the dataset from which they are derived to micro-parameters (i.e., specific, detailed factors that influence a system) which define mechanistic processes based on in vitro and in vivo experiments and can ultimately inform on system-level behavior beyond a single dataset. Consequently, many parameters in a QSP model, including fluxes, volumes, binding constants, cell and protein half-lives, can be fixed based on prior knowledge. However, some elements of the system may be unknown or may need to be simplified or lumped together to avoid model inflation. The associated parameters then need to be estimated based on the available data, usually in the form of time-series measurements following a perturbation, such as PK and PD data. This constitutes the model calibration procedure, which is essentially fitting the model or sub-models to data, as briefly described below in general terms (additional details can be found in the appendix).

Consider a model defined by the following general equation:

$$Y_j(t) = f\left(X_j, \Theta\right) + g\left(X_j, \Theta\right) * \varepsilon_j, \tag{11}$$

where f is the structural model, X_j are independent variables (time, dose, covariates, etc), Θ is a vector of model parameters, g is the residual error model, and $\varepsilon_j \sim N(0, \sigma^2)$, $j = 1 \ldots m$.

The first step in model calibration would consist of finding the estimates $\widehat{\Theta}$ of unknown parameters Θ which correspond to an extremum (minimum or maximum) of an objective function (OF). The most general OF to consider would be the extended least squares (ordinary least squares (OLS) or weighted least squares (WLS) are described in the Appendix):

$$O_{\mathrm{ELS}}(\Theta) = \sum_{j=1}^{m} \left[\frac{\left(Y_j^{obs} - f\left(X_j, \Theta\right)\right)^2}{\mathrm{VAR}_j} + \log\left(\mathrm{VAR}_j\right) \right], \tag{12}$$

where Y_j^{obs} is the observed value at time point j and VAR_j is the variance of the Jth residual. This OF in turn equates to the -2*logarithm of the likelihood function (-2LL) for the normally distributed and independent $\varepsilon_j \sim N(0, \sigma^2)$:

$$-2LL(\Theta) = m * \log(2\pi) + \sum_{j=1}^{m} \left(\frac{\left(Y_j^{\text{obs}} - f(X_j, \Theta) \right)^2}{VAR_j} + \log(\text{VAR}_j) \right), \qquad (13)$$

where VAR_j is defined as $g(X_j, \Theta) * \sigma^2$.

As such, the estimation of parameters in QSP modeling is no different from the maximum likelihood approach typically used for conventional PK, PK/PD, and pharmacometrics models.

Some key challenges in parameter optimization specific to QSP modeling are associated with the nature of the data used in the procedure and the complexity of the mechanistic model. As briefly discussed in Sect. 2.1, the data used in QSP model development usually originate from multiple sources and present themselves in different forms with the most common being aggregated datasets that are provided as a measure of the central tendency (mean, median) and variation (standard deviation (SD), standard error (SE), etc.). Consequently, the application of normally distributed random effects, which is conventionally used in PK/PD and population modeling, is usually not an option for QSP modeling for the following reasons: firstly, random effects applied to aggregated data will reflect a combination of residual/measurement error, between subject variability, and between cohort or between-study variability, as opposed to separable between-subject variability (BSV) and measurement error available with individual subject-level data; secondly, the reliable estimation of random effects will be challenged by the scarcity of the data; thirdly, the assumption of normality for such random effects cannot be always justified. As a result, the quantification of heterogeneity between populations and BSV in QSP model development often requires alternative solutions to non-linear mixed effects modeling. If subject-level data are available for model calibrations, methods may be applied to accommodate for random effects (see Appendix for additional details and supporting mathematics). We refer the reader to the recent model of glucose-insulin dynamics with dapagliflozin treatment by Shah et al. (2021) for an example.

A QSP modeler will usually work with sparse, heterogeneous, and/or aggregated data. The most direct approach to account for heterogeneity between cohorts or studies would be to calibrate different sets of data separately, thereby producing multiple sets of parameter estimates. The range of the estimates can then be interpreted as physiologically/biologically plausible (at cohort- or study level), and the 95% CIs of the estimates (see also the discussion on *practical identifiability* below) may be compared to make an inference on the significance of differences between the datasets. This approach, however, is confronted by the problem of data scarcity, whereby the use of partial datasets may impact the practical identifiability of parameters. Moreover, while QSP model calibration would ideally exploit the data from several dependent variables, such as multiple biomarkers, PD data, or endpoints, it would be optimistic to expect all these variables to have corresponding measurements within each data source used for model development. One solution consists of identifying covariates in a broad sense, as independent variables that mechanistically describe between-study or between-cohort differences affecting any rate, parameter, or dependent variable in the model and taken directly from the data. Thus, no formal covariate search procedure is applied in this scenario: covariates are essentially part of a structural model introduced primarily based on an understanding of the data and underlying mechanisms that contribute to between-subject variability, rather than based on statistical inference.

Knowing realistic distributions of parameters in a population that can capture BSV or between-animal variability is a powerful tool which may substantially increase inferential capabilities of a mathematical model. For most drug development applications, it is indeed important to predict a distribution of responses rather than a single meta-response. However, and as mentioned above, quantification of variability through an implementation of random effects is feasible only when subject- or animal-level data are available. To overcome this limitation, alternative solutions, which are Bayesian in nature, have been developed. For example, if several point estimates are known for a parameter, they can be considered as a plausible range for this parameter during model development procedures and sensitivity analyses. Another approach is to approximate distributions of individual parameter values (i.e., the generation of virtual patient population (VPP)) based on known distributions of selected outcomes rather than individual datapoints (Allen et al. 2016; Rieger et al. 2018; Cheng et al. 2017, 2022). The general workflow of these algorithms typically consists of two stages. Firstly, values of unknown parameters are sampled from a selected (often uniform) distribution within physiologically plausible ranges (in Bayesian terms: a prior) and corresponding model predictions are matched with the plausible range of observed outcomes. Only those parameter sets that result in plausible outcomes are kept for the second stage, at which the probability of a set of parameters to be included into a VPP is estimated based on the observed probability density of an outcome (usually by using a Kolmogorov-Smirnov test-statistic as an OF). This second stage constitutes the key difference separating a sensitivity analysis (covered in detail in Sect. 4.2) from model-based predictions of a population response. One downside of this approach is that the resultant combinations of parameters, which are always estimated based on a limited number of outcomes and timepoints, such as the distributions of biomarkers at baseline that are matched with steady-state model predictions, may cause unrealistic behavior in other model variables within a fraction of simulations, especially in convoluted systems with identifiability issues and correlations between parameters (Duffull and Gulati 2020). In such cases, an understanding of a model's phase plane (see appendix) may represent a significant advantage in controlling the full scope of model outcomes, based on the VPPs generated.

The modular principle of QSP model development described in the previous section becomes essential when integrating different types of data into the model through a sequential approach to model calibration and evaluation (Zhang et al. 2003). While this requires the optimization problem to be solved multiple times per module (sub-model), it allows for the integration of aggregated and subject-level data into a single system. Moreover, the number of parameters which need to be estimated may quickly rise to a few dozen, for example, when multiple sets of PK parameters for different drugs and several systems parameters must be estimated, in addition to residual error parameters and random effects. A modular modeling approach significantly reduces the parameter space per optimization procedure, improving run times and convergence.

The modular modeling approach is a pragmatic one to address technical issues, but the overall sparsity of the available quantitative data remains a challenge in the development of mechanistic models. For example, with typical datasets it often becomes problematic to assess the variance of the residual error. To overcome this challenge, an exploratory analysis may be performed to evaluate the degree of variability in the data (e.g., SE of the mean observations) and its proportionality relative to the dependent variable value by calculating the coefficient of variation (CV = SE/mean). The results of these evaluations may then be used to guide selection of the residual error model and to fix respective parameters with reasonable values. Furthermore, state variables and micro-parameters in QSP models that originate from biology can be constrained within plausible ranges, which helps in model calibration. Physiologically relevant initial values and constraints are often applied to ease the process of parameter optimization and in subsequent "sanity checks" of the resulting estimates. Likewise, random effects, if present, may be assigned only to those parameters expected to be the main, biologically meaningful drivers of variability in the observed responses, provided relevant individual-level data are available (Shah et al. 2021; Kosinsky et al. 2018). Otherwise, the systematic introduction of random effects will lead to a significant increase in free parameters and will make the reliable estimation of parameter values improbable.

The estimation of the unknown or free parameters itself can be performed via different local and global optimization algorithms, such as Nelder-Mead, Broyden-Fletcher-Goldfarb-Shanno (BFGS), Levenberg-Marquardt, simulated annealing, particle swarm, differential evolution, and other algorithms (Nelder and Mead 1965; Watson 1978; Bélisle 1992; Kennedy and Eberhart 1995; Mullen et al. 2011; Fletcher 1987). If random effects are present in a non-linear set of ODEs, the likelihood needs to be approximated. Special local algorithms, such as first-order (FO), first-order conditional estimation (FOCE), or stochastic approximation expectation maximization (SAEM), are designed to handle such cases (Bauer 2019a; Kuhn and Lavielle 2005). Intuitively, global algorithms may be the preferred method for parameter optimization in QSP models; in practice, however, they are associated with high computational costs and are not guaranteed to provide the best solution (Moles et al. 2003). Local non-gradient methods such as the Nelder-Mead and first-order gradient methods are fast and easy to implement; however, they can get trapped into local minima. This risk may be partially mitigated by using physiological priors and constraints on the parameters. Second-order gradient methods such as the Levenberg-Marquardt and BFGS methods may be considered as a satisfactory middle ground, although they also tend to get trapped in local minima, especially in highly nonlinear problems (Bard 1970). A multi-start parameter optimization procedure from different initial sets of model parameters is generally warranted for non-global optimization algorithms to avoid local minima.

Calibration of a model is not necessarily a matter of convergence of a single "best" algorithm, but rather an assessment of a model using multiple criteria. These criteria form the basis of model qualification and validation standards which are necessary to enable sound decision-making based on model-based predictive simulations.

The primary metrics for model discrimination are the values of -2LL objective function and its associated scores, including the Akaike information criterion (AIC, Eq. 14) and the Bayes information criterion (BIC, Eq. 15). The difference in -2LL between two nested models is approximately χ^2-distributed, and for one degree of freedom a difference of 3.84 corresponds to a significance level of $p = 0.05$, provided that the models are hierarchical (Bonate 2011). In the case of non-nested models, AIC or BIC is used for model comparison purposes:

$$\mathrm{AIC} = -2LL + 2 * P, \tag{14}$$

$$\mathrm{BIC} = -2LL + P * \log(M), \tag{15}$$

where P is the number of estimated parameters and M is the number of measurements in the dataset. The model with the lowest AIC or BIC has the advantage.

A second key indicator of the quality and predictive power of a model is practical identifiability of the estimated parameters. Even with optimal values of -2LL, AIC, or BIC, a model may still be overparametrized, that is, the values for the unknown parameters cannot be determined unambiguously from the available data and are, therefore, unidentifiable. Large mechanistic models are particularly prone to this issue considering the overall sparsity of the available data. Several methods may be used to quantify parameter identifiability. The gold standard would be to calculate the SE of the point estimates of the parameters, based on the Fisher Information Matrix (FIM) (Bard 1974) (see Appendix for additional details and supporting equations).

In addition to calculating the FIM, it is recommended to compute the CI of the parameter estimates using bootstrap methods (non-parametric and parametric) and likelihood function profiling (Efron and Tibshirani 1986; Venzon and Moolgavkar 1988), since the FIM-derived variance-covariance matrix provides a lower-bound estimate of the covariance. Non-parametric bootstrap is based on a repeated parameter optimization process with artificially reduced datasets compiled by sampling measurements from the original data, each point having an equal probability of getting into the set. In contrast, parametric bootstrap involves multiple estimations of the parameters, with the original data being permutated using a parametric function. For general use, it is recommended to perform at least 600 iterations of bootstrap to achieve adequate accuracy; however, bootstrap methods work poorly with small sample sizes, which is often the case in QSP modeling (Wilcox 2010).

Reconstruction of the CIs of the parameters using the profile likelihood approach also requires multiple runs of the model calibration procedure per estimated parameter and is based on the difference in the log-likelihood (LL) between a model where a parameter θ_k is estimated and a model where θ_k is fixed (usually within the neighborhood of the optimal value of the former model). Thus, CIs based on the profile likelihood correspond to the interval $\left[\theta_k^{\min}, \theta_k^{\max}\right]$ of the maximum width containing $\widehat{\theta}_k$ such that:

$$\mid LL(\dot{\Theta}) - LL\left(\widehat{\Theta}\right) \mid \le \frac{1}{2}\chi^2_{1,0.95}, \tag{16}$$

where $LL\left(\widehat{\Theta}\right)$ is the log-likelihood when all unknown parameters are estimated, $LL(\dot{\Theta})$ is the log-likelihood when $\widehat{\theta}_k$ is fixed within the set of pre-defined values, and $\chi^2_{1,0.95} = 3.84$. Compared to the FIM approach, the profile likelihood can reveal one-sided identifiability issues in parameter values. On the other hand, a CI based on the LL curvature for $\widehat{\theta}_k$ does not fully account for the uncertainty in other parameters (McCullagh and Nelder 1998). A simple illustrative example benchmarking different parameter identifiability methods is shown in Fig. 4. This

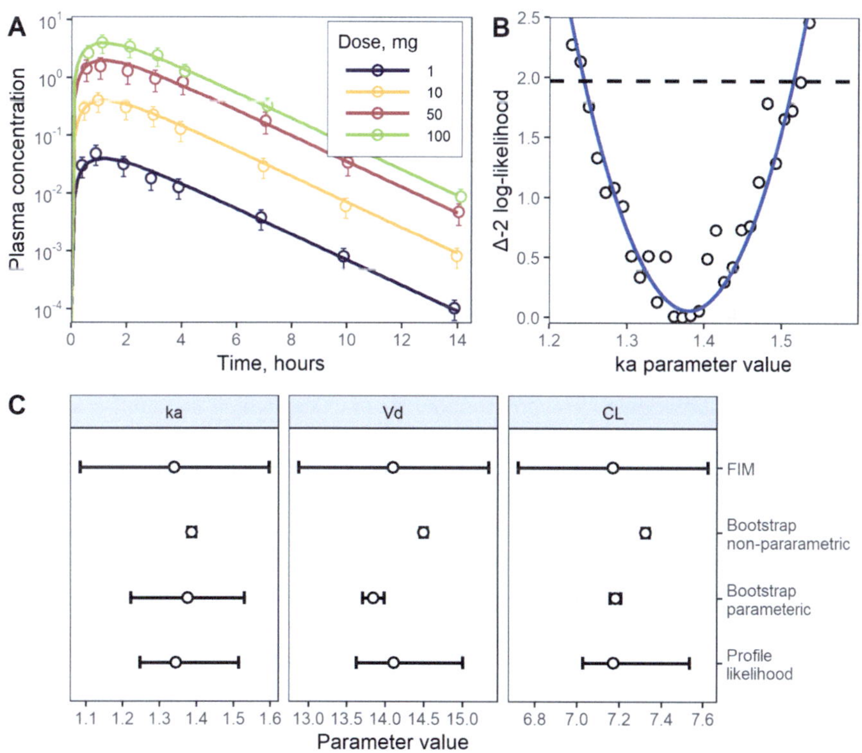

Fig. 4 Illustration of practical identifiability analyses of parameters produced with different methods based on aggregated PK data. (Simulated aggregated data were used to calibrate a one-compartment PK model. (**a**) Curves: model predictions; points with error bars: observed means ± SE; (**b**) Points: calculated likelihood values; curve: quadratic spline; (**c**) Points with error bars: parameter estimates with 95% CIs. FIM Fisher information matrix, SE standard error, CI confidence interval)

example can be interpreted as an attempt to calibrate a PK module of a larger QSP model based on aggregated study-level data; the reader is referred to (Luzyanina et al. 2009) as an advanced example of a thorough identifiability evaluation of a systems model.

Both bootstrapping and profile likelihood methods may not be optimal for computationally intensive QSP models. As an alternative, a multi-start parameter optimization procedure may be applied, which represents a series of model calibration iterations with the original unmodified dataset but different starting parameter values, to confirm the uniqueness of a computed solution. The starting values are selected from a uniform distribution within physiologically plausible limits. The 95% CIs of point estimates calculated by the FIM method are expected to overlap between runs that result in the same computed solution. Typically, 10–15 runs are sufficient to reveal the presence of other local minima or identifiability issues.

A model should also be assessed using goodness-of-fit plots designed to assess the quality of the data descriptions by the model (Bonate 2011; Nguyen et al. 2017). Overall, the types of plots which may be used in QSP modeling are the same as those used in pharmacometrics and include the visualization of observed versus predicted values, weighted residuals against time and against model predictions, distributions of weighted residuals, etc. (Nguyen et al. 2017). If random effects are present, most of the graphs may be prepared in two versions: (i) one using only typical parameter values and (ii) another one using individual parameter values for predictions, with additional diagnostics being performed to assess distributions of random effects/ individual parameters, their correlations, as well as normalized prediction distribution error (NPDE) and visual predictive check (VPC) plots. Systemic bias in predictions, skewed distributions, heteroscedasticity in residuals, and poor simulation-based diagnostics (VPC, NPDE) would then require revisions of the structural and/or statistical models.

In summary, as may be noted from the content of this section, many elements of QSP model calibration follow good modeling practices which are similar to those used in well-established empirical PK/PD modeling (Population Pharmacokinetics 2022) and pharmacometrics. The "learn-and-confirm" cycles of data mining, handling of equations and parameter estimation procedures lie at the core of QSP model development. A simultaneous implementation of all techniques and methods listed in Sects. 4.2 and 4.3, though desirable, is hardly feasible for the diverse variety of situations where QSP modeling should be applied and is, therefore, contextual. Moreover, as of today, several items such as structured systematic literature reviews, extensive identifiability analyses, and (hierarchical) Bayesian modeling may still be considered aspirational in the field of mechanistic modeling in drug development. Ultimately, the applicability of the model (i.e., how fit-for-purpose a model may be) is defined by a balance of complexity versus limitations to unambiguously answer the relevant questions posed within a COU. Furthermore, an adequate inclusion of mechanistic hypotheses into the model structure requires up-to-date knowledge in biology and medical domains and is therefore best achieved through highly collaborative efforts across diverse teams of biology and medical researchers and proximal stakeholders.

4 Model Simulations

4.1 A Taxonomy of Simulations

The main goal of a theoretical construct, such as a mathematical model, is to provide a framework to make inferences with regard to its object of study. While some inferences can be made following model calibration and evaluation (Sect. 3), such as identification of knowledge gaps and optimal functional relationships between variables, further inferences may require forward simulations. From a technical perspective, performing simulations using an ODE-based QSP model, no matter how complex, only requires numerical solutions of the ODE system. Many algorithms have been developed for the task, for both stiff and non-stiff equations, including Runge-Kutta methods, Adams methods, the backward differentiation formula, etc. (Butcher 1996, 2000).

The main challenge of model-based simulations resides in the nuances of the interpretation associated with different components that make up the simulation scenario (Figs. 5 and 6). Firstly, model simulations may include uncertainty in model parameters that reflect confidence in model predictions. Secondly, variability can also be a part of a simulation, in different shapes and forms, be it random effects or virtual patients. In addition, a model can incorporate covariates in a broad sense, i.e., as independent variables associated with any model function, state variable, or parameter, which adds to the complexity of subsequent predictions. However, it should be noted that QSP modeling often strives to integrate intrinsic factors driving pharmacodynamic response as a part of the model structure and not as external regressors. Finally, a simulation can include residual variability or elements of trial execution: missing or wrong doses, observations, subjects, etc. (Holford 1999). Depending on the type(s) of data used to calibrate model parameters and the combination of the abovementioned components used when running a model-

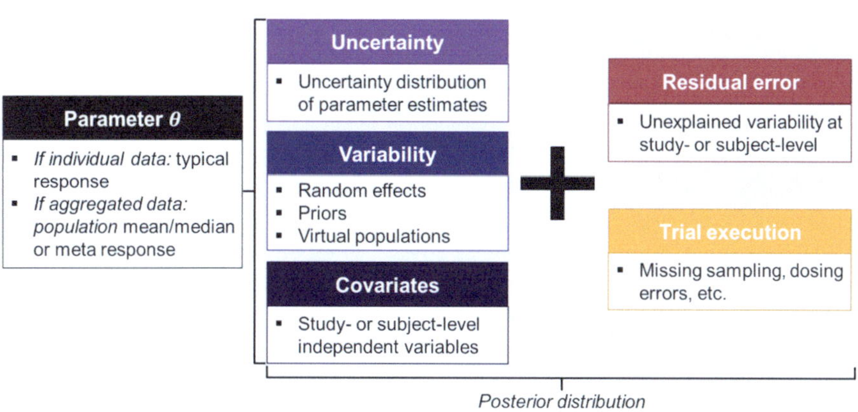

Fig. 5 Key components of model simulation

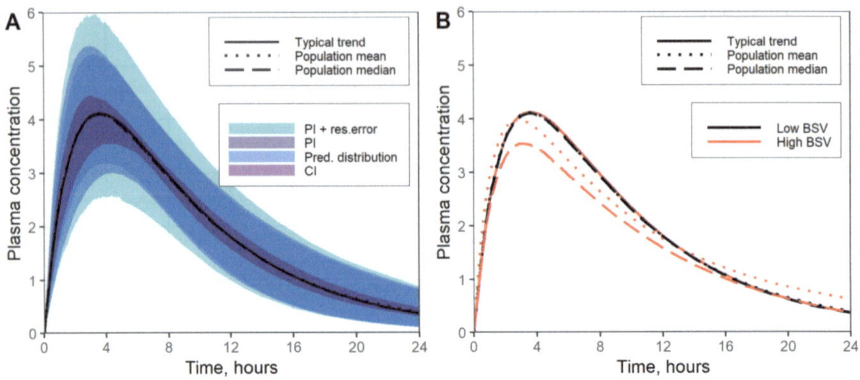

Fig. 6 Illustrative examples of forward simulations. (Simulations using a one-compartment PK model (PI prediction interval; res error: residual error, Pred distribution: prediction distribution, CI confidence interval, BSV between-subject variability). For low BSV simulations, typical trend, mean, and median overlap, thus, can be considered interchangeable, but with caution)

based prediction, the interpretation of the in silico experiment may vary substantially.

The most frequently used simulations are based on the population parameter values (θ *i* n Fig. 5). This type of prediction is used to analyze the behavior of a system as a whole and to analyze the sensitivity of model variables to parameter values. Since QSP models are often calibrated based on aggregated data, θ will represent a meta-parameter reminiscent of the mean derived in a model-based meta-analysis, although not weighted by sample size and variance *per* each datapoint (Chan et al. 2022). Similarly to fixed-effects meta-analyses, the residual error associated with such predictions will reflect the residual variance of the mean (and not its confidence) (Borenstein et al. 2010). If individual-level data were used to calibrate a parameter, similarly to conventional population modeling, then θ would correspond to the system's response of a typical subject, and the associated predictions would represent typical trends in the population. Importantly, these trends are numerically different from population mean and median and should not be interpreted as such; if a model, however, is relatively simple in structure and variability is moderate, trends and population mean and median are expected to be numerically close (Fig. 6). Simulations with residual error only based on a model with random effects are hard to interpret and generally should be avoided.

Overall, covariates may be included in nearly any type of simulation. As discussed in Sect. 3, in QSP models the covariates should be interpreted as independent predictors taken directly from the data into the model structure. If these are indeed present, simulations with specific covariate values may be implemented and benchmarked, or a family of curves may be drawn to evaluate the impact of a covariate on θ -based model predictions, by using fixed covariate values per simulation, in a fashion similar to a local sensitivity analysis (Sect. 4.2). As in the case of θ, it should be remembered that the covariates defined using study-level data are study-level covariates and reflect aggregated and not individual characteristics of a population.

To perform statistical inference, a measure of confidence can be derived for the simulations, i.e., a confidence interval (CI) around model predictions. To build a CI around a central tendency, such as population mean, median, typical trend, aggregated mean trend, or something else, θ should be sampled from a joint distribution defined by the variance-covariance matrix. In most simulation cases, it is recommended to visualize uncertainties around predictions (Musuamba et al. 2021). In general, small relative standard errors of the parameter estimates are the hallmark of a narrow CI.

While QSP rarely handles variability in the form of random effects, theoretical distributions of individual parameters may still be reconstructed using the virtual population approach, as described in Sect. 3, to generate prediction distributions. Alternatively, random effects may be applied, if the data allow it.

Finally, a prediction interval (PI) can be calculated by accounting for both variability and uncertainty, reflective of the complete theoretical variance in the measurements minus the residual variability. Accordingly, by introducing a residual error on top of variability and uncertainty, the PI will represent an interval in which a future observation will fall, with a certain probability. To assess mean trends, it is recommended to sample at least 200 parameters, and at least 1,000 to calculate the PI (Mould and Upton 2012). An illustrative example of various types of intervals as well as central tendencies is shown in Fig. 6.

At first glance, introducing all the above-mentioned statistical components into a simulation may seem to be the best approach to cover the broadest spectrum of posterior distributions that can be made from a single set of simulations. In practice, however, it may not always be possible: a model may not appropriately capture the variability; it may have been calibrated on aggregated data; or the inference may be based on a variable not used in model calibration and, therefore, may lack residual error estimates. Furthermore, "*more*" does not imply "*better,*" since "*more*" will bring additional assumptions and the potential for errors. It is recommended that two mutually complementary rules be followed when exploring possibilities offered by model simulations: (1) perform a correct interpretation of trends and bands and (2) compare what is comparable.

Both rules are best illustrated when considering the model validation exercise. For example, to compare observed mean trends in the data using model predictions, it is sufficient to sample population parameter values only from the uncertainty distribution, if said parameters were identified using aggregated, study-level data. However, if variability was accounted for in parameter estimation, the confidence interval for the population mean should be built by calculating multiple means from multiple virtual populations (based on variability), with θ sampled from the uncertainty distribution. Moreover, statistical inference might not always be directly associated with clinical implications. Consider comparing mean trends of observed data with theoretical predictions made based on a robust QSP model (i.e., with high precision parameter estimates); both data and model predictions may lie close to each other, yet feature narrow, non-overlapping CIs. While this would signify a technical discrepancy between the two and motivate a search for factors explaining the difference, the data characterization by the model could be considered acceptable provided that the discrepancy is within an acceptable clinical or physiological margin.

Mechanistic, dynamic QSP models provide a great framework to derive critical inferences based on evaluations of the system's behavior. Two instruments leveraging forward simulations are often implemented to evaluate the system's behavior: sensitivity analysis and VPP simulations. Both techniques offer similarities and distinctions, yet one is often mistaken for the other. This warrants separate discussions in the following Sects. 4.2 and 4.3.

4.2 Sensitivity Analyses

Sensitivity analysis is an essential component of the QSP model evaluation process. It represents the analysis of changes in model outputs relative to changes in model inputs. It can be performed in a variety of ways, including local vs. global methods (Zhang et al. 2015).

Local sensitivity analysis is arguably the more straightforward approach and represents the degree of response in model variables relative to a change in a single parameter. Numerically, local sensitivity indices (LSI) are defined as:

$$\frac{dY}{d\theta} = \lim_{\Delta\theta \to 0} \frac{Y(t, \theta + \Delta\theta) - Y(t, \theta)}{\Delta\theta}, \tag{17}$$

where $Y(t, \theta)$ is a dependent variable, θ is a model parameter, and $\Delta\theta$ represents a change in θ. To calculate the LSI, one needs to assume a linear relationship between $Y(t, \theta)$ and θ in the neighborhood of θ and set $\Delta\theta$ to an arbitrary small number (e.g., $0.001 * \theta$).

Graphically, the analysis can be assessed by drawing a family of curves, e.g., tornado or forest plots, for a set of θ values varied over a user-defined range (Fig. 7). This range can be arbitrary, e.g., $\pm 0.5 * \theta$, or it may map to the RSE or BSV of the parameters. However, since QSP model parameters usually carry a physiological or biologic reality, varying their values uniformly within a pre-specified range may not necessarily be biologically plausible. An alternative – and arguably more informative – approach would be to perform a visual inspection of model outputs under parameter values that are constrained by physiological boundaries.

Regardless, by calculating sensitivity *per* one parameter at a time, the effect of parameter interactions gets ignored. This is not the case with global sensitivity analysis methods such as partial rank correlation coefficients (PRCC), the Sobol method, the Fourier amplitude sensitivity test (FAST) method with its modifications, and others (Marino et al. 2008; Sobol′ 2001; Saltelli et al. 1999).

Correlation, as a measure of the linear relationship between a parameter and a variable, can be defined by:

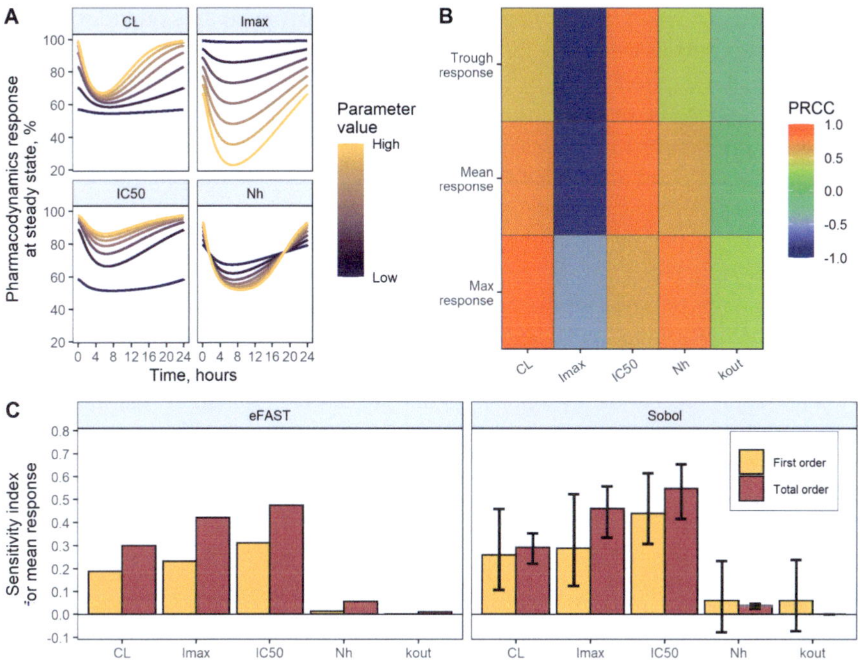

Fig. 7 Typical outputs of local (**a**), PRCC (**b**), eFAST, and Sobol (**c**) sensitivity analyses. (Calculations are based on an indirect response PK/PD model, with the drug concentration affecting the k_{in} parameter in a Hill-type equation (see Eqs. 1 and 3))

$$CC_{\theta_k, Y(t)} = \frac{\mathrm{Cov}(\theta_k, Y(t))}{\sqrt{\mathrm{Var}(\theta_k) * \mathrm{Var}(Y(t))}} = \frac{\sum_{i=1}^{N}\left(\theta_{ik} - \overline{\theta}_k\right)\left(Y(t)_i - \overline{Y}(t)\right)}{\sqrt{\sum_{i=1}^{N}\left(\theta_{ik} - \overline{\theta}_k\right)^2 * \sum_{i=1}^{N}\left(Y_i - \overline{Y}(t)\right)^2}}, \tag{18}$$

where θ_k is the Kth parameter ($k = 1, 2, \ldots, K$) of the system sampled N times, $Y(t)$ is a dependent variable, Var and Cov are variance and covariance, and $\overline{\theta}_k$ and $\overline{Y}(t)$ are average parameter and variable values across all samples. The partial correlation will then correspond to the correlation between the two residuals $\left(\theta_k - \widehat{\theta}_k\right)$ and $\left(Y - \widehat{Y}\right)$, with:

$$\widehat{\theta}_k = c_0 + \sum_{p=1, p \neq k}^{P} c_p \theta_p, \tag{19}$$

$$\widehat{Y} = b_0 + \sum_{p=1, p \neq k}^{P} b_p \theta_p, \tag{20}$$

where c_0, c_p, b_0, b_p are coefficients of a multiple linear regression model.

Calculating the partial correlation for the rank-transformed data provides PRCC, which varies between −1 and 1. The association between a variable and a model parameter increases as PRCC converges to ±1. Parameter samples designated as i in Eq. 18 are typically sampled from a uniform distribution with predefined ranges based on prior knowledge. Since the parameter space in mechanistic models is usually large, it is recommended to apply a Latin hypercube sampling technique to decrease the number of computations while still maintaining a broad range of tested combinations of parameter values (McKay et al. 1979).

The PRCC method can be applied to non-linear relationships between model inputs and outputs; however, these relationships need to be of a monotonic nature. In contrast, both the Sobol and FAST methods, which are based on the decomposition of the model variance and the estimation of the contribution of the input parameter variance to the output data variance, do not present this limitation. The Sobol and extended FAST (eFAST) methods provide first-order (Eq. 17) and total-order sensitivity indices *per* parameter of interest; they require the range of parameters to sample from and differ only in their multi-dimensional integration algorithms. The output of executing these methods given a model with two inputs and one output would be a number assigned per input in the range from 0 to 1. First-order indices show the proportion of output variance caused by each input, second-order indices show their interaction, and the total-order indices show both. No standard cut-off value exists, to differentiate important parameters from less important ones, but a value of 0.05 is frequently accepted as a reference threshold.

Figure 7 illustrates the application of various sensitivity analyses using a relatively simple model example. Figure 7a shows how different parameter values affect the trajectories and the magnitude of changes in a PD marker. The heat map in Fig. 7b indicates how strongly and in which direction measures of PD responses are affected vs. parameter values. For example, trough PD values are significantly (and inversely) more affected by the I_{max} parameter as compared to maximal response values. Figure 7c illustrates that eFAST and Sobol are qualitatively identical; according to these graphs, the sensitivity of the mean PD response gradually increases with CL, I_{max}, and IC_{50} parameters, whereas the Hill coefficient and the turnover constant have minimal effects in this type of a model.

By applying sensitivity analyses to a QSP model, a modeler may better understand the relationship between model parameters and outputs, identify bottlenecks in an intricate, physiologically driven mathematical network of processes, and optimize subsequent in vivo, in vitro, and in silico research. Furthermore, the power of mechanistic models to address various scenarios of interest can be expanded even further by performing VPP simulations and in silico clinical trials – the subject of Sect. 4.3.

4.3 Virtual Clinical Trials

At first glance, the line between global sensitivity analysis and VPP simulations might look thin. Both methodologies invoke parameter sampling from a prescribed range of physiologically plausible boundaries, with a goal of describing the range of

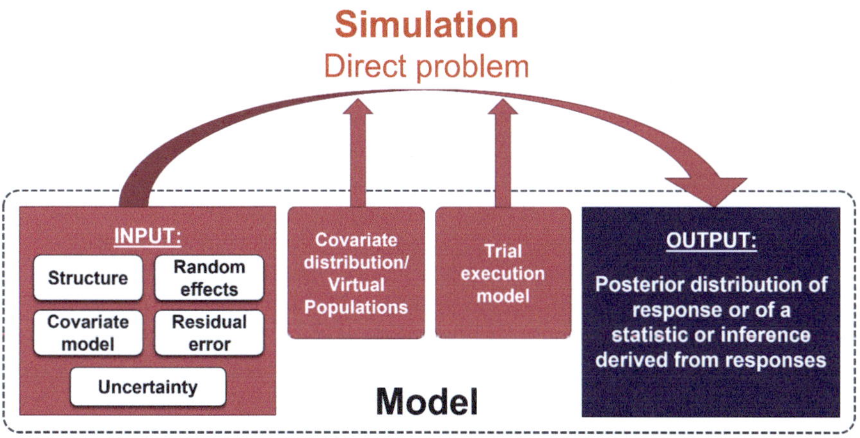

Fig. 8 Elements of an in silico clinical trial simulation

subject-level responses. This is, however, all the two methods have in common – and even then, it is reasonable to assume that the uniform distribution applied in sensitivity analyses (with or without Latin hypercube sampling or other modifications) does not correspond to the distribution of parameters in an actual population, which is akin to using the uninformative prior in Bayesian statistics as a posterior distribution. Furthermore, sampling of parameters without considering their full joint distribution (i.e., both marginal and conditional) will inevitably generate non-physiological combinations of individual parameters. Conditional distribution is a vital component of inter-individual heterogeneity of individual responses and is necessary to analyze biological systems, in contrast to the analysis of a mathematical system via global sensitivity.

A VPP simulation is a procedure that generates individual characteristics, such as parameters and covariates, from the multivariate distribution reflective of the actual population of real-life subjects. It goes hand-in-hand with in silico clinical trials defined as digital evidence that can serve to complement or replace in vivo clinical trials for the development and regulatory evaluation of medical therapies (Musuamba et al. 2021; Viceconti et al. 2016). In silico clinical trials rely on three elements: the input/output model; the covariate model; and the trial execution model (Fig. 8) (Holford et al. 2000, 2010). In pharmacometrics language, the first element would correspond to the "final model." In addition to the structural model, the "final model" includes all components described in detail in the classification of simulation scenarios: parameter uncertainty, between-subject variability, residual error, and covariates (Sect. 4.1). By default, uncertainty in parameter estimates should be included in any VPP and in silico clinical trial simulation. Random effects are used to sample individual parameter values from the estimated (joint) normal distribution under selected transformation (e.g., Appendix Eqs. A.16, A.17 and A.18) reflective of the actual distribution of parameters in the population. The residual error can sometimes be excluded, since it is not always available for the outputs of interest, or the posterior distribution is not required to contain residual variability. Finally, the input/output model may contain covariates as mechanistic features affecting model predictions.

Unlike model-based predictions for individual subjects with single covariate values, a VPP simulation requires a realistic covariate distribution for a target population. Moreover, this distribution is expected to consider both variance and covariance between the covariates, since many covariates are expected to correlate with each other, e.g., weight correlating with sex, or age with renal function. The task of generating such a distribution has been extensively developed in physiologically based PK (PBPK) modeling methodologies and is supported by databases such as NHANES (Willmann et al. 2007; National Center for Health Statistics 1994).

Finally, the execution of any clinical trial is prone to errors that contribute to the incremental heterogeneity in subject-level responses. Numerous deviations from a protocol are possible; these can be roughly categorized as subject-related reasons (e.g., wrong inclusion), doses (wrong/missing doses, errors in dose schedule), and observations (e.g., missing observations, wrong measurement times) (Girard 2005). To avoid uninformative inflation of study power, these components can also be considered in an in silico clinical trial.

In conclusion, valid individual parameter values for a viable mechanistic structural model and a valid covariate distribution represent the core of a VPP simulation, which in turn forms the basis of in silico clinical trials. While QSP modeling techniques may in many ways parallel population PK/PD models built around subject-level data, unlike with empirical models, mechanistic models consider the general subject-level variance in systemic parameters, estimated or assumed based on physiological priors, and are not specific to a particular treatment. Furthermore, even if no random effects were quantified in a model, a full joint distribution of parameter values can be elegantly reproduced by sampling from the plausible ranges for model parameters to achieve physiologically realistic outcomes with adjustments for the probability of being in a virtual population (Allen et al. 2016). Alternatively, a shift from a frequentist to a Bayesian approach might "re-purpose" the use of physiological priors and inference from posterior distribution in QSP model development and may represent a worthy maturation step in the evolution of the discipline.

4.4 External Validation

Finally, an external validation should be performed, as an ultimate step in QSP model development (Singh et al. 2023). This benchmarking exercise typically entails running model-based predictions of scenarios previously not used in model calibration, for which the outcome data is available (Kirouac 2018). As discussed in Sect. 2.2, introducing mechanistic elements of biology and physiology into a QSP model makes it, to some degree, generalizable and suitable for extrapolation. Since the model structure is, in itself, a hypothesis and relies on a number of assumptions, a successful external validation exercise can be thought of as evidence in support of the model hypothesis. Conversely, an unsuccessful external validation should motivate further evaluation of the model structure, model assumptions, data mining, and hypothesis generation.

Multiple attempts have been made to categorize validation exercises in QSP modeling (Singh et al. 2023; Ramanujan et al. 2019; Braakman et al. 2022), with most being classified from either statistical or biological viewpoints. The use of statistical considerations increases the range of quantitative tools available for external validation. For example, if BSV is accounted for in a model, more advanced comparisons can be made such as evaluating observed versus predicted proportions of subjects matching a desired response. Including uncertainty into model simulations enables generation of confidence intervals and an appropriate benchmarking of observed and predicted trends (also discussed in Sect. 4.1). From a biological perspective, external validation may be more qualitative in nature, within target, within pathway, or across pathways (Singh et al. 2023). The qualitative yet critical evaluation of biological plausibility of model outputs should be performed preferably in collaboration with biology experts. Within-target, within-pathway, and across-pathway validation are indicators as to how one may successfully make extrapolation using the model, perhaps with another drug acting on the same target, or on a different target within the same pathway, or on a different pathway altogether. Additional external validation exercises may provide greater confidence in a QSP model. For example, if virtual populations were constructed based on baseline values of a biomarker, external validation using longitudinal data for that biomarker under treatment conditions would be considered essential for demonstrating the predictive potential of the model.

5 Software

The QSP modeling workflow proposed in this chapter can be implemented in a wide range of software solutions using various programming languages. The majority of numerical methods for model calibration, evaluation, and simulation are already available in different environments, including MATLAB, R, Python, etc. Given the usual time constraints and the necessity for robust model assessments, it is recommended to use existing, well-established programming solutions actively used by expert modeling scientists in quantitative pharmacology across the industry, academia, and regulatory health authorities. A number of software products exist, including, but not limited to, Berkeley Madonna, COPASI, Monolix, NLMIXR, NONMEM, MATLAB Simbiology, IQRtools, mrgsolve, and Simurg, to perform model calibration and simulation and, as such, enable key elements of the QSP modeling workflow described in this chapter. The various aforementioned software solutions offer extensive and diverse functionalities, making an objective and comprehensive comparison nearly impossible. Consequently, this section serves primarily to highlight the strengths and focus points of the various software tools suitable for QSP modeling. Software packages specific for PBPK modeling (e.g., GastroPlus,

Simcyp) and agent-based modeling (e.g., NetLogo, Morpheus) are not considered here.

Historically, NONMEM was the first program designed for non-linear mixed effects (NLME) calculations in population PK/PD modeling. Initially developed for IBM360/50 hardware, NONMEM has since (1984) been implemented in Fortran90/95 (Bauer 2019b). The program has been updated over several decades, with greatly expanded functionality, and focused primarily on model calibration of population PK/PD models. It includes more than five optimization algorithms for NLME tasks and remains a gold standard in the industry (Bauer 2019a). NONMEM is a flexible software for mathematical modeling with broad functionality, yet it is associated with a steep learning curve and requires extensive manual coding and additional software add-ons in order to perform all steps proposed in the present QSP modeling workflow; it is also slow, computationally speaking, making it arguably a less practical tool for QSP.

Similar to NONMEM, Monolix was built to perform NLME calculations using the SAEM algorithm within a MATLAB environment (Lavielle and Mentré 2007). It rapidly evolved as a standalone software with user-friendly interface and functionality to match those of NONMEM. Monolix uses structural model files encoded in the MLXTRAN language and features built-in comprehensive data exploration, model diagnostics, and model simulation capabilities. The software is complemented with lixoftConnectors and RsSimulx R packages, allowing Monolix to be called via R programming scripts.

As of today, NLMIXR is the first and only open-source free R package with extensive NLME modeling capabilities (Fidler et al. 2019). NLMIXR uses another package, RxODE, as a numerical solver of ODEs and offers a complete integration within the R environment; it uses a convenient script-based language. Mrgsolve is another, open-source R-based package; it is primarily focused on model-based simulations with a syntax that is directly associated with NONMEM; this allows to perform efficient model-based predictions, through R, based on NONMEM control files (Elmokadem et al. 2019).

Berkeley Madonna, COPASI, and MATLAB Simbiology (www.mathworks.com/products/simbiology.html/) represent a class of QSP software tools with a long history stemming from systems biology modeling (Hoops et al. 2006; Keating et al. 2020). These computer programs feature sophisticated graphical user interfaces designed to compile dynamical networks of interactions between multiple model variables in a convenient manner. All three are able to process structural models encoded in the systems biology markup language (SBML) format and feature a wide array of tools, absent in other software solutions, such as built-in steady-state and sensitivity analysis capabilities (Keating et al. 2020). MATLAB Simbiology arguably has the broadest functionality, including NLME algorithms, built-in sensitivity analyses, and more. The scripting language is also available within MATLAB Simbiology (including command-line/m-file coding), yet it is code-heavy as compared to the syntax of other packages such as NLMIXR or Monolix.

IQR is another QSP modeling package developed within the R environment and features a unique mixture of modeling tools (https://iqrtools.intiquan.com/). While having its own syntax, the software offers the functionality to automatically transition models and control files between NONMEM, Monolix, and NLMIXR; it can also read and write SBML files. IQR does not have its own NLME algorithms but can calibrate models without random effects using a gradient descent-based parameter optimization scheme. It also offers versatile functionality for exploratory data analyses, identifiability analyses, and model diagnostics, complemented with a user interface and a tool for automatic reporting. Among the newly emerging programming solutions, Simurg software (https://www.simurgpharm.com/) may be highlighted as an attempt to implement and refine the "umbrella" workflow for various types of commonly occurring model-based analyses to provide a single environment seamlessly supportive of MIDD. It consists of several modules designed for ODE and delay differential equation–based NLME tasks, time-to-event analyses, meta-regression, and more. Another modeling package is PK-Sim/MoBi (https://www.open-systems-pharmacology.org/), an open-source modeling tool, which aims at linking two complementary mechanistic modeling approaches such as PBPK and QSP and supports the standardization and clarity of spatial components in QSP models.

Bayesian modeling solutions are not yet commonly used among pharmacometricians and QSP modelers, and applications in the industry are not yet widespread. Among the available tools, the Stan probabilistic programming language represents, arguably, the lead option for Bayesian QSP model development in the future, along with other options such as WinBUGS/OpenBUGS, JAGS, and approximate Bayes computation methods that are available in various pieces of software (https://mcmc-jags.sourceforge.io/) (Carpenter et al. 2017; Lunn et al. 2009; Sunnåker et al. 2013).

In summary, multiple software tools and environments supporting key elements of the proposed QSP modeling workflow exist, each with their advantages and challenges. The inclusion of NLME algorithm(s) is a key factor when selecting a QSP software of choice; the availability of additional built-in functionalities and graphical interfaces is also helpful in facilitating the development of QSP models.

6 Conclusions

A QSP model is a mathematical system that integrates temporal and spatial aspects of biological and pharmacological processes to predict molecular, cellular, and/or physiological responses. Technically, a QSP model must address the challenge of maintaining a balance between quantitative and qualitative descriptors of the underlying system, while providing a degree of generalizability which cannot be reached using empirical models. QSP modeling is a powerful approach to address a multitude of questions and support decision-making in the context of MIDD, including clinical trial design optimization, dose and biomarker selection, optimization of combination treatments, species-to-species translation, and more (Rogers et al. 2013; Lemaire et al. 2023). However, developing a QSP model is a challenging and time-consuming endeavor requiring comprehensive knowledge of a therapeutic area, extensive data mining, thorough calibration and validation, and cross-discipline expertise (Musuamba et al. 2021).

In this chapter, we proposed a QSP modeling workflow inspired by Musuamba et al., Cucurull-Sanchez et al., Gadkar et al. (Musuamba et al. 2021; Cucurull-Sanchez et al. 2019; Gadkar et al. 2016), and similar works designed to help address many of the above-mentioned challenges. While designed primarily for QSP models with differential equations, many of the practical aspects and principles described in the present workflow are transposable to other mathematical modeling methods. The chapter also contains a few references to Bayesian theory, which may be useful in the future since a full-fledged QSP modeling workflow in a Bayesian paradigm remains to be established. In summary, the workflow consists of three major parts: structural model selection; model calibration and evaluation; and model-based simulations. Details of the methodology and QSP-specific challenges were presented, with companion technical solutions, including reproducibility and completeness of the data mining process, tools for model quality assessments, and the ability to perform inferences based on forward simulations. This information is meant to serve as a resource to both newcomers and experienced QSP modelers, offering an introduction to the field as well as operating procedures and references for routine analyses.

Appendix

A1. Steady-State Evaluation

Consider a modified version of Eq. 1 from the main text:

$$\frac{dX}{dt} = k_{\text{in}} - k_{\text{out}} * X(t) - \frac{V_{\max} * X(t)}{k_m * X(t)}, \tag{A.1}$$

where k_{in}, k_{out}, $V_{\max}$, and Km are parameters and $X(t)$ is a state variable. Trying to compute $X(SS)$ results in the quadratic equation:

$$X^2 + \left(k_m + \frac{V_{\max} - k_{\text{in}}}{k_{\text{out}}}\right) * X - k_m * \frac{k_{\text{in}}}{k_{\text{out}}} = 0. \tag{A.2}$$

Then,

$$X(0) = X(SS) = \frac{-b \pm \sqrt{b^2 - 4c}}{2}, \tag{A.3}$$

where $b = \left(Km + \frac{V_{\max} - k_{\text{in}}}{k_{\text{out}}}\right)$, $c = -\frac{k_{\text{in}}}{k_{\text{out}}} * Km$.

If $X(t)$ is, for instance, a pharmacodynamic marker, a negative solution to Eq. A.2 may be disregarded, since it would not be considered physiologically plausible. Equation A.3 thus provides a baseline, steady-state solution for $X(t)$. While this approach avoids the burn-in period, defining the initial conditions has the additional benefit of reducing the degrees of freedom in the model, since parameter interdependencies can be formally defined.

As suggested, it is advisable to analyze the corresponding steady states, in order to understand potential behaviors of the system, at least on theoretical grounds. A steady state is defined as $dX_k/dt = 0$, for all variables $X_k(t)$, $k = 1, \ldots, K$. The steady-state value of $X_k(t)$ thus corresponds to $X_k(SS) = \lim_{t \to +} X_k(t)$ or to the value a variable will take given an infinite simulation time. A steady state is said to be Lyapunov-stable when for each $\varepsilon > 0$, there exists a $\delta > 0$ such that if $|X(SS) - \overline{X}| < \delta$, $|X(t) - \overline{X}| < \varepsilon$ for $t_{SS} \leq t < +\infty$; i.e., a steady state is stable when, under small perturbations and as defined by δ, it remains in the neighborhood of ε. To illustrate different kinds of steady states, we will use a version of the Verhulst equation (Eq. 8 in the main text), with an added Allee effect (Allee and Bowen 1932):

$$\frac{dX}{dt} = q * \left(1 - \frac{X(t)}{K}\right) * X(t) * (X(t) - a), \tag{A.4}$$

where q, K, and a are parameters ($q, K, a > 0$) and $X(t)$ is the state variable.

Three steady states exist in this model, two being stable ($X(t) = 0$ and K) and one being unstable ($X(t) = a$) (Fig. 9a). A powerful illustration of how a model can behave relative to a single variable and its steady states is its phase trajectory diagram: a graphical visualization of the time derivative of a variable over initial values for this variable (Fig. 9b). Values for which the trajectory crosses 0 on the Y-axis correspond to steady states. Furthermore, evaluating the sign of the derivative in the neighborhood of these values will reveal whether the steady state is Lyapunov-stable or not. If the derivatives on both sides of a steady-state point "move away"

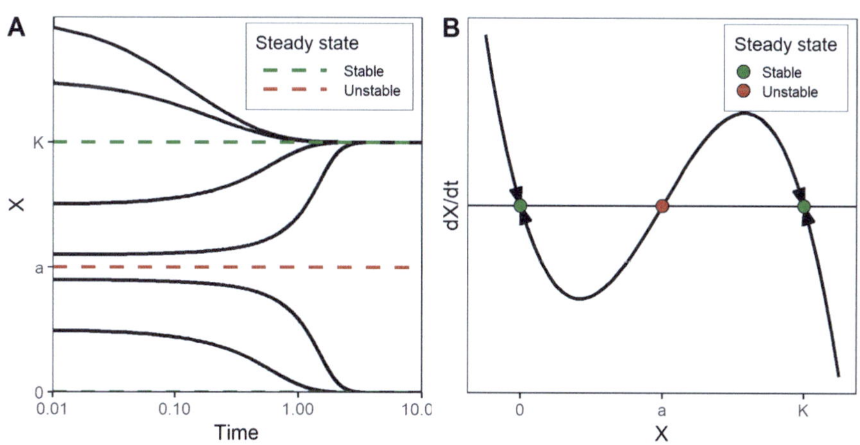

Fig. 9 Change over time in a state variable (**a**) and its phase trajectory (**b**). (Lines – simulations based on Eq. A.4; dashed lines (**a**) or colored points (**b**) model steady states; arrows – sign of the derivative; a and K – model parameters)

from that point, the model will remain in this steady state only if this state also is its initial condition. Thus, in Eq. A.4, parameter a represents an unstable steady state, in contrast to two stable steady states, 0 and K.

A set of phase trajectories for different initial values forms an easily interpretable phase portrait of a system, which allows then to draw conclusions about the nature of changes in the model as a whole (Jones and Sleeman 2003; Khailaie et al. 2013; Sahoo et al. 2020; Balti et al. 2021; Zugaj et al. 2024). The methodology offers the benefit of not requiring an analytical solution of the system, which is typically not obtainable for complex non-linear networks of ODEs, and it allows for the identification of complex patterns such as unstable steady states and oscillators (e.g., Lotka-Volterra equations (Wangersky 1978)).

A2. Structural Identifiability

The general purpose of a structural identifiability analysis is to understand which parameters in the model can be, in theory, unambiguously estimated, provided that the input and output data are noise-free and as exhaustive as possible (Walter and Pronzato 1997). Mathematically, this can be formulated in the following way: parameter θ_i is structurally identifiable, if, for all $\theta \in \Theta \subset \mathbb{R}^l$, the following is true:

$$Y(m(t), \theta) = Y\left(m(t), \widehat{\theta}\right) \Rightarrow \widehat{\theta}_i = \theta_i, \tag{A.5}$$

for a non-linear, continuous-time system

$$\sum_{\theta} \begin{cases} \dot{X}(t) = f_\theta(X(t), m(t)) \\ Y(t) = h_\theta(X(t), m(t)) \end{cases}, \tag{A.6}$$

where $X(t) \in X \subseteq \mathbb{R}^n$ are state variables, $m(t) \in M \subseteq \mathbb{R}^m$ represents input data, and $Y(t) \in Y \subseteq \mathbb{R}^p$ represents model output.

Multiple methods exist to evaluate this proposition, including Taylor expansion, generating series, differential algebra, etc. (Chis et al. 2011). Some of these methods are implemented in specialized toolboxes, e.g., GenSSI, DAISY, COMBOS, and EAR (Ligon et al. 2018; Bellu et al. 2007; Meshkat et al. 2014; Anguelova et al. 2012). However, many methods require symbolic calculations and may be difficult to implement, especially for complex non-linear systems with multiple feedback loops. Consequently, practical identifiability coupled with global sensitivity analysis, as described in the main text of this chapter, may represent a more pragmatic approach.

A3. Model Calibration

Consider a model defined by the following general equation:

$$Y_j(t) = f(X_j, \Theta) + g(X_j, \Theta) * \varepsilon_j, \tag{A.7}$$

where f is the structural model, X_j are independent variables (time, dose, covariates, etc), Θ is a vector of model parameters, g is the residual error model, and $\varepsilon_j \sim N(0, \sigma^2)$, $j = 1 \ldots m$.

The first step in model calibration would consist of finding the estimates $\widehat{\Theta}$ of unknown parameters Θ which correspond to an extremum (minimum or maximum) of an objective function (OF), i.e., $\widehat{\Theta} = \underset{\Theta}{\text{argmax}}(OF)$ or $\widehat{\Theta} = \underset{\Theta}{\text{argmin}}(OF)$, depending on the OF. The simplest OF to consider would be ordinary least squares (OLS) or weighted least squares (WLS):

$$O_{\text{OLS}}(\Theta) = \sum_{j=1}^{m} \left(Y_j^{\text{obs}} - f(X_j, \Theta) \right)^2, \tag{A.8}$$

$$O_{\text{WLS}}(\Theta) = \sum_{j=1}^{m} \left[\left(Y_j^{\text{obs}} - f(X_j, \Theta) \right)^2 * w_j \right], \tag{A.9}$$

where Y_j^{obs} is the observed value at time point j and w_j is the weight of the jth residual.

However, OLS is applicable to models with constant variance, which is not always appropriate for the variety of datasets encountered in practice, while the manual assignment of weights in WLS can be hard to justify (Peck et al. 1984). Both methods are isolated cases of extended least squares (ELS):

$$O_{\text{ELS}}(\Theta) = \sum_{j=1}^{m} \left[\frac{\left(Y_j^{obs} - f(X_j, \Theta) \right)^2}{\text{VAR}_j} + \log(\text{VAR}_j) \right], \tag{A.10}$$

which in turn equates to the -2*logarithm of the likelihood function (-2LL) for the normally distributed and independent $\varepsilon_j \sim N(0, \sigma^2)$:

$$-2LL(\Theta) = m * \log(2\pi) + \sum_{j=1}^{m} \left(\frac{\left(Y_j^{\text{obs}} - f(X_j, \Theta) \right)^2}{\text{VAR}_j} + \log(VAR_j) \right), \tag{A.11}$$

where VAR_j is defined as $g(X_j, \Theta) * \sigma^2$.

As such, the estimation of parameters in QSP modeling is no different from the maximum likelihood approach typically used for conventional PK, PK/PD, and pharmacometrics models, when making frequentist inferences. The residual error model g is usually selected based on the following three options – constant, proportional, and combined:

$$g(X_j, \Theta) = a, \tag{A.12}$$

$$g(X_j, \Theta) = f(X_j, \Theta), \tag{A.13}$$

$$g(X_j, \Theta) = a + f(X_j, \Theta), \tag{A.14}$$

where a is a scalar coefficient.

A4. Model Parameterization with Subject-Level Data

If subject-level data are available for model calibration, Eq. A.7 may be modified to also accommodate for random effects, the way it is done in population modeling:

$$Y_{ij}(t) = f(X_j, \Theta, \Phi) + g(X_j, \Theta, \Phi) * \varepsilon_{ij}, \tag{A.15}$$

where Φ is a vector of random effects, $i = 1, \ldots, n$. Individual parameter values θ_i are then defined, most commonly, by either one of the three types of equations – normal, log-normal, or logit:

$$\theta_i = \theta_{\mathrm{TV}} + \eta_i, \tag{A.16}$$

$$\theta_i = \theta_{\mathrm{TV}} * \exp(\eta_i), \tag{A.17}$$

$$\theta_i = \frac{u * \theta_{\mathrm{TV}} * \exp(\eta_i) + l}{1 + \theta_{\mathrm{TV}} * \exp(\eta_i)}, \tag{A.18}$$

where θ_{TV} is the typical (population) parameter value, η_i is the i*th* value of a random effect, u and l are constants corresponding to the upper and lower ranges of the logit distribution, $\eta_i \sim N(0, \omega^2)$.

A5. Practical Identifiability of the Estimated Parameters

$$I(\widehat{\Theta}) = -\frac{d^2 LL(\widehat{\Theta})}{d\Theta^2}, \tag{A.19}$$

where Θ is a vector of model parameters and $\widehat{\Theta}$ is a vector of maximum likelihood estimates of the parameters. The variance-covariance matrix may then be defined by the following formula:

$$C(\hat{\Theta}) = I(\hat{\Theta})^{-1} = \begin{pmatrix} \mathrm{var}(\theta_1) & \mathrm{cov}(\theta_1, \theta_2) & \ldots & \mathrm{cov}(\theta_1, \theta_n) \\ \mathrm{cov}(\theta_2, \theta_1) & \mathrm{var}(\theta_2) & \ldots & \mathrm{cov}(\theta_2, \theta_n) \\ \ldots & \ldots & \ldots & \ldots \\ \mathrm{cov}(\theta_n, \theta_1) & \mathrm{cov}(\theta_n, \theta_2) & \ldots & \mathrm{var}(\theta_n) \end{pmatrix}, \quad \text{(A.20)}$$

where var(θ_n) is the variance of the nth parameter, cov(θ_{k1}, θ_{k2}) is the covariance between parameters θ_{k1} and θ_{k2}, $1 \leq k1,\ k2 \leq n$.

Consequently, the SE of a parameter is defined as the square root of the diagonal elements of the variance-covariance matrix:

$$\mathrm{SE}\left(\widehat{\theta_k}\right) = \sqrt{C_{k,k}\left(\widehat{\Theta}\right)}, \quad \text{(A.21)}$$

and the confidence interval can be calculated from Eq. A.22 as follows:

$$\mathrm{CI}\left(\widehat{\theta_k}\right) = \widehat{\theta_k} \pm t_{1-\frac{\alpha}{2}} * \mathrm{SE}\left(\widehat{\theta_k}\right), \quad \text{(A.22)}$$

where t represents the Student's t-distribution quantile, α is the significance level, and $\widehat{\theta_k}$ is the point estimate of the θ_k parameter.

The relative standard error (RSE, %) is then defined by the formula:

$$\mathrm{RSE}\left(\widehat{\theta_k}\right) = \frac{SE\left(\widehat{\theta_k}\right)}{\widehat{\theta_k}} * 100\%. \quad \text{(A.23)}$$

Considering altogether Eqs. A.22 and A.23, and the quantiles of the standard normal distribution, it is noted that the 95% confidence interval (CI) of a parameter's point estimate will include zero if the RSE of that parameter exceeds 51%. In other words, an RSE greater than 51% does not allow to reject the hypothesis that this parameter is different from zero, at a significance level of $p = 0.05$, and the parameter must therefore be considered as unidentifiable.

An additional benefit from computing the variance-covariance matrix is the ability to calculate the linear correlation coefficient between two parameter estimates, using the following transformation:

$$\mathrm{corr}\left(\widehat{\theta_{k1}}, \widehat{\theta_{k2}}\right) = \frac{C_{k1,k2}}{\mathrm{SE}\left(\widehat{\theta_{k1}}\right) * \mathrm{SE}\left(\widehat{\theta_{k2}}\right)}. \quad \text{(A.24)}$$

If $\left| \mathrm{corr}\left(\widehat{\theta_{k1}}, \widehat{\theta_{k2}}\right) \right| \geq 0.5$, more than 25% of the observed data variance is equally well described by either one of the two parameters, and the hypothesis that the two parameters follow the same distribution cannot be rejected.

References

Allee WC, Bowen ES (1932) Studies in animal aggregations: mass protection against colloidal silver among goldfishes. J Exp Zool 61:185–207. https://doi.org/10.1002/jez.1400610202

Allen LJS (2011) An introduction to stochastic processes with applications to biology, 2nd edn. Chapman & Hall/CRC, Boca Raton

Allen RJ, Rieger TR, Musante CJ (2016) Efficient generation and selection of virtual populations in quantitative systems pharmacology models. CPT Pharmacometrics Syst Pharmacol 5:140–146. https://doi.org/10.1002/psp4.12063

Anguelova M, Karlsson J, Jirstrand M (2012) Minimal output sets for identifiability. Math Biosci 239:139–153. https://doi.org/10.1016/j.mbs.2012.04.005

Bai JPF, Schmidt BJ, Gadkar KG, Damian V, Earp JC, Friedrich C et al (2021) FDA-Industry Scientific Exchange on assessing quantitative systems pharmacology models in clinical drug development: a meeting report, summary of challenges/gaps, and future perspective. AAPS J 23:60. https://doi.org/10.1208/s12248-021-00585-x

Balti A, Zugaj D, Fenneteau F, Tremblay P-O, Nekka F (2021) Dynamical systems analysis as an additional tool to inform treatment outcomes: the case study of a quantitative systems pharmacology model of immuno-oncology. Chaos 31:023124. https://doi.org/10.1063/5.0022238

Bard Y (1970) Comparison of gradient methods for the solution of nonlinear parameter estimation problems. SIAM J Numer Anal 7:157–186. https://doi.org/10.1137/0707011

Bard Y (1974) Nonlinear parameter estimation. Academic, New York

Bauer RJ (2019b) NONMEM tutorial part I: description of commands and options, with simple examples of population analysis. CPT Pharmacometrics Syst Pharmacol. https://doi.org/10.1002/psp4.12404

Bauer RJ (2019a) NONMEM tutorial part II: estimation methods and advanced examples. CPT Pharmacometrics Syst Pharmacol 8:538–556. https://doi.org/10.1002/psp4.12422

Bélisle CJP (1992) Convergence theorems for a class of simulated annealing algorithms on $\mathbb{R}^d$. J Appl Probab 29:885–895. https://doi.org/10.2307/3214721

Bellu G, Saccomani MP, Audoly S, D'Angiò L (2007) DAISY: a new software tool to test global identifiability of biological and physiological systems. Comput Methods Prog Biomed 88:52–61. https://doi.org/10.1016/j.cmpb.2007.07.002

Bonate PL (2011) Pharmacokinetic-pharmacodynamic modeling and simulation, 2nd edn. Springer, New York

Bonate PL (2014) Be a model communicator: and sell your models to anyone. Peter Bonate. ISBN 0692323813

Borenstein M, Hedges LV, Higgins JPT, Rothstein HR (2010) A basic introduction to fixed-effect and random-effects models for meta-analysis. Res Synth Method 1:97–111. https://doi.org/10.1002/jrsm.12

Box GEP (1979) Robustness in the strategy of scientific model building. In: Robustness in statistics. Elsevier, pp 201–236. https://doi.org/10.1016/B978-0-12-438150-6.50018-2

Braakman S, Pathmanathan P, Moore H (2022) Evaluation framework for systems models. CPT Pharmacometrics Syst Pharmacol 11:264–289. https://doi.org/10.1002/psp4.12755

Bradshaw EL, Spilker ME, Zang R, Bansal L, He H, Jones RDO et al (2019) Applications of quantitative systems pharmacology in model-informed drug discovery: perspective on impact and opportunities. CPT Pharmacometrics Syst Pharmacol 8:777–791. https://doi.org/10.1002/psp4.12463

Butcher JC (1996) A history of Runge-Kutta methods. Appl Numer Math 20:247–260. https://doi.org/10.1016/0168-9274(95)00108-5

Butcher JC (2000) Numerical methods for ordinary differential equations in the 20th century. J Comput Appl Math 125:1–29. https://doi.org/10.1016/S0377-0427(00)00455-6

Carpenter B, Gelman A, Hoffman MD, Lee D, Goodrich B, Betancourt M et al (2017) Stan: a probabilistic programming language. J Stat Softw 76:10.18637/jss.v076.i01

Chan P, Peskov K, Song X (2022) Applications of model-based meta-analysis in drug development. Pharm Res 39:1761–1777. https://doi.org/10.1007/s11095-022-03201-5

Cheng Y, Thalhauser CJ, Smithline S, Pagidala J, Miladinov M, Vezina HE et al (2017) QSP toolbox: computational implementation of integrated workflow components for deploying multi-scale mechanistic models. AAPS J 19:1002–1016. https://doi.org/10.1208/s12248-017-0100-x

Cheng Y, Straube R, Alnaif AE, Huang L, Leil TA, Schmidt BJ (2022) Virtual populations for quantitative systems pharmacology models. In: Bai JPF, Hur J (eds) Systems medicine, vol 2486. Springer US, New York, pp 129–179. https://doi.org/10.1007/978-1-0716-2265-0_8

Chis O-T, Banga JR, Balsa-Canto E (2011) Structural identifiability of systems biology models: a critical comparison of methods. PLoS One 6:e27755. https://doi.org/10.1371/journal.pone.0027755

Cleland WW (1963) The kinetics of enzyme-catalyzed reactions with two or more substrates or products. Biochim Biophys Acta (BBA) – Specialized Section on Enzymological Subjects 67:188–196. https://doi.org/10.1016/0926-6569(63)90227-X

Cornish-Bowden A (2012) Fundamentals of enzyme kinetics. 4th, completely revised and greatly enlarged edition. Wiley-Blackwell, Weinheim

Cucurull-Sanchez L, Chappell MJ, Chelliah V, Amy Cheung SY, Derks G, Penney M et al (2019) Best practices to maximize the use and reuse of quantitative and systems pharmacology models: recommendations from the United Kingdom quantitative and systems pharmacology network. CPT Pharmacometrics Syst Pharmacol 8:259–272. https://doi.org/10.1002/psp4.12381

Dayneka NL, Garg V, Jusko WJ (1993) Comparison of four basic models of indirect pharmacodynamic responses. J Pharmacokinet Biopharm 21:457–478. https://doi.org/10.1007/BF01061691

Duffull S, Gulati A (2020) Potential issues with virtual populations when applied to nonlinear quantitative systems pharmacology models. CPT Pharmacometrics Syst Pharmacol 9:613–616. https://doi.org/10.1002/psp4.12559

Efron B, Tibshirani R (1986) Bootstrap methods for standard errors, confidence intervals, and other measures of statistical accuracy. Stat Sci 1. https://doi.org/10.1214/ss/1177013815

Elmokadem A, Riggs MM, Baron KT (2019) Quantitative systems pharmacology and physiologically-based pharmacokinetic modeling with mrgsolve: a hands-on tutorial. CPT Pharmacometrics Syst Pharmacol 8:883–893. https://doi.org/10.1002/psp4.12467

Fahrmeir L, Kneib T, Lang S, Marx BD (2021) Regression models. Regression. Springer Berlin Heidelberg, Berlin/Heidelberg, pp 23–84. https://doi.org/10.1007/978-3-662-63882-8_2

Fidler M, Wilkins JJ, Hooijmaijers R, Post TM, Schoemaker R, Trame MN et al (2019) Nonlinear mixed-effects model development and simulation using nlmixr and related R open-source packages. CPT Pharmacometrics Syst Pharmacol 8:621–633. https://doi.org/10.1002/psp4.12445

Finlay DB, Duffull SB, Glass M (2020) 100 years of modelling ligand–receptor binding and response: a focus on GPCRs. Br J Pharmacol 177:1472–1484. https://doi.org/10.1111/bph.14988

Fletcher R (1987) Practical methods of optimization, 2nd edn. Wiley, Chichester/New York

Gadkar K, Kirouac D, Mager D, Van Der Graaf P, Ramanujan S (2016) A six-stage workflow for robust application of systems pharmacology. CPT Pharmacom Syst Pharm 5:235–249. https://doi.org/10.1002/psp4.12071

Gibiansky L, Gibiansky E, Kakkar T, Ma P (2008) Approximations of the target-mediated drug disposition model and identifiability of model parameters. J Pharmacokinet Pharmacodyn 35:573–591. https://doi.org/10.1007/s10928-008-9102-8

Girard P (2005) Clinical trial simulation: a tool for understanding study failures and preventing them. Basic Clin Pharmacol Toxicol 96:228–234. https://doi.org/10.1111/j.1742-7843.2005.pto960313.x

Gong C, Ruiz-Martinez A, Kimko H, Popel AS (2021) A spatial quantitative systems pharmacology platform spQSP-IO for simulations of tumor-immune interactions and effects of checkpoint inhibitor immunotherapy. Cancers (Basel) 13:3751. https://doi.org/10.3390/cancers13153751

Goutelle S, Maurin M, Rougier F, Barbaut X, Bourguignon L, Ducher M et al (2008) The Hill equation: a review of its capabilities in pharmacological modelling. Fundam Clin Pharmacol 22: 633–648. https://doi.org/10.1111/j.1472-8206.2008.00633.x

Helbing D (2012) Agent-based modeling. In: Helbing D (ed) Social self-organization. Springer Berlin Heidelberg, Berlin/Heidelberg, pp 25–70. https://doi.org/10.1007/978-3-642-24004-1_2

Helmlinger G, Al-Huniti N, Aksenov S, Peskov K, Hallow KM, Chu L et al (2017) Drug-disease modeling in the pharmaceutical industry – where mechanistic systems pharmacology and statistical pharmacometrics meet. Eur J Pharm Sci 109:S39–S46. https://doi.org/10.1016/j.ejps.2017.05.028

Helmlinger G, Sokolov V, Peskov K, Hallow KM, Kosinsky Y, Voronova V et al (2019) Quantitative systems pharmacology: an exemplar model-building workflow with applications in cardiovascular, metabolic, and oncology drug development. CPT Pharmacometrics Syst Pharmacol 8:380–395. https://doi.org/10.1002/psp4.12426

Holford N (1999) Simulation in drug development: good practices. http://holford.fmhs.auckland.ac.nz/docs/simulation-in-drug-development-good-practices.pdf

Holford NHG, Kimko HC, Monteleone JPR, Peck CC (2000) Simulation of clinical trials. Annu Rev Pharmacol Toxicol 40:209–234. https://doi.org/10.1146/annurev.pharmtox.40.1.209

Holford N, Ma SC, Ploeger BA (2010) Clinical trial simulation: a review. Clin Pharmacol Ther 88: 166–182. https://doi.org/10.1038/clpt.2010.114

Hoops S, Sahle S, Gauges R, Lee C, Pahle J, Simus N et al (2006) COPASI—a COmplex PAthway SImulator. Bioinformatics 22:3067–3074. https://doi.org/10.1093/bioinformatics/btl485

Hsieh N-H, Reisfeld B, Bois FY, Chiu WA (2018) Applying a global sensitivity analysis workflow to improve the computational efficiencies in physiologically-based pharmacokinetic modeling. Front Pharmacol 9:588. https://doi.org/10.3389/fphar.2018.00588

Jafarnejad M, Gong C, Gabrielson E, Bartelink IH, Vicini P, Wang B et al (2019) A computational model of neoadjuvant PD-1 inhibition in non-small cell lung cancer. AAPS J 21:79. https://doi.org/10.1208/s12248-019-0350-x

Jones DS, Sleeman BD (2003) Differential equations and mathematical biology. Chapman & Hall/CRC, Boca Raton

Keating SM, Waltemath D, König M, Zhang F, Dräger A, Chaouiya C et al (2020) SBML level 3: an extensible format for the exchange and reuse of biological models. Mol Syst Biol 16: 10.15252/msb.20199110

Kennedy J, Eberhart R (1995) Particle swarm optimization. In: Proceedings of ICNN'95 – international conference on neural networks, vol 4. IEEE, Perth, pp 1942–1948. https://doi.org/10.1109/ICNN.1995.488968

Khailaie S, Bahrami F, Janahmadi M, Milanez-Almeida P, Huehn J, Meyer-Hermann M (2013) A mathematical model of immune activation with a unified self-nonself concept. Front Immunol 4. https://doi.org/10.3389/fimmu.2013.00474

Kirouac DC (2018) How do we "validate" a QSP model? CPT Pharmacometrics Syst Pharmacol 7: 547–548. https://doi.org/10.1002/psp4.12310

Klipp E, Liebermeister W (2006) Mathematical modeling of intracellular signaling pathways. BMC Neurosci 7:S10. https://doi.org/10.1186/1471-2202-7-S1-S10

Koolman J, Röhm K-H (2013) Color atlas of biochemistry, 3rd ed., Rev updated edn. Thieme, Stuttgart

Kosinsky Y, Dovedi SJ, Peskov K, Voronova V, Chu L, Tomkinson H et al (2018) Radiation and PD-(L)1 treatment combinations: immune response and dose optimization via a predictive systems model. J Immunother Cancer 6:17. https://doi.org/10.1186/s40425-018-0327-9

Kuhn E, Lavielle M (2005) Maximum likelihood estimation in nonlinear mixed effects models. Comput Stat Data Anal 49:1020–1038. https://doi.org/10.1016/j.csda.2004.07.002

Lavielle M, Mentré F (2007) Estimation of population pharmacokinetic parameters of Saquinavir in HIV patients with the MONOLIX software. J Pharmacokinet Pharmacodyn 34:229–249. https://doi.org/10.1007/s10928-006-9043-z

Lemaire V, Bassen D, Reed M, Song R, Khalili S, Lien YTK et al (2023) From cold to hot: changing perceptions and future opportunities for quantitative systems pharmacology modeling in cancer immunotherapy. Clin Pharmacol Ther 113:963–972. https://doi.org/10.1002/cpt.2770

Ligon TS, Fröhlich F, Chiş OT, Banga JR, Balsa-Canto E, Hasenauer J (2018) GenSSI 2.0: multi-experiment structural identifiability analysis of SBML models. Bioinformatics 34:1421–1423. https://doi.org/10.1093/bioinformatics/btx735

Lo-Thong O, Charton P, Cadet XF, Grondin-Perez B, Saavedra E, Damour C et al (2020) Identification of flux checkpoints in a metabolic pathway through white-box, grey-box and black-box modeling approaches. Sci Rep 10:13446. https://doi.org/10.1038/s41598-020-7295-5

Lunn D, Spiegelhalter D, Thomas A, Best N (2009) The BUGS project: evolution, critique and future directions: the Bugs project. Statist Med 28:3049–3067. https://doi.org/10.1002/sim.3680

Luzyanina T, Roose D, Bocharov G (2009) Distributed parameter identification for a label-structured cell population dynamics model using CFSE histogram time-series data. J Math Biol 59:581–603. https://doi.org/10.1007/s00285-008-0244-5

Ma H, Wang H, Sové RJ, Wang J, Giragossian C, Popel AS (2020) Combination therapy with T cell engager and PD-L1 blockade enhances the antitumor potency of T cells as predicted by a QSP model. J Immunother Cancer 8:e001141. https://doi.org/10.1136/jitc-2020-001141

Mager DE, Jusko WJ (2001) General pharmacokinetic model for drugs exhibiting target-mediated drug disposition. J Pharmacokinet Pharmacodyn 28:507–532. https://doi.org/10.1023/a:1014414520282

Mager DE, Krzyzanski W (2005) Quasi-equilibrium pharmacokinetic model for drugs exhibiting target-mediated drug disposition. Pharm Res 22:1589–1596. https://doi.org/10.1007/s11095-005-6650-0

Marcoline FV, Furth J, Nayak S, Grabe M, Macey RI (2022) Berkeley Madonna version 10–A simulation package for solving mathematical models. CPT Pharmacometrics Syst Pharmacol 11:290–301. https://doi.org/10.1002/psp4.12757

Marino S, Hogue IB, Ray CJ, Kirschner DE (2008) A methodology for performing global uncertainty and sensitivity analysis in systems biology. J Theor Biol 254:178–196. https://doi.org/10.1016/j.jtbi.2008.04.011

Marzban S, Han R, Juhász N, Röst G (2021) A hybrid PDE–ABM model for viral dynamics with application to SARS-CoV-2 and influenza. R Soc Open Sci 8:210787. https://doi.org/10.1098/rsos.210787

McCullagh P, Nelder JA (1998) Generalized linear models, 2nd edn. Chapman & Hall/CRC, Boca Raton

McKay MD, Beckman RJ, Conover WJ (1979) A comparison of three methods for selecting values of input variables in the analysis of output from a computer code. Technometrics 21:239. https://doi.org/10.2307/1268522

Meshkat N, Kuo CE, DiStefano J (2014) On finding and using identifiable parameter combinations in nonlinear dynamic systems biology models and COMBOS: a novel web implementation. PLoS One 9:e110261. https://doi.org/10.1371/journal.pone.0110261

Milberg O, Gong C, Jafarnejad M, Bartelink IH, Wang B, Vicini P et al (2019) A QSP model for predicting clinical responses to monotherapy, combination and sequential therapy following CTLA-4, PD-1, and PD-L1 checkpoint blockade. Sci Rep 9:11286. https://doi.org/10.1038/s41598-01947802-4

Moise N, Friedman A (2021) A mathematical model of the multiple sclerosis plaque. J Theor Biol 512:110532. https://doi.org/10.1016/j.jtbi.2020.110532

Moles CG, Mendes P, Banga JR (2003) Parameter estimation in biochemical pathways: a comparison of global optimization methods. Genome Res 13:2467–2474. https://doi.org/10.1101/gr.1262503

Mould D, Upton R (2012) Basic concepts in population modeling, simulation, and model-based drug development. CPT Pharmacometrics Syst Pharmacol 1:6. https://doi.org/10.1038/psp.2012.4

Mullen K, Ardia D, Gil D, Windover D, Cline J (2011) DEoptim: an *R* package for global optimization by differential evolution. J Stat Softw 40. https://doi.org/10.18637/jss.v040.i06
Murad MH, Asi N, Alsawas M, Alahdab F (2016) New evidence pyramid. Evid Based Med 21: 125–127. https://doi.org/10.1136/ebmed-2016-110401
Musuamba FT, Skottheim Rusten I, Lesage R, Russo G, Bursi R, Emili L et al (2021) Scientific and regulatory evaluation of mechanistic *in silico* drug and disease models in drug development: building model credibility. CPT Pharmacometrics Syst Pharmacol 10:804–825. https://doi.org/10.1002/psp4.12669
National Center for Health Statistics (1994) The Third National Health and Nutrition Examination Survey, 1988–94 (NHANES III)
Nelder JA, Mead R (1965) A simplex method for function minimization. Comput J 7:308–313. https://doi.org/10.1093/comjnl/7.4.308
Nguyen THT, Mouksassi M, Holford N, Al-Huniti N, Freedman I, Hooker AC et al (2017) Model evaluation of continuous data Pharmacometric models: metrics and graphics. CPT Pharmacometrics Syst Pharmacol 6:87–109. https://doi.org/10.1002/psp4.12161
Nijsen MJMA, Wu F, Bansal L, Bradshaw-Pierce E, Chan JR, Liederer BM et al (2018) Preclinical QSP modeling in the pharmaceutical industry: an IQ consortium survey examining the current landscape: preclinical QSP in pharmaceutical industry. CPT Pharmacometrics Syst Pharmacol 7:135–146. https://doi.org/10.1002/psp4.12282
Novère NL, Hucka M, Mi H, Moodie S, Schreiber F, Sorokin A et al (2009) The systems biology graphical notation. Nat Biotechnol 27:735–741. https://doi.org/10.1038/nbt.1558
Page MJ, McKenzie JE, Bossuyt PM, Boutron I, Hoffmann TC, Mulrow CD et al (2021) The PRISMA 2020 statement: an updated guideline for reporting systematic reviews. BMJ 372:n71. https://doi.org/10.1136/bmj.n71
Peck CC, Beal SL, Sheiner LB, Nichols AI (1984) Extended least squares nonlinear regression: a possible solution to the "choice of weights" problem in analysis of individual pharmacokinetic data. J Pharmacokinet Biopharm 12:545–558. https://doi.org/10.1007/BF01060132
Peck CC, Barr WH, Benet LZ, Collins J, Desjardins RE, Furst DE et al (1994) Opportunities for integration of pharmacokinetics, pharmacodynamics, and toxicokinetics in rational drug development. J Clin Pharmacol 34:111–119. https://doi.org/10.1002/j.1552-4604.1994.tb03974.x
Peleg M, Corradini MG, Normand MD (2007) The logistic (Verhulst) model for sigmoid microbial growth curves revisited. Food Res Int 40:808–818. https://doi.org/10.1016/j.foodres.2007.01.012
Population Pharmacokinetics: Guidance for industry 2022
Ramanujan S, Chan JR, Friedrich CM, Thalhauser CJ (2019) A flexible approach for context-dependent assessment of quantitative systems pharmacology models. CPT Pharmacometrics Syst Pharmacol 8:340–343. https://doi.org/10.1002/psp4.12409
Rieger TR, Allen RJ, Bystricky L, Chen Y, Colopy GW, Cui Y et al (2018) Improving the generation and selection of virtual populations in quantitative systems pharmacology models. Prog Biophys Mol Biol 139:15–22. https://doi.org/10.1016/j.pbiomolbio.2018.06.02
Rihan FA (2021) Delay differential equations and applications to biology. Springer, Singapore. https://doi.org/10.1007/978-981-16-0626-7
Rogers M, Lyster P, Okita R (2013) NIH support for the emergence of quantitative and systems pharmacology. CPT Pharmacometrics Syst Pharmacol 2:e37. https://doi.org/10.1038/psp.2013.13
Sahoo P, Yang X, Abler D, Maestrini D, Adhikarla V, Frankhouser D et al (2020) Mathematical deconvolution of CAR T-cell proliferation and exhaustion from real-time killing assay data. J R Soc Interface 17:20190734. https://doi.org/10.1098/rsif.2019.0734
Saltelli A, Tarantola S, Chan KP-S (1999) A quantitative model-independent method for global sensitivity analysis of model output. Technometrics 41:39–56. https://doi.org/10.1080/00401706.1999.10485594

Schauer M, Heinrich R (1983) Quasi-steady-state approximation in the mathematical modeling of biochemical reaction networks. Math Biosci 65:155–170. https://doi.org/10.1016/0025-5564(83)90058-5

Segel IH (2001) Enzyme kinetics: behavior and analysis of rapid equilibrium and steady-state enzyme systems, Wiley Classics Library ed., [10. print.]. Wiley, New York

Shah M, Stolbov L, Yakovleva T, Tang W, Sokolov V, Penland RC et al (2021) A model-based approach to investigating the relationship between glucose-insulin dynamics and dapagliflozin treatment effect in patients with type 2 diabetes. Diabetes Obes Metab 23:991–1000. https://doi.org/10.1111/dom.14305

Sheiner LB (1997) Learning versus confirming in clinical drug development. Clin Pharmacol Ther 61:275–291. https://doi.org/10.1016/S0009-9236(97)90160-0

Singh FA, Afzal N, Smithline SJ, Thalhauser CJ (2023) Assessing the performance of QSP models: biology as the driver for validation. J Pharmacokinet Pharmacodyn. https://doi.org/10.1007/s10928-023-09871-x

Sobol′ IM (2001) Global sensitivity indices for nonlinear mathematical models and their Monte Carlo estimates. Math Comput Simul 55:271–280. https://doi.org/10.1016/S0378-4754(00)00270-6

Sokolov V, Yakovleva T, Chu L, Tang W, Greasley PJ, Johansson S et al (2020) Differentiating the sodium-glucose cotransporter 1 inhibition capacity of Canagliflozin vs. Dapagliflozin and Empagliflozin using quantitative systems harmacology modeling. CPT Pharmacometrics Syst Pharmacol 9:222–229. https://doi.org/10.1002/psp4.12498

Sokolov V, Yakovleva T, Stolbov L, Penland RC, Boulton D, Parkinson J et al (2023) A mechanistic modeling platform of SGLT2 inhibition: implications for type 1 diabetes. CPT Pharmacom Syst Pharm 12:831–841. https://doi.org/10.1002/psp4.12956

Sunnåker M, Busetto AG, Numminen E, Corander J, Foll M, Dessimoz C (2013) Approximate Bayesian computation. PLoS Comput Biol 9:e1002803. https://doi.org/10.1371/journal.pcbi.1002803

The American Society of Mechanical Engineers (ASME) (n.d.) Assessing credibility of computational modeling through verification and validation: application to medical devices VV40 – 2018. https://www.asme.org/products/codes-standards/v-v-40-2018-assessing-credibility-computational

Venzon DJ, Moolgavkar SH (1988) A method for computing profile-likelihood-based confidence intervals. Appl Stat 37:87. https://doi.org/10.2307/2347496

Viceconti M, Henney A, Morley-Fletcher E (2016) In silico clinical trials: how computer simulation will transform the biomedical industry. Int J Clin Trials 3(37). https://doi.org/10.18203/2349-3259.ijct20161408

Walter E, Pronzato L (1997) Identification of parametric models from experimental data. Springer/Masson, Berlin/New York/Paris

Wang R-S, Saadatpour A, Albert R (2012) Boolean modeling in systems biology: an overview of methodology and applications. Phys Biol 9:055001. https://doi.org/10.1088/1478-3975/9/5/055001

Wangersky PJ (1978) Lotka-Volterra population models. Annu Rev Ecol Syst 9:189–218. https://doi.org/10.1146/annurev.es.09.110178.001201

Watson GA (1978) Numerical analysis: proceedings of the biennial conference. Springer-Verlag, Berlin

Wendt JF, Anderson JD, Von Karman Institute for Fluid Dynamics (eds) (2008) Computational fluid dynamics: an introduction, 3rd edn. Springer, Berlin/London

Wilcox RR (2010) Fundamentals of modern statistical methods: substantially improving power and accuracy, 2nd edn. Springer, New York

Willmann S, Höhn K, Edginton A, Sevestre M, Solodenko J, Weiss W et al (2007) Development of a physiology-based whole-body population model for assessing the influence of individual variability on the pharmacokinetics of drugs. J Pharmacokinet Pharmacodyn 34:401–431. https://doi.org/10.1007/s10928-007-9053-5

Yakovleva T, Sokolov V, Chu L, Tang W, Greasley PJ, Peilot Sjögren H et al (2019) Comparison of the urinary glucose excretion contributions of SGLT2 and SGLT1: a quantitative systems pharmacology analysis in healthy individuals and patients with type 2 diabetes treated with SGLT2 inhibitors. Diabetes Obes Metab 21:2684–2693. https://doi.org/10.1111/dom.13858

Zhang L, Beal SL, Sheiner LB (2003) Simultaneous vs. sequential analysis for population PK/PD data I: best-case performance. J Pharmacokinet Pharmacodyn 30:387–404. https://doi.org/10.1023/b:jopa.0000012998.04442.1f

Zhang X-Y, Trame MN, Lesko LJ, Schmidt S (2015) Sobol sensitivity analysis: a tool to guide the development and evaluation of systems pharmacology models. CPT Pharmacometrics Syst Pharmacol 4:69–79. https://doi.org/10.1002/psp4.6

Zugaj D, Fenneteau F, Tremblay P-O, Nekka F (2024) Dynamical behavior-based approach for the evaluation of treatment efficacy: the case of immuno-oncology. Chaos: an interdisciplinary. J Nonlinear Sci 34:013142. https://doi.org/10.1063/5.0170329

Role of Data in Development and Application of Quantitative Systems Pharmacology Models

Jason R. Chan, Abhishek Gulati, Mary E. Spilker, and Erica L. Bradshaw

Contents

J. R. Chan (✉)
Global PK/PD & Pharmacometrics, Eli Lilly and Company, Indianapolis, IN, USA
e-mail: jrchan@lilly.com

A. Gulati
Quantitative Pharmacology & Pharmacometrics, Merck & Co., Inc., West Point, PA, USA
e-mail: abhishek.gulati@merck.com

M. E. Spilker
Pharmacokinetics, Dynamics and Metabolism, Pfizer Inc., La Jolla, CA, USA
e-mail: mary.spilker@pfizer.com

E. L. Bradshaw
Biomosaic Consulting, LLC, Carlsbad, CA, USA

E. L. Bradshaw, M. E. Spilker (eds.), *Quantitative Systems Pharmacology*,
Handbook of Experimental Pharmacology 289, https://doi.org/10.1007/164_2025_752

Abstract

Quantitative systems pharmacology (QSP) is an interdisciplinary approach that integrates mathematical modeling, biological knowledge, and experimental data to gain insights into the complex dynamics of drug action and disease progression on the human physiological system. Different types of data from multiple and often disparate sources are used throughout all phases of QSP model development. Herein, we explore the various types of data used to inform the development and application of QSP models, their sources, and how the quality of data impacts the interpretation of model-derived results.

Keywords

Assay systems · Biological variability · Data curation · Data integration · Measurement variability · Systems pharmacology

1 Introduction

Quantitative systems pharmacology (QSP) is a mechanistic modeling approach that combines computational and experimental methods with the goal of understanding the complexities of drug action on disease progression (Sorger et al. 2011a). QSP models play a pivotal role in drug discovery through clinical development by providing a comprehensive knowledge-based framework to assess the efficacy and safety of potential therapeutics. A typical QSP model incorporates drug information (pharmacokinetic and pharmacodynamic) as input into a mechanistic biology model with the goal of extrapolating changes in the modeled biology to outcomes. Figure 1 provides a high-level illustration of the model-building process and highlights some examples of model uses. At the heart of QSP models lies the critical role of data, which serves as the foundation for model development, calibration, and validation. The integration of diverse data types informed by natural history studies, preclinical in vitro and in vivo experiments, and clinical trial data spanning Phases 1, 2, and 3 enables QSP models to capture the intricate relationships between drug molecules, cellular processes, and physiological responses related to clinical outcomes. Through integration of the different types of data arising from these studies, QSP models facilitate informed decision-making and guide the development of optimized treatment strategies (Bradshaw et al. 2019).

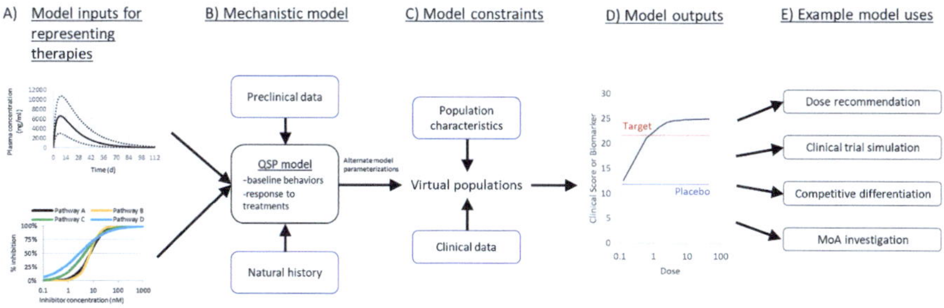

Fig. 1 General schematic of a QSP model with application to drug development is illustrated. Various types of data may be applied to QSP models to represent therapies (**a**), to develop the underlying mechanistic model (**b**), and to provide model constraints (**c**). QSP models can be calibrated and qualified against the dose/response relationship for existing therapies and used to predict outcomes for drug candidates (**d**). This information can then be applied for various purposes relevant to drug development (**e**)

The value of the QSP approach is that it uses information from biological systems for extrapolation, which allows for knowledge-based predictions into spaces where no data exists (Hasegawa and Duffull 2018). The scope of biology modeled is a critical consideration when building a QSP model (Gadkar et al. 2016). The model should contain sufficient biological detail to answer the question(s) of interest and, given the multidisciplinary nature of QSP, should include input from biologists, biomarker experts, clinicians, and modelers. Given the diverse datasets that are often leveraged in QSP models, it is critical to partner with data experts that can facilitate an understanding of the measurement methodology and how best to represent the data in a QSP model. This includes providing an assessment of the qualitative/quantitative nature of the data, translatability, and the measurement error associated with it.

This chapter will step through some types of data underpinning QSP model development and application. Initially, QSP model development begins with the creation of a visual map or flow diagram, akin to a figure one may find in a review article on disease pathophysiology. This qualitative, visual model schematic is quite useful to communicate model scope and is easily understood by all stakeholders. This has been illustrated in "What Is QSP and Why Does It Exist?: A Brief History" (Chapter 1), with a pathophysiology schematic of atopic dermatitis and the corresponding QSP model in Figs. 1 and 2, respectively. The end goal of this initial phase of QSP model development is to have all key stakeholders agree on the relevant biology and to put the key elements (e.g., cells, mediators, and analytes) together into a single framework, much like pieces of a puzzle, that represents the current understanding of the pathophysiology. While it is debatable how many mechanisms are required for a model to be considered a QSP model, the data considerations needed to develop or validate a model are the same, regardless of whether the model has two mechanisms or hundreds.

Once the key elements have been defined, teams will seek out datasets to develop the basic biological mechanisms in the model. This effort can leverage natural history data as well as preclinical and existing clinical datasets. These datasets may encompass diverse data types representing biomarkers, biomeasures, and disease endpoint data relevant to the natural progression of the disease or following a perturbation. The dynamic, temporal response of these data to perturbations is often critical for development of a QSP model. Equally important is the incorporation of the pharmacologic intervention, which is an essential element of a QSP model. This is often represented in QSP models as therapeutic concentrations that are linked to pharmacodynamic (PD) effects. With these general elements, a QSP model can describe and predict individual or average responses. Depending on the QSP model context of use, the model may need to further incorporate variability within and between subjects to develop virtual patients and populations. The interplay between different datasets that inform on the structure and performance of a QSP model has the potential to confirm mechanistic understanding of a biological phenomenon or further inform on this understanding. This chapter will highlight a few of the more common data types used in the development and application of QSP models, their sources, and how the quality of data impacts the interpretation of model-derived results.

2 Data Used in QSP Model Development

Data obtained from natural history studies, preclinical and clinical experimentation, and "omic" studies are useful for informing the structure and dynamics of QSP models. Natural history studies elucidate the temporal evolution of biology associated with disease progression in humans. This can identify critical biological components which can then be further interrogated with controlled in vitro and in vivo studies. These latter studies can provide data describing the baseline/steady-state conditions of the unperturbed system as well as those following perturbations that generate concentration- and time- dependent data. While data from both conditions are important, it is often the dynamics of a system that are most useful for QSP and mechanistic modeling purposes.

2.1 Natural History

Natural history studies follow a group of people over time who have, or are at risk of developing, a specific medical condition or disease. These studies are observational and non-interventional in nature and may be retrospective or prospective. This type of data can be invaluable for QSP model development. Importantly, they characterize how biology relevant to disease changes over time, providing mechanistic information that can be used for model development and calibration. The TrialNet Natural History study for type 1 diabetes is a good example (Mahon et al. 2009). This study consists of three phases: screening, baseline risk assessment, and follow-up risk assessments to understand factors that may characterize diabetes onset. Of particular interest is the collection of oral glucose tolerance tests every six months as well as samples for future tests of T-cell function, autoantibody isotypes, RNA gene expression, and proteomics. Another example is the Alzheimer's Disease Neuroimaging Initiative, which is a longitudinal multicenter study whose objective is to develop clinical imaging, genetic, and biochemical biomarkers for the early detection and tracking of Alzheimer's disease (Mueller et al. 2005). Data such as these have the potential to provide baseline disease progression details within a QSP model that can be subsequently perturbed with therapeutic interventions. Confidence in simulations coming from QSP models that are developed using natural history studies would intuitively be higher than QSP models that are based on preclinical data alone.

2.2 Preclinical Data

Preclinical data play a crucial role in the development of QSP models. Preclinical in vitro and in vivo data are essential for model parameterization and for describing the dynamics of the biological processes to be modeled. These experiments can decouple the system for detailed study of a particular mechanism and can often provide data that simply cannot be sufficiently measured in humans. Furthermore, this data can inform on a range of diverse biological elements such as receptor numbers per cell, baseline protein concentrations, protein turnover rates, receptor internalization rates, and partition coefficients in a manner that is sufficient for QSP model parameterization. Preclinical datasets are often used to inform on a drug's mechanism of action and provide initial potency or pharmacology information. They can also be used to assess how in vitro data and parameters may extrapolate to in vivo measures and ultimately to humans. In this regard, they may inform on potential species differences that would need to be addressed for incorporation into clinical QSP models.

2.2.1 In Vitro Data

In QSP model development, in vitro systems and measurements are valuable for the elucidation of specific model components and associated parameter values that can be difficult to assess through in vivo experiments or from clinical data. Cell-free, cell-based, and tissue-based systems are the most common in vitro systems used to study how endogenous and exogenous molecules interact with receptors, proteins, and enzymes to help us understand normal physiological functions, disease mechanisms, and the effects of therapeutic agents. Within these systems, several different types of assay formats may be used to obtain data such as binding measurements, enzyme activity, gene expression, protein levels, signaling function, and cell growth, and viability. As with QSP models, assay systems and methods can have a specific context of use; therefore, it is important for modelers to be aware of what an assay is designed to measure and their potential limitations. Here, we briefly discuss strengths and limitations of in vitro systems and some key considerations for a few of the frequently employed methods. The authors direct the readers to *The Assay Guidance Manual* which is an eBook managed by the National Center for

Advancing Translational Sciences (NCATS) on the National Library of Medicine website (AGM eBook) for a comprehensive review of the design, development, and use of assays and assay systems to support quantitative biology and pharmacology . (Markossian et al. 2004; Coussens et al. 2018)

Cell-Free, Cell-Based, and Tissue-Based Assay Systems and Data
Cell-free assays examine biochemical reactions outside of living cells, using isolated enzymes, proteins, nucleic acids, or other biomolecules to study enzyme activity, protein-protein interactions, ligand binding, and metabolic pathways without the complexity of cellular systems. A key strength of cell-free assays is their simplicity and precision, offering a high degree of control over experimental conditions and allowing for the isolation of specific variables. Binding measurement, specifically the association and dissociation rates of drug-target, ligand-target, or other protein-protein interactions, and enzyme kinetics to understanding protein activity and how they respond to binding components (i.e., an inhibitor or a substrate) are often key measurements made in cell-free environments. Several different methods can be employed to obtain these data such as radioactive ligand binding assays and surface plasmon resonance (SPR) for binding measurements, and substrate conversion, bioluminescence- and fluorescence-based imaging assays to measure enzyme activity. (Hoare 2004; Brooks et al. 2004; Auld et al. 2004; Wu et al. 2003; Zegzouti et al. 2009; Ciuffreda et al. 2024) Some key considerations that may limit the direct in vitro to in vivo translation of results from these types of assays include the following: (1) The use of cofactors or competing interactions may alter binding kinetics which do not reflect the in vivo conditions. (2) Protein immobilization can alter the conformation or accessibility and can affect binding affinity and kinetics compared to behavior in cellular context. For example, proteins which require multimeric formation, such as tumor necrosis factor alpha (TNFa), can be susceptible to ligand immobilization effects on conformation, which directly impact understanding of binding affinity (McMillan et al. 2021). Additionally, when receptor dynamics are influenced by membrane fluidity, receptor clustering or lateral mobility assays with immobilized proteins may not provide information that can directly be incorporated into a model. (3) Lastly, the use of defined buffer systems with controlled pH, ionic strength, and temperature may not reflect the in vivo condition of interest and limit the utility of direct readouts from an in vitro assay.

Cell-free biochemical assays are powerful tools for generating precise and detailed biochemical data; however, they come with several key limitations that can impact the applicability of the data, especially in the context of systems modeling. The lack of cellular context removes cellular architecture (presence of membranes, organelles, etc.) and does not capture key components of a "system" such as regulatory networks and post-translational modifications. Additionally, simplified assay systems typically use defined buffers and purified systems that may not accurately recapitulate the cellular environment. For example, in cell-free biochemical assays, IC50 is measured using ATP concentrations at or below Km. However, physiological ATP concentrations typically are in the millimolar range (Smyth and Collins 2009). Proteins can also lose structure and activity in these artificial environments, especially over time, which may affect the reliability and translatability of kinetic measurements. When assessing pharmacologic interactions, cell-free systems are often a first step in assessing potency parameters followed by testing in relevant cell-based systems.

Cell-based systems are useful for the assessment of functional responses such as receptor activation, signal transduction, gene expression, protein synthesis, secretion, degradation, or cell viability in response to endogenous or exogenous agents. As a simplified example, ligand stimulation of a receptor, such as the ligand epidermal growth factor (EGF) binding to its receptor, EGFR, results in activation of EGFR, which in turn activates downstream signaling events such as activation of pathway proteins (e.g., Ras, RAF, MEK, ERK, PI3K, and AKT), ultimately leading to phenotypic changes in cellular growth and proliferation. (Tito et al. 2025; Wee and Wang 2017) In oncology, cellular systems (i.e., cancer cell lines grown in 2D culture) are commonly used to examine the effects of a therapeutic intervention on signal transduction and how that affects cellular proliferation. Data generated from these experiments can include measurements of proteins in the pathway, for instance EGFR, ERK, and PI3K, their activated states, phospho-EGFR, phospho-ERK and phospho-AKT, and cellular proliferation. (Toulany et al. 2007; Renshaw et al. 2013) When measuring phosphorylated proteins, the results are often reported as phosphorylated protein relative to total protein levels (e.g., phospho-AKT/total AKT). In these cases, it is important to review the dynamics of the total protein to ensure changes are appropriately allocated to the phosphorylated and/or total protein levels. As another example, TNFa (ligand) binding to its receptor (TNFR1) results in activation of the NF-kB pathway leading to transcription, expression, and secretion of cytokines (i.e., TNFa, IL-6, IL-8, and IL-10.) (Liu et al. 2017), which can be evaluated in plated (i.e., HEK293) (Takashima et al. 2013; Razonable et al. 2005) or suspension (i.e., Jurkat, THP-1) cell systems (Khalaf et al. 2010). The quantification of these types of data is central to QSP model development and parameterization.

A thorough review of the numerous methods available to enable measurements of proteins, genes, transcripts, and cell viability is beyond the scope of this chapter and book. Here we highlight some considerations and provide a table (Table 1) capturing a few key methods frequently employed. It is important for the QSP modeler to have an appreciation for the strengths and weaknesses of different assay formats to facilitate an understanding of potential data limitations. For instance, Western blot data is often characterized through digitization and image processing methods and presented numerically/graphically. But it is important to evaluate how protein levels of interest are assessed, e.g., if levels are benchmarked to a known sample control. Even with a conserved protein control, data is generally viewed as semi-quantitative and therefore presented as a fold or percent change relative to a benchmark. Despite Western blot data being acknowledged as semi-quantitative, it is often used as a building block in QSP models of protein networks given its broad use in biology literature. More quantitative datasets may be generated using other methods such as enzyme-linked immunosorbant assays (ELISA), Meso Scale Discovery (MSD), or mass spectrometry assays, where reagents, dilution methods, and standard curves are optimized for delivery of quantitative protein values. Standard curves and internal

Table 1 Sampling of in vitro assay systems used in QSP models

Method	Sensitivity	Specificity	Quantitative?	Sample Type	Measurements (uses)	Key strengths	Limitations
SPR	High	Limited	Quantitative	Purified chemicals, DNA, proteins, cells, biofluids	Kinetics of biomolecule interactions	Label-free, high sensitivity, real-time measurements	Highly sensitive to pH, temperature and sample preparation conditions. Limited specificity.
BCA Assay	High (~0.5 μg/mL)	No	Approximate	protein extracts	Total protein	Stable signal, detergent-compatible	Total protein measure (no specificity), incompatible with some reducing agents
Western Blot	High	Yes (antibody-based)	Semi-quantitative	Complex samples	Protein and size	High specificity and can assess protein size and post-translational modifications	Qualitative, requires normalization, susceptible to signal saturation, time consuming, and labor intensive
JESS / WES (ProteinSimple)	Very High (pg level)	Yes	Quantitative	Lysates, serum, etc.	Protein and size	High sensitivity and good reproducibility, low sample requirement, multiplexing possible	Dependent upon antibody quality and specificity; potential for cross-reactivity. Need for normalization
ELISA	High (ng–pg/mL)	Yes	Quantitative	Biofluids, lysates	Protein, antibodies	High sensitivity and specificity, highly customizable, & scalable	Dependent upon antibody quality and specificity; potential for cross-reactivity & matrix effects
MSD (Meso Scale)	Ultra-high (fg–pg)	Yes	Quantitative	Biofluids, tissue lysates	Protein, antibodies	Ultra-sensitive, wide dynamic range, multiplexed detection	Dependent upon antibody quality and specificity; potential for cross-reactivity

LC-MS/MS	Ultra-high (fmol)	Yes	Quantitative	Any	Small molecules, peptides, proteins, antibodies	Deep proteome coverage; targeted or discovery-based	Sample prep is highly important, instrumentation is expensive and requires expertise
AlphaLISA/ AlphaScreen	High (pg/mL)	Yes	Quantitative	Cell culture, serum		High sensitivity, good for low-abundance proteins	Prone to interference
Flow cytometry/ CyTOF	Moderate	Yes	Quantitative	Single cells	Proteins, cell populations, DNA cycle, cell death/ health	Evaluate multiple parameters, cell-specific	Requires high-quality antibodies and calibration, single cell suspensions, and intracellular targets requires membrane permeabilization
qPCR	High (single-copy detection)	Yes (primer/ probe specific)	Quantitative	Extracts from cells & tissues	DNA	Highly sensitive, specific, broad dynamic range	Requires clean samples, risk of contamination
RT-qPCR	High (1–10 copies RNA)	Yes (primer/ probe specific)	Quantitative	Extracts from cells & tissues	mRNA	Quantitative mRNA detection, gene expression analysis	RNA quality dependent, reverse transcription efficiency

controls are of particular importance when using an assay to compare compounds, such as when prioritizing candidates during lead optimization, a typical QSP application.

Functional endpoints such as the assessment of cell growth, viability, and mechanisms of cells death provide essential functional and mechanistic data for QSP modeling. When carefully selected and interpreted within their appropriate context of use, they allow for representation of biological processes and dynamic behaviors relevant to disease pathophysiology and therapeutic intervention. Commonly used cell growth and viability assays for the assessment of cell health and proliferative capacity include assays that measure cell number, cell size, metabolic activity, or DNA replication. Assays such as MTT, XTT, WST-1, and resazurin (Alamar Blue) that assess cellular metabolic activity and ATP-based assays (e.g., CellTiter-Glo) that provide a more direct measure of cellular energy status are well suited for concentration-response studies and may be high throughput and robust; however, results may be influenced by metabolic state rather than absolute cell number, and may not distinguish between cytostatic and cytotoxic effects. (Riss et al. 2004a, b) Mechanisms of cell death can be interrogated through a variety of assays that distinguish apoptosis from necrosis or other forms of cell death (e.g., pyroptosis, ferroptosis). For example, Annexin V staining (often paired with propidium iodide) can be used to detect early apoptotic changes in the plasma membrane, while caspase activity assays (e.g., Caspase-Glo) measure key enzymatic events in the apoptotic cascade (Riss et al. 2004b, c). TUNEL assays can assess DNA fragmentation, while LDH release assays provide a measure of membrane integrity commonly associated with necrotic cell death (Riss et al. 2004b, c) These assays, while informative, may require orthogonal validation as some features (e.g., membrane permeabilization) are not exclusive to a single mode of death. The diversity of measurement features and their interpretation highlight the utility of QSP modeler collaboration with assay specialists and biology experts when leveraging in vitro data for QSP modeling.

A key strength of cell-based assays is the maintenance of native cellular architecture and intracellular signaling machinery, allowing for the capture of time-dependent processes and the dynamic interplay of signaling pathways, receptor trafficking, and intracellular feedback mechanisms. This makes them especially useful for capturing complex cellular behaviors relevant to disease biology or drug response. However, cell-based systems also have several limitations. Data from cell-based systems may be more variable than cell-free assays due to differences in cell line origin, passage number, and culture conditions, which can impact reproducibility and translatability. Additionally, the response observed may be influenced by off-target effects or compensatory signaling pathways, complicating the interpretation of results. It is also important to note that overexpression systems, frequently used to enhance signal detection, may not reflect endogenous receptor or protein levels and can bias downstream readouts. Immortalized cell lines are frequently used and may differ significantly from primary cells. While primary cells may be viewed as a more translatable system, or at least more closely related to in vivo or disease state, these cells are prone to batch-to-batch variability. Moreover, differences in protein expression, pathways, and signaling may be different across species and cell systems, which can play an important role if in vitro to in vivo understanding of data is required or for extrapolation of a model system from a preclinical model to humans.

Tissue-based in vitro and ex vivo assay systems, along with physiologically relevant blood-derived assays, provide an increasing level of biological complexity and are useful for generating mechanistic and quantitative data to support QSP modeling. These systems bridge the gap between molecular targets and whole-organism responses by preserving essential aspects of structural and spatial organization, cell-cell interactions, and micro-environmental context. Precision cut tissue (or tumor) slices are ex vivo systems that preserve native tissue architecture, cell heterogeneity, and stromal interactions, enabling functional assessments of drug response in a patient-specific context (Centenera et al. 2018; Bigaeva et al. 2020). In oncology, patient-derived tumor slices have been used to test checkpoint inhibitors and kinase-targeted therapies, with outputs such as immune cell infiltration and cytokine release used to calibrate immune-oncology models (Dimou et al. 2022; Sivakumar et al. 2019; Roelants et al. 2020). In immunology, lymphoid tissue slices and whole blood stimulation assays (Müller et al. 2024; Belanger et al. 2021) can support dynamic analyses of cytokine networks (e.g., TNF-α, IL-1β, IL-6), which may inform immune response modeling.

3D organoid cultures, which are miniaturized organ-like structures derived from stem cells or patient tissues, reproduce aspects of tissue-specific architecture, differentiation, and function (Kijanska and Kelm 2004). For example, intestinal organoids have been used to model inflammatory cytokine effects in IBD (Kandilogiannakis et al. 2021), and airway organoids have supported SARS-CoV-2 and other viral infection studies (Han et al. 2022; Sachs et al. 2019). These models offer prolonged culture times, patient specificity, and experimental manipulability, though they may lack vascular and immune elements unless designed to include these aspects. Another tissue-based system are the organ-on-a-chip (OoC; also termed microphysiological systems (MPS)) which integrate microfluidic flow and mechanics to recapitulate tissue-tissue interfaces and dynamic physiology (Wu et al. 2020). Notable examples include lung-on-a-chip (Huh et al. 2010), liver-on-a-chip (Mehta et al. 2024), and multi-organ chips for ADME-tox prediction (Sung 2021; Shroff et al. 2023). These systems allow readouts of PK/PD behavior and mechanistic responses to therapeutics, but are often limited by complexity, cost, and lower scalability. Advanced technologies such as multiplexed immunohistochemistry, RNAscope, and spatial transcriptomics enable spatially resolved quantification of biomarker expression, cell-cell interactions, and pathway activity (Eng et al. 2019; Zhu et al. 2024; Cohn et al. 2023; Goltsev et al. 2018). These high-dimensional datasets are valuable tools to help bridge molecular signaling with tissue-level behavior. While challenges remain, including variability, standardization, and data integration, these physiologically relevant systems form a foundational part of QSP workflows, enhancing translational modeling and predictive pharmacology in complex disease settings.

Data Interpretation and Considerations in QSP Modeling

In vitro preclinical data, analysis, and interpretation are fundamental for QSP modeling and enable the translation of assay results into predictive and mechanistic systems-level insights. When evaluating preclinical assay results and/or data a few important considerations to bear in mind are factors such as assay conditions, assay systems, nonlinearity and saturation, variability, and scaling. Assay factors such as substrate, inhibitor and ligand concentrations, receptor expression levels, and buffer/media components can all impact data generated and influence in vitro to in vivo translation. The generation of kinetic parameters should consider potential nonlinear behavior and whether protein inhibition is competitive, uncompetitive, or non-competitive which may affect how assays are designed and analyzed. For example, in competitive inhibition, as the concentration of the inhibitor increases, the observed reaction rate will decrease, but this effect can be overcome by increasing the substrate concentration. Therefore, the substrate concentration can have a profound impact on the interpretation of enzyme Km for the inhibitor. As mentioned above, the assay system used is very important to understand elements such as intracellular and extracellular proteins having different interactions and function, distribution (such as a molecules ability to cross membranes or be taken up into cells), and potential feedback mechanisms, as these may impact or influence in vitro to in vivo translation of assay data. Variability is a critical aspect of assay data interpretation and use. Experimental variability can exist as part of the assay measurement and as part of the system. When assays are developed, precision and accuracy of measurements as well as linearity of assays and upper and lower limits of quantification should be assessed. System or assay variability may arise from the proteins or cell systems being used, with assay variability increasing generally with increasing systems complexity (proteins > cellular > multi-cellular > tissue-based). Additionally, primary cell and tissue-based systems may be prone to higher levels of overall experimental variability due to biological variability in the system itself. In QSP, it is necessary to scale these in vitro–derived data to predict in vivo systems. Where possible, it is ideal to document any appropriate physiological and biological assumptions, and variability should be captured with confidence intervals or population distributions. This information can then be communicated to key stakeholders to achieve alignment and facilitate the interpretation of QSP simulations.

2.2.2 In Vivo Data

Building upon in vitro knowledge, preclinical in vivo datasets can further elucidate biological mechanisms and confirm in vitro–derived knowledge of a therapeutic's potency or mechanism of action within the context of a whole organism. This data can provide valuable information for QSP models with applications in the preclinical, translational, and clinical spaces. Preclinical and clinical in vivo data share many similarities, some of which are discussed in Sect. 3. This section will highlight aspects of preclinical datasets that cannot be easily generated clinically and provide valuable knowledge for use in QSP models.

Preclinical Models for Mechanistic Understanding

While clinical manifestations and phenotypes often define a disease, circulating biomarkers and tissue histology may provide clues to disease pathophysiology. Unfortunately, this data can be costly and difficult to acquire in humans in a manner that can inform on components of QSP models. For example, histology data provides a snapshot of the disease at a given time point and if derived from a biopsy may reflect only a small portion of the tissue. This data is often insufficient to fully understand disease pathogenesis and potential response to new therapies. To fill this knowledge gap, animal models are used for mechanistic understanding and hypotheses testing of pathways, endogenous molecules, and/or cell types in a disease setting. Moreover, they allow for interrogation of dynamic-, cellular-, and organ-level interactions that cannot be easily recapitulated in vitro. Animal model disease manifestations typically occur on a faster time scale than in humans, which facilitates data acquisition relevant to disease progression. They also provide an opportunity to sample tissues in a manner that is not feasible in humans including the potential to interrogate entire organs. For example, in oncology, human-derived cell lines can be implanted in immune-compromised mice to enable the study of tumor growth (and evaluation of response to therapeutics) over the course of a few weeks, rather than months or years. Furthermore, throughout the study, the entire tumor can be retrieved and examined using many of the same assay formats as described for in vitro cell-based assays.

Various tools exist to probe biological mechanisms, including but not limited to genetic manipulation (knockout, knock-in), adoptive transfer, siRNA, and pharmacological intervention with small and large molecule antagonists, inhibitors, or agonists. The method used to induce the disease pathology in preclinical species should be carefully considered since it may result in pathophysiology that is reflective of only a subset of patients or a specific phase of a disease (Rydell-Törmänen and Johnson 2019). For example, genetically modified mouse models to study cystic fibrosis have exhibited gastrointestinal effects similar to the disease in humans but could not recapitulate the clinical pulmonary effects (Davidson and Rolfe 2001). It is important to highlight that while individual studies may be limited in scope, when taken together, conclusions from disparate experiments can be combined to create a more wholistic view of the disease, linking different mechanisms together in a manner consistent with the current understanding of the biology to form the backbone of a QSP model. For example, data for various treatments in vivo were used to develop a QSP model of type 1 diabetes (Shoda et al. 2010).

In Vivo Drug Concentrations

When a valid animal model is available to mechanistically study a disease, it can support QSP model development in several ways. First, it can be used to evaluate the exposure at the site of action (i.e., in the tissue(s) of interest) and relate those concentrations to measurements in an accessible tissue, such as blood or plasma. Access to tissue exposure data can be used to evaluate the potential for asymmetric tissue distribution of a therapeutic. These results can be used to either confirm the future use of blood/plasma as a surrogate for tissue concentrations or establish the need for a more sophisticated relationship between the blood/plasma compartment and the tissue(s) of interest. If the latter is needed, the data may be used to build the blood/plasma to tissue relationship that is ultimately used in a QSP model. The nuances of the tissue and blood or plasma measurements should be considered when preparing the data for use in a model. For example, tissue measurements are often derived from tissue that is homogenized before concentrations are measured. Thus, the concentration reflects that of the whole tissue and any spatial information is lost. If spatial details or various tissue sub-compartments are critical to the QSP model, additional sampling methods, such as imaging (Thurber et al. 2013; Cunha et al. 2014; Bai et al. 2023) or microdialysis methods (Hammarlund-Udenaes 2017) may be required. Protein binding correction is another consideration for small molecule therapeutics that can be important for alignment of concentrations across in vitro, preclinical and clinical contexts. This is discussed in detail in Sect. 3.1.

Beyond direct measurement of the therapeutic concentrations, it is also important to probe preclinical biology in concentration ranges that are relevant to the clinical exposures. Often clinical therapeutics are used in preclinical studies as benchmarks and points of validation for modeling efforts. Consideration of the species cross-reactivity of a therapeutic as well as exploring biology in clinically relevant concentration ranges is important to derive meaningful preclinical knowledge that can be repurposed to inform on clinical questions.

In Vivo Biomarkers and Biomeasures

An advantage of preclinical biomarker studies is that datasets can be acquired with more frequent sampling and can encompass a wholisitc organ view given the ability to sample entire organs which cannot be achieved in humans. This helps establish quantitative relationships between drug concentrations and biomarker responses and permits the reassessment of these relationships in a reproducible system with less variability than is seen in humans. In vivo preclinical biomarker data will often be measured with similar assays as those described in the *in vitro* section and can reflect information that is qualitative or quantitative depending on the study requirements and resources.

Homogenization of tissue and subsequent analysis of the relative quantification of activated and total protein concentrations is a common type of biomarker for kinase inhibitors and is often achieved by ELISA or MSD assays (Table 1). Sample handling and processing, evaluation of the detection reagents, and the linear range can all affect the assay performance. However, once optimized, these assays can provide quantitative biomarker information in hundreds of samples in a reasonable time frame to impact preclinical drug discovery projects and QSP modeling. This information is related to, but distinct from, quantification of proteins with IHC methods that are commonly used in the clinic. To relate these preclinical and clinical datasets, bridging studies may be needed to transfer knowledge across these two common PD measurement methods.

The quality of information derived from biomarker studies will depend on the measurement system as well as the study design. When considering in vivo biomarker data, the dynamic exposures and appropriately sampled PD biomarkers can provide important information about the kinetics of the biomarker, delays between a drug's target engagement and biomarker response, and the occurrence of compensation mechanisms that may induce transients (e.g., rebounds) in the biomarker. It is also advised to evaluate the dynamics of the unperturbed biomarker over time to capture any baseline changes that need to be included in the model. This provides steady state conditions, which need to be properly accounted for in the QSP model.

Biomeasures is a recently coined term that reflects the measurement of the biological components that contribute to drug-independent preclinical and clinical physiology (Agoram and van der Graaf 2012). This can include concentrations of items such as cell surface receptors, soluble receptors, cytokines, and cells to name a few. Dynamic factors such as protein turnover levels or receptor internalization rates are also important biomeasures for incorporation into QSP models. The measurement of these values may be determined in healthy and disease states to best capture relevant differences as needed for the QSP model's context of use.

Translational Considerations

As touched upon earlier, the main concern with animal models relates to their translatability and recapitulation of human disease(s). Thus, a systematic approach should be taken to uncover data that may influence translation. This could include understanding the impact of species differences in genetics, anatomy, cellular function, and pharmacologic cross-reactivity/potency differences. For example, there are known differences between mice and humans for both innate and adaptive immunity (Mestas and Hughes 2004). Th17 cells, which have been demonstrated to have a pathogenic role in many autoimmune diseases, are differentiated from naïve CD4 T cells under the influence of cytokines that can differ between mouse and humans. (de Jong et al. 2010) Differences in anatomy may also contribute to failure of animal models to successfully translate (Wang et al. 2012). One advantage of using a QSP model is that known differences between preclinical species and humans can be accounted for in the model, which has the potential to improve translation. A simple example of this is the correction for small molecule drug's plasma protein binding, which can vary between species and is routinely measured. This scale factor can be easily used to align in vitro, in vivo preclinical and clinical unbound concentrations for use in QSP models. Incorporation of species differences through model modifications or reparameterizations enables QSP models to better approximate the human condition, offering a model-based approach to translation.

2.3 (Multi)-omic Data

QSP models provide an excellent vehicle for utilizing large-scale datasets such as gene expression (genomics), transcriptomics, proteomics, lipidomics, metabolomics, and peptidomics which can provide crucial insights into key biological processes that define the disease state. This information can focus the scope of biology so that only the most impactful pathways are represented, and physiologic characteristics are appropriately defined when developing new QSP models. Omic datasets are increasingly being generated from in vitro systems, preclinical species, and humans, which, when paired with modeling approaches can facilitate mechanistic understanding and translation (Uatay et al. 2024). As the breadth of applications continues to expand for omic technologies, the technologies themselves also continue to advance. For example, genomic data has progressed from bulk RNA samples to single-cell RNA sequencing which can provide additional insights for QSP model development (Zhang et al. 2019, 2023) and can now start to inform on the spatial heterogeneity in tissue microenvironments. (Arulraj et al. 2024)

Information from omic data can be used for de novo model development and can also be of value when used to qualify the scope of biology in preexisting QSP models. For example, gene expression datasets have been examined to understand the key biological processes that define rheumatoid arthritis, in comparison to osteoarthritis and normal synovial tissue (You et al. 2012). This analysis gives an opportunity to assess the biology represented in QSP models for rheumatoid arthritis (Rullmann et al. 2005; Thalhauser et al. 2015; Channavazzala et al. 2019; Macfarlane et al. 2019). Omic datasets, in combination with natural history studies and basic physiology information, provides ideal data for describing disease pathophysiology in QSP models and have the potential to increase confidence in the underlying biology compared to those QSP models that are based on preclinical data alone. Recent publications have demonstrated this blending of data in the area of immuno-oncology, where the translation of syngeneic mouse models is still debated (Arulraj et al. 2024; Lazarou et al. 2020). These works are addressing the main drawback with omic datasets, which is that they are typically qualitative measures reflecting statistical fold changes or pathway up–/downregulation rather than quantitative measures of change at the gene, protein, or cellular level. Thus, efforts in the modeling community are blending omics data with quantitative measures of absolute cell counts, measurable protein concentrations, and tissue microenvironment knowledge to bring omic knowledge into the QSP framework. These efforts further highlight a lack of basic quantitative physiologic information to blend with omics data regarding quantitative measures of protein concentrations, tissue cell numbers, turnover rates, and other basic parameters needed to merge omic data for use in QSP models (Lazarou et al. 2020) However, when it is possible to bring omics datasets together with quantitative measures of the biological system, these blended models have the potential to improve QSP model calibration and the development of virtual populations. This area is expected to advance in the coming years through collaborations between QSP scientists and bioinformatic experts where pathway analysis tools or unsupervised machine learning (Marriott et al. 2023; Zhang et al. 2022) may be applied to omic datasets for diseases where pathophysiology is emerging to create QSP models, virtual populations, and digital twins.

2.4 Clinical Outcomes

Representation of clinical outcomes is a key component of QSP models for application to drug discovery and development. A model that can adequately represent outcomes that are measured in the clinic can be used to make human dose projections early in drug discovery that are benchmarked to clinical success criteria. When multiple therapies can be represented by QSP models, the model can be used to understand potential competitive differentiation. In some cases, representation of clinical outputs may be very straightforward and obvious. For example, a model of diabetes typically will track blood glucose levels (Bosch et al. 2022). The regulation of blood glucose levels is fairly well understood, and blood glucose itself is measured clinically. By contrast, in autoimmune diseases, clinical endpoints are often composite measures that include subjective components. For example, the ACR20 score for rheumatoid arthritis (RA) measures the percentage of patients who experience at least a 20% improvement in tender and swollen joints as well as a 20% improvement in three of five other criteria which includes patient and physician global assessments, pain, disability, and an acute-phase reactant (erythrocyte sedimentation rate or C-reactive protein). (Felson et al. 1995) For these models, abstractions from model variables need to be made in order to represent the outcomes. Ideally, this would be done in a manner that is consistent with current understanding of the physiology and supported by data. Currently, there is no best practice for representing endpoints with subjective components in QSP models and different approaches may be equally viable. For example, a pathobiology-based approach could be used, where in the case of inflammatory diseases, like RA, the QSP model would characterize the disease by cellular infiltrates. One could infer that reductions in the cellular infiltrate would correlate with improved disease activity. Another approach could be to look at the literature for correlations that have been previously noted and then incorporate an optimization function of the correlates to represent the clinical output. In any case, it is important to document the approach for transparency to stakeholders.

3 Data Used to Represent Therapies

As proposed in a 2011 NIH white paper by the QSP Workshop Group, a key goal of QSP models is to understand how drugs modulate biology (e.g., cellular networks) in space and time to impact human pathophysiology (Sorger et al. 2011b). Drugs are represented in QSP models according to their concentration profiles and/or perturbation of relevant pathways. In this regard, pharmacokinetic and pharmacodynamic data are the key data types that define therapeutic behavior and drug action in QSP models.

3.1 Pharmacokinetic (PK) Data

Pharmacokinetic data refers to information about how a drug is absorbed, distributed, metabolized, and eliminated by the body over time. These data describe drug concentrations in plasma and tissues over time for different doses, frequency, and routes of administration. PK data for small molecules is often measured using mass spectrometry with available standards of the therapeutic agent and any relevant metabolites. Biologic molecules, such as antibodies, use antibody-based assay methods such as ELISA assays (Table 1) to reflect unbound or total (antigen bound and unbound) concentrations. Both datasets can be useful for PK models and knowledge of the assay construct is important when modeling biological therapies. Additional information about biological PK models can be found in "Breadth of Pharmacology Modeling: Fundamentals of Pharmacokinetics, Pharmacodynamics and Mechanistic Modeling" (Chapter 2).

PK data provides crucial insights into the drug's behavior within the physiological system and forms a fundamental component of QSP models which aim to connect drug concentrations to biological pathways in the model. Indeed, QSP models have utility in assessing the feasibility of different modalities by predicting drug dosing and affinity or potency requirements for targets with varying properties (Bansal et al. 2022). Methods to model PK data have been well established and guidance for evaluating PK models has been put forward by regulators (Population Pharmacokinetics | FDA). For QSP, a typical approach to represent drug PK is to simply use PK models for the drug of interest that have been published or developed directly from acquired preclinical or clinical data. For clinical QSP applications, population PK models are typically preferred as they are built from extensive patient data, describe the observed variability, and often include covariates that may be leveraged in QSP models. If PK models have not been disclosed, new models can be derived by fitting a model to reported mean PK profiles. While clearly not as rigorous as a population PK model, this approach is typically sufficient since the goal of QSP models is to understand the impact of uncertainty in the biology on therapeutic efficacy and not necessarily uncertainty in PK. PK data are also used in QSP models of novel compounds that have not yet been tested in humans. Since these candidates have not entered Ph1, it is necessary to predict the human PK. This can be accomplished via allometry from preclinical PK studies or in vitro to in vivo extrapolation (IVIVE). Methods for these approaches have been described elsewhere (Wang et al. 2016; Sodhi and Benet 2021).

While variability in PK may be incorporated in QSP models, for applications that aim to elucidate mechanistic insight, the introduction of variability in PK may be unnecessarily obscuring. However, variability in PK can be important for human dose projection with QSP models, where it may lead to an understanding of an effective dose range from which to understand the impact of developability concerns like solubility, tablet sizes, and volume used in autoinjectors.

PK data, in combination with pharmacodynamic data, are used in QSP models to simulate drug responses for both model calibration and model qualification. When using PK data from the literature for QSP models, it is important to note the patient population. Often the PK data are from healthy volunteers; however, PK parameters can be different in patient populations compared to healthy subjects. For example, drugs that are renally eliminated may clear more slowly in patients with impaired kidney function, leading to elevated circulating plasma concentrations, compared to healthy volunteers, and may result in exaggerated pharmacology. Understanding the data that goes into building the PK model helps determine the level of confidence in the drug's representation in the QSP model.

Another important consideration for PK data is the target tissue where the drug is expected to function for therapeutic benefit. For small molecules, the free drug hypothesis assumes that only unbound drug is available to interact with its target. (Smith et al. 2010) Measured small molecule drug concentrations are often corrected for protein binding using a value derived from equilibrium dialysis methods (Di 2021). This correction is applied to plasma, tissue, and in vitro cellular samples. In the latter case, it is important to know the composition of the culture media for the cellular assays that are used to connect drug concentrations to the target biology within a QSP model as an adjustment to the IC50 may be required (Austin et al. 2002). To elaborate, PK models predict plasma concentrations, but cellular assays often use 10% fetal bovine serum. For highly bound drugs, the IC50 derived from cellular assays may reflect a higher potency than if the assay were conducted in 100% plasma, which could have implications for projecting the effective human dose. Another approach is to use cellular assays in whole blood. In this case, the adjustment can be made if the blood to plasma ratio is known. This method is preferred when the unbound fraction is very low as the percent error in this measurement can be quite high. PBPK models, which predict free drug concentrations in various tissue compartments, may also be relevant as input into QSP models, especially if there are predicted drug-drug interactions that may impact plasma drug concentrations or if a particular organ with active drug transporters is the target tissue since free drug concentrations that differ from plasma measurements may have significant impact on predicted efficacy (Polak et al. 2019). This may be of particular importance to small molecule drug candidates and brings additional information to the model rather than assuming unbound drug concentrations are similar between plasma and tissue compartments. For large molecules, other methods have been described to estimate tissue concentrations from plasma measurements. (Shah and Betts 2013; Evans et al. 2014) Special considerations for the PK and other QSP model components may be necessary for more recent modalities. For example, the PK and pharmacodynamic effects may be significantly separated in time for siRNA, virus vectors, cell, gene, and vaccine therapies necessitating mechanisms in the QSP model that account for these longer timescales. (Mc Laughlin et al. 2023; Singh et al. 2020) Antibody drug conjugates (ADCs) have their own considerations including internalization rates, payload deconjugation, and efflux of free payload back into circulation (Scheuher et al. 2023; Lam et al. 2022). This is an area with active QSP model development that leverages a wide range of datasets to inform on the mechanisms associated with ADC delivery and pharmacological activity (Scheuher et al. 2023; Chang and Shah 2025).

3.2 Pharmacodynamic (PD) Data

PD data characterizes the effects of drugs on target molecules, cells, tissues, and whole organisms. These data can be obtained from in vitro assays, animal models, and clinical trials using techniques such as enzyme assays, cell-based assays, and physiological measurements (see Sects. 2.2.1 and 2.2.2). They provide a means to connect drug concentrations to the biology represented in QSP models. As such, assays that directly inform on pathways in the QSP model are prioritized and conceptually provide more confidence in model predictions. It's important to mention that the terms PD and biomarkers are frequently used interchangeably, so many of the observations made in Sect. 2 are relevant here as well.

While relevant PD data is generated in drug discovery phases, the first available human biomarker data that can be incorporated into QSP models becomes available from Phase 1 studies. These studies are performed across a range of escalating doses and provide data that can inform on the therapeutic's exposure-response relationship. A QSP model can recapitulate Phase 1 PD data relevant to disease patients, and (when necessary) healthy patients would provide a high level of confidence for Phase 2 dose selection. While general markers of disease activity may be informative, like C-reactive protein in autoimmune disease, the ideal biomarker in Phase 1 would clearly demonstrate expression of pharmacology relevant to the disease state. An example of this is the use of ex vivo stimulation assays in Phase 1 studies for JAK inhibitors. These assays measure inhibition of cytokine signaling, where the validated targets (or their surrogates) are used for stimulation – for example, IL-6 for rheumatoid arthritis (Mohamed et al. 2020) and IL-23 for psoriasis (Catlett et al. 2023).

However, often there are no suitable biomarker data available in Phase 1 to connect PK to the pathways in QSP models. This can be the case for Phase 1 studies conducted in healthy volunteers where disease and/or target-specific biomarkers are not expressed. In these cases, clinical observations can be augmented with preclinical knowledge from relevant in vitro and in vivo data. Ideally, the in vitro assays would be reformatted to an ex vivo or clinically aligned version that can be measured in Phase 1. This provides an opportunity not only to increase confidence in using the in vitro data for translation, but also an opportunity to update the model with Phase 1 clinical data that can then be applied to Phase 2 dose selection. The aforementioned assays for measuring JAK inhibition are examples where the in vitro data correlate well with the ex vivo data. In these studies, the IC50 derived from in vitro cytokine stimulation assays where whole blood samples are simulated with cytokines in the presence of various concentrations of drug correlates well with the IC50 derived from Phase 1 studies where blood samples are drawn at various timepoints and stimulated ex vivo (Traves et al. 2021).

Ideally, when using in vitro assays, drug concentrations should cover a broad range to adequately characterize IC50, hill coefficient, and maximal inhibition. This is true for in vivo preclinical studies too, which should incorporate a range of doses and potentially dose regimens to fully elucidate the exposure-response relationship. This can also be a limitation of datasets that are derived from clinical trials that may not cover the full exposure-response range. In this case, the QSP model, which can bring together in vitro, preclinical, and clinical datasets may provide the most complete picture of the exposure-response relationship.

Unfortunately, sufficient clinical dose response data are often difficult to find in the public domain, and when located, they often only represent a single dose, necessitating assumptions for inclusion in a QSP model. For example, the dose used may be assumed to be at IC90 with the Imax and hill coefficient both assumed to be equal to 1. IC50 is clearly a key parameter, but changes in the hill coefficient can be quite impactful on the derived IC50 value from this single dose. In this regard, a steep hill curve may result in a candidate drug failing to reach target efficacy at lower doses, which can affect selection of a minimally active dose for testing in dose ranging Phase 2b trials. The assumption on Imax may also be critical. A full dose response curve may reveal that the drug is unable to reach 100% inhibition in a critical pathway. Thus, an assumption of 100% inhibition may result in an overprediction for efficacy of the candidate compound. When faced with having to make these assumptions, it should be documented and communicated to project teams as they clearly can affect how model simulations are interpreted. A sensitivity analysis may highlight the impact of these assumptions and lead the project team to conduct additional experiments to verify the literature and resolve these uncertainties.

3.3 Other Considerations

In addition to the data itself, documentation of the measurement assays, data sources, and similar factors should be communicated to provide context for interpretation of the simulation results within the QSP model's context of use. For example, where there are data from multiple sources, a weighted average may be calculated or a particular study could be prioritized, which may affect the interpretation of simulations. Clearly, the source of data and how they are collected are a key consideration. If text mining approaches are applied to literature-based data extraction, some level of quality control must be applied to ensure that the data are reliable and accurate. Similarly, if data are extracted by digitizing figures, then additional documentation would be recommended to ensure that the values are correctly extracted. Lack of confidence in model predictions could result if data sources or model assumptions have not been documented appropriately and communicated to stakeholders.

4 Data Used to Develop Virtual Patients

Virtual patients are alternate parameterizations of QSP models that attempt to capture the biomarker and outcome responses of individual patients. The goal of virtual patient generation is to ensure that the model is able to capture the breadth of response to key therapies for calibration and validation, which facilitates the generation of virtual populations. A cohort of virtual patients is derived by sampling a distribution around nominal parameter values in a manner similar to simulations with PKPD and pharmacometrics models (Trang et al. 2017). However, given the mechanistic nature of QSP models, many of its parameters are based on measurable quantities from experimental data where there is observed variability. As such, sampling can be based on the known parameter variability derived from experimental data, rather than by some arbitrary amount. If variability around a parameter or biological species is unknown, assuming a lognormal distribution may be a good start. However, there should be a strategy in place on how the value will be updated as more data is obtained. A thorough discussion with biologists or subject matter experts may also lead to consensus on an approximate range or distribution. For smaller QSP models, it may be possible to explore variability in all parameters. However, this can become increasingly impractical due to computational constraints and thus, as the size of the model increases, any parameter search to create virtual patients may cover only a small fraction of the parameter space. Faster computer processors, parallelization, and other sampling approaches (e.g., Latin Hypercube Sampling, or Sobol sequences) can help, but for larger QSP models, thoughtful selection of parameters may be necessary. While sensitivity analyses may be conducted to prioritize the most sensitive parameters, another complementary approach is to vary parameters based on biologic considerations around targets of interest. The former increases the probability of observing generally variable responses in a cohort of virtual patients, while the latter increases the probability of observing variable responses to particular treatments, including both known clinical therapies and drug candidates that have yet to enter the clinic. Selection of parameters that are not sensitive, but relevant to a target of interest, ensures that simulations adequately explore the most relevant parameter space and may result in unexpected behavior that could lead to novel insights due to the nonlinear nature of many QSP models.

Parameter sampling in PKPD and pharmacometrics models typically assumes that all combinations of parameter values are reasonable. By contrast, alternate parameterizations of QSP models may result in a virtual patient that has unrealistic behavior. Checks are typically put in place to filter out these virtual patients from further analysis. Some checks are obvious from first principles. For instance, some virtual patients may have fluxes that are impossible to become negative and must be screened out. For others, data can also be used to define reasonable baseline behavior. For example, virtual patients generated in a QSP model to be hyperlipidemic must have LDL and HDL levels that are aligned with real-world data as indicated in the NHANES database (Allen et al. 2016; NHANES Data 2011–2012). If supported by data, correlations between parameters may be enforced for building virtual patients, effectively reducing the parameter space and constraining the model. Other data types may also be used to put constraints on model behavior when creating virtual patients, including histology or imaging. For example, in auto-immune/inflammation QSP models, the number of cells present per unit tissue at baseline can be estimated from histological studies and used as a screen for unrealistic virtual patients. The concept of using data as constraints to model behavior at baseline may also be extended to response to therapy, if available. For larger models, checking each state variable for plausible behavior may not be practical and approaches such as model-order reduction may be considered. (Duffull and Gulati 2020)

5 Data Used to Develop Virtual Populations

Depending on the resulting set of parameter values, virtual patients can exhibit different behaviors ranging from no response to complete response for a particular intervention with possibly differential responses occurring across therapies. Virtual populations are derived from a large cohort of virtual patients by selecting a subset of patients such that they are consistent with clinical data for both baseline measures and response to treatment. Various methods have been described for creating virtual populations, including the application of prevalence weights or algorithm-directed sampling (Schmidt et al. 2013; Rieger et al. 2018). Data considerations are the same regardless of which method is used. Development of virtual populations can become progressively more difficult as the number of therapies used for model fitting increases. Failure to find suitable virtual populations suggests several possibilities. First, there may be an inconsistency in the clinical data used for fitting, making it impossible to fit two trials simultaneously. Alternatively, the model may be missing some key pieces of biology, leading to review of the literature, and creation of additional virtual patients or a change in the base model. Virtual population building can therefore be an iterative process, providing a unique approach to understanding how basic biology can impact clinical response. Theoretically, a QSP model could be used to evaluate the likelihood of competing hypotheses. For example, if implementation of an initial hypothesis resulted in the inability to develop virtual populations compared to the competing hypothesis, then it could be reasonable to surmise that the initial hypothesis may be less likely.

5.1 Clinical Data

Clinical data are essential for developing virtual populations as they are used for model calibration and qualification. These data may be continuous or non-continuous, represent individual responses or population level responses, and represent mechanistic or descriptive endpoints. The data may also include correlations between biomarkers and/or clinical measures, which could be revealed by having individual patient-level data. Typically, the QSP model is fit to select clinical trial data and then qualified by predicting clinical trial data that were not used for calibration. There are some key considerations for assessing clinical trial data for developing virtual populations. First is an assessment of the number of trials that may be used. Phase 3 trial data are prioritized as they are considered definitive and used for registration. Phase 2 data can also be used and are valuable as the drug being tested often targets a novel mechanism of action and is tested over a dose range. Fitting a model to these data effectively adds constraints to the model, which is particularly important for QSP models with many parameters. This is manifested as a shrinking of the parameter space for which virtual patients may be viable. Second, the trials used should have similar entry criteria and baseline patient characteristics. While not always possible, this facilitates using QSP models for competitive differentiation. Lastly, trials used for calibration should be distinct from an MOA perspective from those used for qualification, as successful qualification provides confidence in the scope of biology represented. If this is not possible, then fitting the model to efficacy of some doses within a treatment and qualifying the model by predicting other doses are acceptable. In this case, qualifying the model against doses outside the range used for calibration is preferred as qualification demonstrates that the mechanisms in the model are suitable for making predictions that are extrapolations. Another approach, though least preferred, would be to qualify the model by predicting efficacy for a therapy where another compound in the same class was used for model calibration. While the process of applying clinical data to the development and qualification of QSP models appears to be straightforward, there are several challenges that may arise that should be addressed before embarking on building virtual populations. A few of these challenges are described below.

5.1.1 Conflicting Data

Assessing clinical data for potential conflicts is necessary for efficient generation of virtual populations. If the clinical data used to fit the model are conflicting, then it may be impossible to achieve an acceptable fit for a virtual population. The classic example is the situation where there are multiple identical Phase 3 studies for the same drug, yet the results on the primary outcome were positive in one study but negative in the other. In this scenario, it is recommended to aggregate the data together rather than to pick one trial over another in order avoid biasing virtual populations toward one trial over another. Another example is where the pharmacokinetics suggest that the clinical doses may not achieve meaningful levels of inhibition, given the IC50 for available PD data to achieve efficacy in the model. In this case, the patients will never be able to respond to the drug in the model and thus make it impossible to create populations. Similarly, when comparing therapies with the same mechanism of action, it is possible that differences in pharmacokinetics and potency will not be sufficient to explain differences in clinical response. These situations warrant a further investigation into the pharmacokinetics and pharmacodynamics data available and suggests that perhaps the mechanism of action of the drug is not fully understood or that other drug properties need to be accounted for (e.g., active metabolites and transporters). As conflicts are possible, assessment of the clinical data, accounting for efficacy, pharmacokinetics, and pharmacodynamics is advised prior to developing virtual populations.

5.1.2 Placebo Response

How to handle placebo responses in a QSP model is an often-overlooked subject that can have significant consequences for interpreting simulations that aim to understand competitive differentiation. Consider the situation where there are two therapies being compared with similar rates of response but dramatically different placebo responses in their corresponding clinical trials. Are these treatments considered to be similarly efficacious or is one demonstrably better than the other? How placebo is integrated in the QSP model affects the calibration and could influence simulations for novel drug targets. Several approaches to addressing the placebo response have been described, ranging from placebo subtraction, statistical transformation, to simply ignoring it (Wang et al. 2019). In some cases, it may be appropriate to explicitly model the placebo response. For example, a model that considers viral dynamics would have a placebo that is based on infection rates. As another example, weight loss in a placebo arm of an obesity trial could directly be attributable to decreased food intake in an obesity model where food intake is a key model input. Which method to use to account for placebo depends on the nature of the clinical data and scope of biology in the QSP model. Modelers should align with key stakeholders on how to address placebo responses to avoid future confusion and unnecessary iteration. Nonetheless, the modeling community could benefit from a systematic evaluation to understand how the different placebo methods may impact simulations of novel drug targets in a QSP model.

5.1.3 Success Criteria

Another consideration for applying clinical data to build virtual populations is to align on what constitutes successful calibration and qualification, as the QSP community has yet to adopt a standardized approach (Chan et al. 2024). This lack of consistency is likely due to context as QSP models can have varied uses and the nature and quality of data may vary from project to project. Here, it is also desirable to align with key stakeholders on how the model calibration and validation will be assessed. For example, for calibrating to clinical data is it necessary for a virtual population to be within the 95% confidence interval of the observed data? Or is it more appropriate for the observed data to be within the 95% prediction interval of the model? Given that the difficulty of developing virtual populations is proportional to the number of therapies that are used for fitting, is it necessary to fit all therapies? Or should there be a prioritization with some therapies being more relevant to the research question than others? Alignment on these types of issues prior to conducting simulations for novel targets will help avoid confusion that may occur if it becomes necessary to change success criteria and the simulation results for novel targets changes meaningfully.

5.2 Biomarkers

Biomarkers can be applied toward virtual populations in various ways, providing another means to constrain QSP models. For example, histological data from patient synovial biopsies have been used to define subtypes of rheumatoid arthritis patients and how these subtypes may respond to different therapies (Dennis Jr et al. 2014). Virtual populations in a RA QSP model can then be tailored to represent these subtypes. In this example, clustering could be used on baseline cell densities for lymphocytes, macrophages, and fibroblast-like synoviocytes to represent subtypes described as low inflammatory, fibroid, lymphoid, or myeloid. Alternatively, model outputs corresponding to measured biomarkers, such as ICAM-1 and CXCL13, may also be used to segregate patients. Response to treatment data can then be used to qualify these biomarker-defined phenotypes or as further calibration criteria. Novel compounds can then be simulated in the context of these phenotypes and may generate testable hypotheses in support of personalized medicine. Biomarker data in QSP models can also be used for model qualification. For example, when simulating a therapy in the model, changes in model outputs not related to efficacy can be compared against reported data to qualify the scope of the biology captured in the QSP model. Importantly, if there are known correlations in biomarkers that correspond to model outputs or parameters, these data can be used in QSP models to qualify virtual populations. Alternatively, these correlations may also be used as design criteria for building virtual populations as mentioned previously.

6 Communication of Data

Communication of the data used in QSP models and for what purposes is essential to provide context for stakeholders to understand the strengths and limitations of simulations. The two key aspects for communicating data are assessment of data quality and documentation. Attention to each ensures transparency and helps limit any potential misinterpretation of the simulations.

6.1 Assessing Data Quality

Assessing the quality of data used to develop, calibrate, and qualify QSP models is necessary for appropriate interpretation of simulation results and for avoiding over-indexing on a particular inference. Generally, models that are built from highly relevant data are likely more reliable than models built with less relevant data. For example, model predictions based on PK data derived from clinical studies in an appropriate patient population should be interpreted with higher confidence than if the PK were projected from preclinical species to humans using allometric scaling. Similarly, PD assays using primary human cells may be more reliable than assays that used transformed cell lines. Moreover, PD assays that evaluate pathways known to be highly relevant for a particular disease indication would result in more confidence in model predictions than if the assays, while indicative of target engagement, affect pathways with unclear relevance. For example, a PD assay that reads out inhibition of TNF and/or IL-6 would be very relevant for rheumatoid arthritis, where both TNF and IL-6 are validated targets with approved therapies that target these cytokines (Humira (adalimumab) label; Actemra (tocilizumab) injection label). By contrast, an assay for inhibition of the chemokine CXCL10 may be less reliable for prediction of efficacy since its role in rheumatoid arthritis is not clear and there is redundancy where other related chemokines (CXCL9, CXCL11) bind to its receptor (CXCR3) (Karin 2020).

While there is a wealth of data in the literature, often quantitative data are not available to inform QSP models. In these instances, it may be appropriate to use qualitative or semi-quantitative data. For example, while an assay may be conducted over a dose range, the output could be gene expression or Western blots. Gene expression often does not correlate with protein levels and there can be post-translational modifications that affect function that are not captured in these data. Densitometry tools can be used to obtain quantitative data from Western blots. While these data are not preferred, they may nonetheless be useful to develop QSP models while more appropriate experimental data are being generated. Alternatively, they may be used for a base parameterization of the model and then allowed to be varied during model calibration to incorporate the uncertainty in those parameters due to their data source.

Given the different uses of data in QSP models, pedigree tables are useful for modelers to communicate the nature and quality of data used. This assessment can increase confidence in the conclusions drawn from modeling. It can also make clear any data gaps or deficiencies that must be considered. Additional experiments can be proposed to close gaps or data that was initially overlooked may come to light. The pedigree tables below may be used as a starting point and can be modified based on the context. Table 2 is useful for understanding the data behind the biological and clinical scope of the model, while Table 3 is particularly relevant for understanding the data behind the representation of drug candidates in the QSP model.

6.2 Documentation

Documentation of the data used to build, calibrate, and validate QSP models is an important task that facilitates transparency and communication with multidisciplinary stakeholders. A recent industry survey (Chan et al. 2024) has indicated that the vast majority of QSP modelers regularly document modeling activities, particularly at key milestones. Somewhat surprisingly there was a preference for less formal documentation practices such as slide presentations (e.g., PowerPoint), comments in modeling code/scripts, and spreadsheets (e.g., Excel). Regardless of the method used for documentation, key information should be recorded that, at a minimum, includes the parameter name, description, value, units, and source. Some example table headers are shown in Table 4.

7 Future Perspectives

As pharmaceutical scientists work toward understanding human physiology better and developing new medications in different therapeutic areas, it is important that newer technologies be adopted to aid in the development of QSP models. Depending on the complexity of the model and available data, QSP models can take significant time and resource to build, qualify, and apply. Data from machine learning approaches and organoid/organ-on-a-chip systems are some of the newer technologies that have been explored for their use in developing QSP models.

Table 2 Illustrative pedigree table for assessment of data used to develop QSP models

Data category	Question	Major weakness	Minor weakness	Neutral	Minor strength	Major strength
Pathophysiology	Is there consensus on biological mechanisms in the model and their mathematical description?	Crude speculation	Embryonic field	Limited consensus	Accepted theory	Well-established theory
	What is the nature of data that supports the structure of the underlying biology in the model?	Pathways are only based on hypothesized mechanisms	Pathways represented are based on preclinical animal data	Pathways supported by data specific to humans	Pathways represented are consistent with prevailing view of pathophysiology	Pathways represented are supported by large-scale datasets
	What data are used to quantify pathways in the underlying biology of the model?	Only data from transformed cell lines or non-human species is used	Most data are not from primary human cells	Some data from primary human cells	Majority of data from primary human cells	Only data from primary human cells is used
	What is the nature of data used for quantification of pathways	None available, parameters need to be estimated	Only one dose reported, necessitating assumptions for EC50 and Hill slope	Only EC50 reported	IC50 and Emax reported	Full dose response available
	Are natural history data used for developing the model?	Model has significant inconsistencies with natural history data	Model has some inconsistencies with natural history data	Disease progression not in scope of model	Model calibrated to natural history data	Model qualified by predicting natural history
		Not described				

	How are clinical outcomes represented in the model?		Abstractions not supported by data, considerable uncertainty in pathophysiology	Abstractions not supported by data, but consistent with understanding of biology	Abstractions supported by data	No abstractions required
Virtual patients	What criteria are used to select parameters for virtual patient creation?	Not described	Arbitrary	Local sensitivity analysis	Global sensitivity analysis	Biological considerations and sensitivity analysis
	How is the range determined?	Not disclosed	Arbitrary	Sensitivity analysis	Expert opinion	Constrained by observed data
	What screening criteria are used to assess virtual patients?	None applied	Spot checks on baseline behavior	Response to treatments	Systematic sanity checks	Observed data as constraints
Virtual populations	What is the nature of clinical trial data in the indication of interest?	No Phase 3 trials yet, no positive Ph2 trials	No Phase 3 trials yet, some positive Ph2 trials	Negative Ph3 trials only, some positive Ph2 trials	Some positive Ph3 trials	Many positive Ph3 trials
	How do key entry criteria compare across trials?	Some differences that may be critical	Some differences that may be material	Similar in majority of trials being compared	Similar across all trials being compared	Identical across all trials being compared
	How do key baseline measures compare across trials?	Some means/medians similar, some outliers certainly will be material	Most means/medians similar, several outliers likely to be material	Most means/medians are similar, some outliers may be material	Most means/medians are similar, outliers not expected to be material	All means/medians are similar
	What is the nature of trials used for calibration and qualification?	No trials for either calibration or validation	Trials only for calibration, none for validation	Trials available for calibration and qualification, but are in the same class (MoA)	Trials available for calibration and qualification have different MoA	Trials for calibration have MOA distinct from trials used for qualification

(continued)

Table 2 (continued)

Data category	Question	Major weakness	Minor weakness	Neutral	Minor strength	Major strength
	How do Ph2 results relate to Ph3 and do duplicate Ph3 trials replicate?	Conflicting Ph3 trials for same compound	Many positive Ph2 trials that fail in Ph3	Insufficient data to assess	Most positive Ph2 trials are positive in Ph3	Duplicate Ph3 trials are always consistent
	What is the nature of the observed placebo responses in clinical trials?	High placebo rates with high variability	High placebo rates with low variability	Moderate placebo responses with moderate variability	Moderate placebo responses with low variability	Low placebo rates with low variability
	What are the criteria for successful calibration?	Poor calibration against clinical data for key trials relevant to the question being asked of the model	Mean/median simulations are within the 95% CI or standard deviation of the observe data, but there are inconsistencies with trends relative to each treatment	Observed data are all within the prediction intervals of the model simulations, and are consistent with trends relative to each treatment	Mean/median simulations are within the 95% CI or standard deviation of the observed data, and are consistent with trends relative to each treatment	Little variation compared to mean/medians of all observed data
	What are the criteria for successful qualification?	Observed data are all within the prediction intervals of the model simulations; but prediction intervals are very wide	Observed data are all within the prediction intervals of the model simulations; prediction intervals are moderately wide	Observed data are all within the prediction intervals of the model simulations; prediction intervals are fairly narrow	Mean/median simulations are within the 95% CI or standard deviation of primary endpoint	Mean/median simulations are within the 95% CI or standard deviation of primary and secondary endpoints

	What is the quality of model qualification?	No qualification done	Qualification is for dose(s) within the dose range used for calibration for a particular therapy	Qualification is for dose(s) outside the dose range used for calibration for a particular therapy	Qualification is for a treatment that has an MOA distinct from treatments used for model calibration	Successful prediction of a Ph3 trial made prior to trial completion (prospective validation)
	How are biomarkers applied to virtual populations?	Model has major inconsistencies that may impact efficacy	Model has minor inconsistencies with biomarker data and are not expected to impact efficacy	Not applied, not assessed	For development of virtual patients or for calibration of virtual populations	For qualification of virtual populations

Table 3 Illustrative pedigree table for assessment of data used to represent a drug candidate in QSP models

Data category	Question	Major weakness	Minor weakness	Neutral	Minor strength	Major strength
Pharmacokinetics	What is the source of PK data for drug candidate?	Completely hypothetical or significant uncertainty/ liability identified	Allometric scaling from preclinical species	IVIVE	Phase 1	Phase 2/3
	What is the source of PK data for key comparator therapies?	Completely hypothetical	Allometric scaling from preclinical species	IVIVE	Phase 1	Phase 2/3
Target engagement	What is the nature of the assay for target engagement?	TE data incompatible with PD assay data (e.g., activity suggested at 0% occupancy)	Insufficient data to infer target tissue TE from plasma measurement where drug concentrations in target tissue are expected to be significantly different	Plasma measurement for TE	Target tissue TE inferred from plasma measurement	Target tissue direct measurement
Pharmacology	What is the relevance of the PD assay to the disease of interest?	No assay available	No clear connection	Some indirect connection	Assay targets broad range of relevant biology	Assay targets pathway that is validated clinically
	What is the nature of the PD data used to quantify drug effects in the model?	In vitro, non-human cells	In vitro, transformed human cell line	In vitro, primary human cells	In vitro, human whole blood	Ex vivo/ in vivo, from clinical studies
	What is the nature of data used for parameterizing the PD?	None available, parameters need to be estimated	Only one dose reported, necessitating assumptions	Only IC50 reported	IC50 and Emax reported	Full dose response available
	Are there many replicates to quantify the PD?	Only single replicate	Few replicates with inconsistent measurements	Few replicates, but consistent measurements	Many replicates but with several outliers	Many replicates with few outliers

Table 4 Illustrative table headers for documenting data used in QSP models

(A) Data supporting initial species values								
Species name	Description	Location	Units	Minimum value	Nominal value	Maximum value	Reference(s)	Notes
Unique identifier	Brief statement, including translation of acronyms	Compartment	Ensure consistency with other species	Lowest plausible value	Most likely value	Highest possible value	Data source (s) for values	Any relevant comments that may affect confidence in modeled pathways
(B) Data supporting parameter values								
Parameter name	Description	Units	Minimum value	Nominal value	Maximum value	Population use	Reference(s)	Notes
Unique identifier	Brief statement, including translation of acronyms	Ensure consistency with other species	Lowest plausible value	Most likely value	Highest possible value	Flag to indicate if parameter is varied for creating virtual patients	Data source (s) for values	Explanations for min and max values
(C) Data supporting therapies represented								
Therapy	Parameter	Description	Value	Units	Reference (s)	Assumption(s)	Notes	
Drug name	PK or PD parameter name	Brief statement	Most likely value	Ensure consistency across model	Data source (s)	Any relevant comments that could affect simulations, e.g., weight assumptions	Any relevant details, e.g., patient population	

(continued)

Table 4 (continued)

(A) Data supporting initial species values								
Species name	Description	Location	Units	Minimum value	Nominal value	Maximum value	Reference(s)	Notes
(D) Data supporting model calibration								
Species name	Therapy	Time	Minimum value	Maximum value	Units	Reference(s)	Notes	
Unique identifier	Drug name, if applicable	Measurement timepoint	Lowest plausible value	Highest plausible value	Ensure consistency across model	Data source(s)	Any relevant comments that could affect simulations, e.g., assay description	
(E) Data for model validation								
Model output	Therapy	Dose	Time	Value	Units	Range	Range detail	Reference(s)
Clinical measure	Drug name	Dose, route, and schedule	Measurement timepoint	Mean or median	Enables comparison with other studies	Acceptable range as suggested by data	IQR, 95%CI, standard deviation, etc.	Data source(s)

Organoid systems mimic the natural physiology and microenvironment of human or animal tissues (Blutt et al. 2017; Rahmani et al. 2019). These systems are a better representation of physiology than traditional in vitro systems which typically isolate a single cell type (or a few different cell types in co-culture systems) and where the culture conditions are two dimensional with non-physiological matrices. For example, endothelial cells are typically grown on polystyrene cell culture plates under static conditions. Physiologically, endothelial cells, together with supporting cells, assemble a laminin-rich basement membrane matrix that provides structural and organizational stability, and microenvironmental cues can affect their differentiation and function (Davis and Senger 2005; Gifre-Renom et al. 2022). Moreover, endothelial cells are constantly exposed to varying degrees of shear stress, which can affect their function (Tsao et al. 1996, 1997). Data from organoid systems are likely more relevant for translation than standard in vitro culture systems. Indeed, experimentally observed effects in human-derived organoids have been translated into predictions of chemotherapy-induced diarrhea in patients (Verma et al. 2023).

Machine learning methods have been used to identify patterns and relationships in data sources upon which QSP models can be based on (Menden et al. 2019; Kolluri et al. 2022). They have also been used to parameterize QSP models (McComb et al. 2022) or identify mechanisms responsible for certain model outcomes (McGillen et al. 2014). Other applications of ML methods include feature selection and optimizing computational burden. Collectively, these applications show promise to speed development and application of QSP models.

As advancements in data sources and technologies emerge, it is important to evaluate their utility to increase the predictive ability of QSP models and/or to speed model development and assessment. Efforts should be made to evaluate their pros and cons relative to current practices.

8 Conclusions

The ultimate goal of applying QSP models to drug development is to increase the likelihood that a candidate drug will become an approved treatment. A retrospective study has shown that three key factors drive successful transition of a candidate drug from Phase 2 to Phase 3. When assessing early phase data, a good understanding of drug exposure, demonstration of target engagement, and expression of relevant pharmacology results in the most Phase 3 starts (Morgan et al. 2012). Missing or weak support for any of these factors can dramatically reduce success in Phase 2. Since QSP models use data relevant to these factors as input for drug candidates, applying QSP models early in development provides an opportunity to assess potential clinical efficacy and these assessments can then be applied toward decision-making. For example, QSP models may be used to project human dose and the corresponding predicted exposure can provide a reference against which safety margins from preclinical toxicology studies may be calculated. Where drug targets may be applicable to multiple indications, QSP models may be used to understand which ones may be more promising. Assessment of the data used to represent drug candidates in QSP models, coupled with an understanding of the data used to develop the underlying mechanistic model, provides an opportunity early in development to understand strengths and weaknesses of a drug candidate. Transparent evaluation of the data used for QSP modeling is critical for appropriate interpretation of QSP simulations, leading to fewer misunderstandings and better decision-making.

References

Actemra (tocilizumab) injection label. https://www.accessdata.fda.gov/drugsatfda_docs/label/2013/125276s092lbl.pdf

Agoram BM, van der Graaf PH (2012) Biomarkers and biomeasures: Key enablers for pharmacokinetic–pharmacodynamic modeling in drug discovery and development. Bioanalysis 4(10):1143–1145. https://doi.org/10.4155/bio.12.88

Allen RJ, Rieger TR, Musante CJ (2016) Efficient generation and selection of virtual populations in quantitative systems pharmacology models. CPT Pharmacometrics Syst Pharmacol 5(3): 140–146. https://doi.org/10.1002/psp4.12063. Epub 2016 Mar 17. PMID: 27069777; PMCID: PMC4809626

Arulraj T, Wang H, Ippolito A, Zhang S, Fertig EJ, Popel AS (2024) Leveraging multi-omics data to empower quantitative systems pharmacology in immuno-oncology. Brief Bioinform 25(3): bbae131. https://doi.org/10.1093/bib/bbae131

Auld DS, Farmen MW, Kahl SD et al (2004) Receptor Binding Assays for HTS and Drug Discovery. 2012 May 1 [Updated 2018 Jul 1]. In: Markossian S, Grossman A, Baskir H et al, editors. Assay Guidance Manual [Internet]. Eli Lilly & Company and the National Center for Advancing Translational Sciences, Bethesda. Available from: https://www.ncbi.nlm.nih.gov/books/NBK91992/

Austin RP et al (2002) The influence of nonspecific microsomal binding on apparent intrinsic clearance, and its prediction from physicochemical properties. Drug Metab Dispos 30(12): 1497–1503

Bai JW, Qiu SQ, Zhang GJ (2023) Molecular and functional imaging in cancer-targeted therapy: current applications and future directions. Sig Transduct Target Ther 8:89. https://doi.org/10.1038/s41392-023-01366-y

Bansal L, Nichols EM, Howsmon DP, Neisen J, Bessant CM, Cunningham F, Petit-Frere S, Ludbrook S, Damian V (2022) Mathematical modeling of complement pathway dynamics for target validation and selection of drug modalities for complement therapies. Front Pharmacol 13:855743. https://doi.org/10.3389/fphar.2022.855743. PMID: 35517827; PMCID: PMC9061988

Belanger MC, Ball AG, Catterton MA, Kinman AWL, Anbaei P, Groff BD, Melchor SJ, Lukens JR, Ross AE, Pompano RR (2021) Acute Lymph Node Slices Are a Functional Model System to Study Immunity Ex Vivo. ACS Pharmacol Transl Sci 4(1):128–142. https://doi.org/10.1021/acsptsci.0c00143. PMID: 33615167; PMCID: PMC7887751

Bigaeva E et al (2020) Exploring organ-specific features of fibrogenesis using murine precision-cut tissue slices. Biochim Biophys Acta 1866(1):165582

Blutt SE, Broughman JR, Zou W, Zeng XL, Karandikar UC, In J, Zachos NC, Kovbasnjuk O, Donowitz M, Estes MK (2017) Gastrointestinal microphysiological systems. Exp Biol Med (Maywood) 242(16):1633–1642. https://doi.org/10.1177/1535370217710638. Epub 2017 May 23. PMID: 28534432; PMCID: PMC5661769

Bosch R, Petrone M, Arends R, Vicini P, Sijbrands EJG, Hoefman S, Snelder N (2022) A novel integrated QSP model of in vivo human glucose regulation to support the development of a glucagon/GLP-1 dual agonist. CPT Pharmacometrics Syst Pharmacol 11(3):302–317. https://doi.org/10.1002/psp4.12752. Epub 2021 Dec 24. PMID: 34889083; PMCID: PMC8923724

Bradshaw EL et al (2019) Applications of quantitative systems pharmacology in model-informed drug discovery: perspective on impact and opportunities. CPT Pharmacometrics Syst Pharmacol 8(11):777–791

Brooks HB, Geeganage S, Kahl SD et al (2004) Basics of enzymatic assays for HTS. 2012 May 1 [Updated 2012 Oct 1]. In: Markossian S, Grossman A, Baskir H et al, editors. Assay Guidance Manual [Internet]. Bethesda (MD): Eli Lilly & Company and the National Center for Advancing Translational Sciences. Available from: https://www.ncbi.nlm.nih.gov/books/NBK92007/

Catlett IM, Aras U, Hansen L, Liu Y, Bei D, Girgis IG, Murthy B (2023) First-in-human study of deucravacitinib: a selective, potent, allosteric small-molecule inhibitor of tyrosine kinase 2. Clin Transl Sci 16(1):151–164. https://doi.org/10.1111/cts.13435. Epub 2022 Nov 22. PMID: 36325947; PMCID: PMC9841305

Centenera MM, Hickey TE, Jindal S, Ryan NK, Ravindranathan P, Mohammed H, Robinson JL, Schiewer MJ, Ma S, Kapur P, Sutherland PD, Hoffmann CE, Roehrborn CG, Gomella LG, Carroll JS, Birrell SN, Knudsen KE, Raj GV, Butler LM, Tilley WD (2018) A patient-derived explant (PDE) model of hormone-dependent cancer. Mol Oncol 12(9):1608–1622. https://doi.org/10.1002/1878-0261.12354. Epub 2018 Aug 16. PMID: 30117261; PMCID: PMC6120230

Chan JR, Allen R, Boras B, Cabal A, Damian V, Gibbons FD, Gulati A, Hosseini I, Kearns JD, Saito R, Cucurull-Sanchez L, Selimkhanov J, Stein AM, Umehara K, Wang G, Wang W, Neves-Zaph S (2024) Current practices for QSP model assessment: an IQ consortium survey. J Pharmacokinet Pharmacodyn 51(5):543–555. https://doi.org/10.1007/s10928-022-09811-1. Epub 2022 Aug 11. PMID: 35953664; PMCID: PMC9371373

Chang and Shah (2025) A translational physiologically-based pharmacokinetic model for MMAE-based antibody-drug conjugates. J Pharmacokinet Pharmacodyn 52:27. https://doi.org/10.1007/s10928-025-09978-3

Channavazzala M et al (2019) Development of Rheumatoid Arthritis QSP model capturing mechanistic pathways and clinical read-outs to enable simulation of novel therapies and trials. ACoP10, Orlando, ISSN:2688–3953, vol 1

Ciuffreda P, Xynomilakis O, Casati S, Ottria R (2024) Fluorescence-based enzyme activity assay: ascertaining the activity and inhibition of endocannabinoid hydrolytic enzymes. Int J Mol Sci 25(14):7693. https://doi.org/10.3390/ijms25147693. PMID: 39062935; PMCID: PMC11276806

Cohn DE et al (2023) Delineating spatial cell-cell interactions in the solid tumour microenvironment through the lens of highly multiplexed imaging. Front Immunol 14:1275890

Coussens NP, Sittampalam GS, Guha R et al (2018) *Assay Guidance Manual*: Quantitative biology and pharmacology in preclinical drug discovery. Clin Transl Sci 11:461–470

Cunha L, Horvath I, Ferreira S et al (2014) Preclinical Imaging: an Essential Ally in Modern Biosciences. Mol Diagn Ther 18:153–173. https://doi.org/10.1007/s40291-013-0062-3

Davidson DJ, Rolfe M (2001) Mouse models of cystic fibrosis. Trends Genet 17(10):S29–S37, ISSN 0168–9525. https://doi.org/10.1016/S0168-9525(01)02452-0

Davis GE, Senger DR (2005) Endothelial extracellular matrix: biosynthesis, remodeling, and functions during vascular morphogenesis and neovessel stabilization. Circ Res 97(11):1093–1107. https://doi.org/10.1161/01.RES.0000191547.64391.e3

de Jong E, Suddason T, Lord GM (2010) Translational mini-review series on Th17 cells: development of mouse and human T helper 17 cells. Clin Exp Immunol 159(2):148–158. https://doi.org/10.1111/j.1365-2249.2009.04041.x. Epub 2009 Nov 11. PMID: 19912248; PMCID: PMC2810383

Dennis G Jr, Holweg CT, Kummerfeld SK, Choy DF, Setiadi AF, Hackney JA, Haverty PM, Gilbert H, Lin WY, Diehl L, Fischer S, Song A, Musselman D, Klearman M, Gabay C, Kavanaugh A, Endres J, Fox DA, Martin F, Townsend MJ (2014) Synovial phenotypes in rheumatoid arthritis correlate with response to biologic therapeutics. Arthritis Res Ther 16(2):R90. https://doi.org/10.1186/ar4555. Epub 2014 Apr 30. PMID: 25167216; PMCID: PMC4060385

Di L (2021) An update on the importance of plasma protein binding in drug discovery and development. Expert Opinion on Drug Discovery 16(12):1453–1465. https://doi.org/10.1080/17460441.2021.1961741

Dimou P, Trivedi S, Liousia M, D'Souza RR, Klampatsa A (2022) Precision-Cut Tumor Slices (PCTS) as an Ex Vivo Model in Immunotherapy Research. Antibodies (Basel) 11(2):26. https://doi.org/10.3390/antib11020026. PMID: 35466279; PMCID: PMC9036232

Duffull S, Gulati A (2020) Potential issues with virtual populations when applied to nonlinear quantitative systems pharmacology models. CPT Pharmacometrics Syst Pharmacol 9(11):613–616. https://doi.org/10.1002/psp4.12559. Epub 2020 Sep 24. PMID: 32970387; PMCID: PMC7679128

Eng CHL, Lawson M, Zhu Q et al (2019) Transcriptome-scale super-resolved imaging in tissues by RNA seqFISH+. Nature 568:235–239. https://doi.org/10.1038/s41586-019-1049-y

Evans CH, Kraus VB, Setton LA (2014) Progress in intra-articular therapy. Nat Rev Rheumatol 10(1):11–22. https://doi.org/10.1038/nrrheum.2013.159. Epub 2013 Nov 5. PMID: 24189839; PMCID: PMC4402210

Felson DT, Anderson JJ, Boers M, Bombardier C, Furst D, Goldsmith C, Katz LM, Lightfoot R Jr, Paulus H, Strand V et al (1995) American College of Rheumatology. Preliminary definition of improvement in rheumatoid arthritis. Arthritis Rheum 38(6):727–735. https://doi.org/10.1002/art.1780380602

Gadkar K et al (2016) A six-stage workflow for robust application of systems pharmacology. CPT Pharmacometrics Syst Pharmacol 5(5):235–249

Gifre-Renom L, Daems M, Luttun A, Jones EAV (2022) Organ-Specific Endothelial Cell Differentiation and Impact of Microenvironmental Cues on Endothelial Heterogeneity. Int J Mol Sci 23(3):1477. https://doi.org/10.3390/ijms23031477. PMID: 35163400; PMCID: PMC8836165

Goltsev Y et al (2018) Deep profiling of mouse splenic architecture with CODEX multiplexed imaging. Cell 174(4):968–981

Hammarlund-Udenaes M (2017) Microdialysis as an important technique in systems pharmacology—a historical and methodological review. AAPS J 19(5). https://doi.org/10.1208/s12248-017-0108-2

Han Y et al (2022) Human organoid models to study SARS-CoV-2 infection. Nat Methods 19(4):418–428

Hasegawa C, Duffull SB (2018) Selection and qualification of simplified QSP models when using model order reduction techniques. AAPS J 20:1–11

Hoare SRJ (2004) Analyzing kinetic binding data. 2021 Apr 1. In: Markossian S, Grossman A, Baskir H et al, editors. Assay Guidance Manual [Internet]. Eli Lilly & Company and the National Center for Advancing Translational Sciences, Bethesda. Available from: https://www.ncbi.nlm.nih.gov/books/NBK569501/

Huh D et al (2010) Reconstituting organ-level lung functions on a chip. Science 328(5986):1662–1668

Humira (adalimumab) label. https://www.accessdata.fda.gov/drugsatfda_docs/label/2011/125057s0276lbl.pdf

Kandilogiannakis L et al (2021) Development of a human intestinal organoid model for in vitro studies on gut inflammation and fibrosis. Stem Cells Int 2021(1):9929461

Karin N (2020) CXCR3 Ligands in Cancer and Autoimmunity, Chemoattraction of Effector T Cells, and Beyond. Front Immunol 11:976. https://doi.org/10.3389/fimmu.2020.00976. PMID: 32547545; PMCID: PMC7274023

Khalaf H, Jass J, Olsson PE (2010) Differential cytokine regulation by NF-κB and AP-1 in Jurkat T-cells. BMC Immunol 11:26. https://doi.org/10.1186/1471-2172-11-26

Kijanska M, Kelm J (2004) In vitro 3D spheroids and microtissues: ATP-based cell viability and toxicity assays. 2016 Jan 21. In: Markossian S, Grossman A, Baskir H et al, editors. Assay Guidance Manual [Internet]. Eli Lilly & Company and the National Center for Advancing Translational Sciences, Bethesda. Available from: https://www.ncbi.nlm.nih.gov/books/NBK343426/

Kolluri S, Lin J, Liu R, Zhang Y, Zhang W (2022) Machine learning and artificial intelligence in pharmaceutical research and development: a review. AAPS J 24:19

Lam I, Pilla Reddy V, Ball K, Arends RH, Mac Gabhann F (2022) Development of and insights from systems pharmacology models of antibody-drug conjugates. CPT Pharmacometrics Syst Pharmacol 11(8):967–990. https://doi.org/10.1002/psp4.12833. Epub 2022 Jul 3. PMID: 35712824; PMCID: PMC9381915

Lazarou G, Chelliah V, Small BG, Walker M, van der Graaf PH, Kierzek AM (2020) Integration of omics data sources to inform mechanistic modeling of immune-oncology therapies: a tutorial for clinical pharmacologists. Clin Pharmacol Ther 107:858–870. https://doi.org/10.1002/cpt.1786

Liu T, Zhang L, Joo D, Sun SC (2017) NF-κB signaling in inflammation. Signal Transduct Target Ther 2:17023. https://doi.org/10.1038/sigtrans.2017.23. Epub 2017 Jul 14. PMID: 29158945; PMCID: PMC5661633

Macfarlane FR, Chaplain MAJ, Eftimie R (2019) Quantitative predictive modelling approaches to understanding rheumatoid arthritis: a brief review. Cells 9(1):74

Mahon JL et al (2009) The TrialNet Natural History Study of the Development of Type 1 Diabetes: objectives, design, and initial results. Pediatr Diabetes 10(2):97–104

Markossian S, Grossman A, Baskir H et al, editors (2004) Assay Guidance Manual [Internet]. Eli Lilly & Company and the National Center for Advancing Translational Sciences, Bethesda. Available from: https://www.ncbi.nlm.nih.gov/books/NBK53196/

Marriott H, Kabiljo R, Hunt GP et al (2023) Unsupervised machine learning identifies distinct ALS molecular subtypes in post-mortem motor cortex and blood expression data. Acta Neuropathol Commun 11:208. https://doi.org/10.1186/s40478-023-01686-8

Mc Laughlin AM, Milligan PA, Yee C, Bergstrand M (2023) Model-informed drug development of autologous CAR-T cell therapy: Strategies to optimize CAR-T cell exposure leveraging cell kinetic/dynamic modeling. CPT Pharmacometrics Syst Pharmacol 12(11):1577–1590. https://doi.org/10.1002/psp4.13011. Epub 2023 Jul 28. PMID: 37448343; PMCID: PMC10681459

McComb M, Bies R, Ramanathan M (2022) Machine learning in pharmacometrics: opportunities and challenges. Br J Clin Pharmacol 88:1482–1499

McGillen JB, Kelly CJ, Martínez-González A, Martin NK, Gaffney EA, Maini PK, Pérez-García VM (2014) Glucose-lactate metabolic cooperation in cancer: insights from a spatial mathematical model and implications for targeted therapy. J Theor Biol 361:190–203

McMillan D, Martinez-Fleites C, Porter J et al (2021) Structural insights into the disruption of TNF-TNFR1 signalling by small molecules stabilising a distorted TNF. Nat Commun 12:582. https://doi.org/10.1038/s41467-020-20828-3

Mehta V, Karnam G, Madgula V (2024) Liver-on-chips for drug discovery and development. Mater Today Biol 101143

Menden MP, Wang D, Mason MJ, Szalai B, Bulusu KC, Guan Y, Yu T, Kang J, Jeon M, Wolfinger R, et al; AstraZeneca-Sanger Drug Combination DREAM Consortium (2019) Community assessment to advance computational prediction of cancer drug combinations in a pharmacogenomic screen. Nat Commun 10:2674

Mestas J, Hughes CCW (2004) Of mice and not men: differences between mouse and human immunology. J Immunol 172(5):2731–2738

Mohamed MF, Beck D, Camp HS, Othman AA (2020) Preferential Inhibition of JAK1 Relative to JAK3 by Upadacitinib: Exposure-Response Analyses of Ex Vivo Data From 2 Phase 1 Clinical Trials and Comparison to Tofacitinib. J Clin Pharmacol 60(2):188–197. https://doi.org/10.1002/jcph.1513. Epub 2019 Aug 25. Erratum in: J Clin Pharmacol. 2020 Apr;60(4):551. PMID: 31448433; PMCID: PMC6973126

Morgan P, Van Der Graaf PH, Arrowsmith J, Feltner DE, Drummond KS, Wegner CD, Street SD (2012) Can the flow of medicines be improved? Fundamental pharmacokinetic and pharmacological principles toward improving Phase II survival. Drug Discov Today 17(9–10):419–424. https://doi.org/10.1016/j.drudis.2011.12.020. Epub 2011 Dec 29

Mueller SG et al (2005) Ways toward an early diagnosis in Alzheimer's disease: the Alzheimer's Disease Neuroimaging Initiative (ADNI). Alzheimers Dement 1(1):55–66

Müller S et al (2024) Whole blood stimulation as a tool for studying the human immune system. Eur J Immunol 54(2):2350519

NHANES Data (2011–2012) Centers for Disease Control and Prevention (CDC). National Health and Nutrition Examination Survey Data. http://www.cdc.gov/nchs/nhanes.htm

Polak S, Tylutki Z, Holbrook M, Wiśniowska B (2019) Better prediction of the local concentration-effect relationship: the role of physiologically based pharmacokinetics and quantitative systems pharmacology and toxicology in the evolution of model-informed drug discovery and development. Drug Discov Today 24(7):1344–1354. https://doi.org/10.1016/j.drudis.2019.05.016. Epub 2019 May 24

Population Pharmacokinetics | FDA https://www.fda.gov/regulatory-information/search-fda-guidance-documents/population-pharmacokinetics

Rahmani S, Breyner NM, Su HM, Verdu EF, Didar TF (2019) Intestinal organoids: A new paradigm for engineering intestinal epithelium in vitro. Biomaterials 194:195–214. https://doi.org/10.1016/j.biomaterials.2018.12.006. Epub 2018 Dec 10

Razonable RR, Henault M, Lee LN, Laethem C, Johnston PA, Watson HL, Paya CV (2005) Secretion of proinflammatory cytokines and chemokines during amphotericin B exposure is mediated by coactivation of toll-like receptors 1 and 2. Antimicrob Agents Chemother 49. https://doi.org/10.1128/aac.49.4.1617-1621.2005

Renshaw J et al (2013) Dual blockade of the PI3K/AKT/mTOR (AZD8055) and RAS/MEK/ERK (AZD6244) pathways synergistically inhibits rhabdomyosarcoma cell growth in vitro and in vivo. Clin Cancer Res 19(21):5940–5951

Rieger TR, Allen RJ, Bystricky L, Chen Y, Colopy GW, Cui Y, Gonzalez A, Liu Y, White RD, Everett RA, Banks HT, Musante CJ (2018) Improving the generation and selection of virtual populations in quantitative systems pharmacology models. Prog Biophys Mol Biol 139:15–22. https://doi.org/10.1016/j.pbiomolbio.2018.06.002. Epub 2018 Jun 15

Riss TL, Moravec RA, Niles AL et al (2004a) Cell viability assays. 2013 May 1 [Updated 2016 Jul 1]. In: Markossian S, Grossman A, Baskir H et al, editors. Assay Guidance Manual [Internet]. Eli Lilly & Company and the National Center for Advancing Translational Sciences, Bethesda. Available from: https://www.ncbi.nlm.nih.gov/books/NBK144065/

Riss T, Niles A, Moravec R et al (2004b) Cytotoxicity assays: In Vitro methods to measure dead cells. 2019 May 1. In: Markossian S, Grossman A, Baskir H, et al, editors. Assay Guidance Manual [Internet]. Eli Lilly & Company and the National Center for Advancing Translational Sciences, Bethesda. Available from: https://www.ncbi.nlm.nih.gov/books/NBK540958/

Riss TL, O'Brien MA, Moravec RA et al (2004c) Apoptosis marker assays for HTS. 2021 Jul 1. In: Markossian S, Grossman A, Baskir H et al, editors. Assay Guidance Manual [Internet]. Eli Lilly & Company and the National Center for Advancing Translational Sciences, Bethesda. Available from: https://www.ncbi.nlm.nih.gov/books/NBK572437/

Roelants C, Pillet C, Franquet Q, Sarrazin C, Peilleron N, Giacosa S, Guyon L, Fontanell A, Fiard G, Long JA, Descotes JL, Cochet C, Filhol O (2020) Ex-Vivo Treatment of Tumor Tissue Slices as a Predictive Preclinical Method to Evaluate Targeted Therapies for Patients with Renal Carcinoma. Cancers (Basel) 12(1):232. https://doi.org/10.3390/cancers12010232. PMID: 31963500; PMCID: PMC7016787

Rullmann JAC et al (2005) Systems biology for battling rheumatoid arthritis: application of the Entelos PhysioLab platform. IEEE Proc-Syst Biol 152(4):256–262

Rydell-Törmänen K, Johnson JR (2019) The applicability of mouse models to the study of human disease. In: Bertoncello I (ed) Mouse cell culture. Methods in molecular biology, vol 1940. Humana Press, New York. https://doi.org/10.1007/978-1-4939-9086-3_1

Sachs N et al (2019) Long-term expanding human airway organoids for disease modeling. EMBO J 38(4):e100300

Scheuher B, Ghusinga KR, McGirr K, Nowak M, Panday S, Apgar J, Subramanian K, Betts A (2023) Towards a platform quantitative systems pharmacology (QSP) model for preclinical to clinical translation of antibody drug conjugates (ADCs). J Pharmacokinet Pharmacodyn. https://doi.org/10.1007/s10928-023-09884-6. Epub ahead of print

Schmidt BJ, Casey FP, Paterson T, Chan JR (2013) Alternate virtual populations elucidate the type I interferon signature predictive of the response to rituximab in rheumatoid arthritis. BMC Bioinform 14:221. https://doi.org/10.1186/1471-2105-14-221. PMID: 23841912; PMCID: PMC3717130

Shah DK, Betts AM (2013) Antibody biodistribution coefficients: inferring tissue concentrations of monoclonal antibodies based on the plasma concentrations in several preclinical species and human. MAbs 5(2):297–305. https://doi.org/10.4161/mabs.23684. Epub 2013 Feb 13. PMID: 23406896; PMCID: PMC3893240

Shoda L, Kreuwel H, Gadkar K, Zheng Y, Whiting C, Atkinson M, Bluestone J, Mathis D, Young D, Ramanujan S (2010) The Type 1 Diabetes PhysioLab Platform: a validated physiologically based mathematical model of pathogenesis in the non-obese diabetic mouse. Clin Exp Immunol 161(2):250–267. https://doi.org/10.1111/j.1365-2249.2010.04166.x. Epub 2010 May 18. PMID: 20491795; PMCID: PMC2909407

Shroff T et al (2023) Studying metabolism with multi-organ chips: new tools for disease modelling, pharmacokinetics and pharmacodynamics. In: Advances in medical imaging, detection, and diagnosis, pp 1273–1315

Singh AP, Zheng X, Lin-Schmidt X, Chen W, Carpenter TJ, Zong A et al (2020) Development of a quantitative relationship between CAR-affinity, antigen abundance, tumor cell depletion and CAR-T cell expansion using a multiscale systems PK-PD model. MAbs 12(1). https://doi.org/10.1080/19420862.2019.1688616

Sivakumar R, Chan M, Shin JS, Nishida-Aoki N, Kenerson HL, Elemento O, Beltran H, Yeung R, Gujral TS (2019) Organotypic tumor slice cultures provide a versatile platform for immuno-oncology and drug discovery. Onco Targets Ther 8(12):e1670019. https://doi.org/10.1080/2162402X.2019.1670019. PMID: 31741771; PMCID: PMC6844320

Smith DA, Di L, Kerns EH (2010) The effect of plasma protein binding on in vivo efficacy: misconceptions in drug discovery. Nat Rev Drug Discov 9(12):929–939. https://doi.org/10.1038/nrd3287

Smyth LA, Collins I (2009) Measuring and interpreting the selectivity of protein kinase inhibitors. J Chem Biol 2(3):131–151. https://doi.org/10.1007/s12154-009-0023-9. Epub 2009 Jun 6. PMID: 19568781; PMCID: PMC2725273

Sodhi JK, Benet LZ (2021) Successful and unsuccessful prediction of human hepatic clearance for lead optimization. J Med Chem 64(7):3546–3559. https://doi.org/10.1021/acs.jmedchem.0c01930. Epub 2021 Mar 25. PMID: 33765384; PMCID: PMC8504179

Sorger PK et al (2011a) An NIH white paper by the QSP workshop group. In: Quantitative and systems pharmacology in the post-genomic era: new approaches to discovering drugs and understanding therapeutic mechanisms

Sorger PK et al (2011b) Quantitative and systems pharmacology in the post-genomic era: new approaches to discovering drugs and understanding therapeutic mechanisms. An NIH white paper by the QSP workshop group. NIH, Bethesda, pp 1–48

Sung JH (2021) Multi-organ-on-a-chip for pharmacokinetics and toxicokinetic study of drugs. Expert Opin Drug Metab Toxicol 17(8):969–986. https://doi.org/10.1080/17425255.2021.1908996

Takashima O, Tsuruta F, Kigoshi Y, Nakamura S, Kim J, Katoh MC, Fukuda T, Irie K, Chiba T (2013) Brap2 regulates temporal control of NF-κB localization mediated by inflammatory response. PLoS One 8(3):e58911. https://doi.org/10.1371/journal.pone.0058911. Epub 2013 Mar 15. PMID: 23554956; PMCID: PMC3598860

Thalhauser CJ et al (2015) Mechanistic predictions of response to combinations of biologic agents in a quantitative systems pharmacology model of rheumatoid arthritis. J Pharmacokinet Pharmacodyn 42(1):S11–S107

Thurber G, Yang K, Reiner T et al (2013) Single-cell and subcellular pharmacokinetic imaging allows insight into drug action *in vivo*. Nat Commun 4:1504. https://doi.org/10.1038/ncomms2506

Tito C, Masciarelli S, Colotti G, Fazi F (2025) EGF receptor in organ development, tissue homeostasis and regeneration. J Biomed Sci 32(1):24. https://doi.org/10.1186/s12929-025-01119-9. PMID: 39966897; PMCID: PMC11837477

Toulany M, Baumann M, Rodemann P (2007) Stimulated PI3K-AKT signaling mediated through ligand or radiation-induced EGFR depends indirectly, but not directly, on constitutive K-Ras activity. Mol Cancer Res 5(8):863–872

Trang M, Dudley MN, Bhavnani SM (2017) Use of Monte Carlo simulation and considerations for PK-PD targets to support antibacterial dose selection. Curr Opin Pharmacol 36:107–113. https://doi.org/10.1016/j.coph.2017.09.009. Epub 2017 Nov 10

Traves PG, Murray B, Campigotto F, Galien R, Meng A, Di Paolo JA (2021) JAK selectivity and the implications for clinical inhibition of pharmacodynamic cytokine signalling by filgotinib, upadacitinib, tofacitinib and baricitinib. Ann Rheum Dis 80(7):865–875. https://doi.org/10.1136/annrheumdis-2020-219012. Epub 2021 Mar 19. PMID: 33741556; PMCID: PMC8237188

Tsao PS, Buitrago R, Chan JR, Cooke JP (1996) Fluid flow inhibits endothelial adhesiveness. Nitric oxide and transcriptional regulation of VCAM-1. Circulation 94(7):1682–1689. https://doi.org/10.1161/01.cir.94.7.1682

Tsao PS, Wang B, Buitrago R, Shyy JY, Cooke JP (1997) Nitric oxide regulates monocyte chemotactic protein-1. Circulation 96(3):934–940. https://doi.org/10.1161/01.cir.96.3.934

Uatay A, Gall L, Irons L, Tewari SG, Zhu XS, Gibbs M, Kimko H (2024) Physiological indirect response model to omics-powered quantitative systems pharmacology model. J Pharm Sci 113(1):11–21. https://doi.org/10.1016/j.xphs.2023.10.032. Epub 2023 Oct 26

Verma M, Gall L, Biasetti J, Di Veroli GY, Pichardo-Almarza C, Gibbs MA, Kimko H (2023) Quantitative systems modeling approaches towards model-informed drug development: Perspective through case studies. Front Syst Biol 2:1063308. https://doi.org/10.3389/fsysb.2022.1063308

Wang X et al (2012) Inter-species differences contributing to poor translatability of sustained glucose lowering effects of glycogen phosphorylase inhibition from rodent to human. Diabetes 61. ALEXANDRIA: AMER DIABETES ASSOC

Wang J, Iyer S, Fielder PJ, Davis JD, Deng R (2016) Projecting human pharmacokinetics of monoclonal antibodies from nonclinical data: comparative evaluation of prediction approaches in early drug development. Biopharm Drug Dispos 37(2):51–65. https://doi.org/10.1002/bdd.1952. Epub 2015 May 15

Wang EB, Shen L, Heathman M, Chan JR (2019) Incorporating placebo response in quantitative systems pharmacology models. CPT Pharmacometrics Syst Pharmacol 8(6):344–346. https://doi.org/10.1002/psp4.12412. Epub 2019 Jun 9. PMID: 30983120; PMCID: PMC6618253

Wee P, Wang Z (2017) Epidermal growth factor receptor cell proliferation signaling pathways. Cancers (Basel) 9(5):52. https://doi.org/10.3390/cancers9050052. PMID: 28513565; PMCID: PMC5447962

Wu G, Yuan Y, Hodge CN (2003) Determining appropriate substrate conversion for enzymatic assays in high-throughput screening. SLAS Discov 8(6):694–700

Wu Q et al (2020) Organ-on-a-chip: recent breakthroughs and future prospects. Biomed Eng Online 19:1–19

You S et al (2012) A systems approach to rheumatoid arthritis. PLoS One 7(12):e51508

Zegzouti H, Zdanovskaia M, Hsiao K, Goueli SA (2009) ADP-Glo: a Bioluminescent and homogeneous ADP monitoring assay for kinases. Assay Drug Dev Technol 7(6):560–572. https://doi.org/10.1089/adt.2009.0222

Zhang F, Wei K, Slowikowski K et al (2019) Defining inflammatory cell states in rheumatoid arthritis joint synovial tissues by integrating single-cell transcriptomics and mass cytometry. Nat Immunol 20:928–942. https://doi.org/10.1038/s41590-019-0378-1

Zhang T, Androulakis IP, Bonate P, Cheng L, Helikar T, Parikh J, Rackauckas C, Subramanian K, Cho CR, Working Group (2022) Two heads are better than one: current landscape of integrating QSP and machine learning: an ISoP QSP SIG white paper by the working group on the integration of quantitative systems pharmacology and machine learning. J Pharmacokinet Pharmacodyn 49(1):5–18. https://doi.org/10.1007/s10928-022-09805-z. Epub 2022 Feb 1. PMID: 35103884; PMCID: PMC8837505

Zhang F, Jonsson AH, Nathan A, Millard N, Curtis M, Xiao Q, Gutierrez-Arcelus M, Apruzzese W, Watts GFM, Weisenfeld D, Nayar S, Rangel-Moreno J, Meednu N, Marks KE, Mantel I, Kang JB, Rumker L, Mears J, Slowikowski K, Weinand K, Orange DE, Geraldino-Pardilla L, Deane KD, Tabechian D, Ceponis A, Firestein GS, Maybury M, Sahbudin I, Ben-Artzi A, Mandelin AM 2nd, Nerviani A, Lewis MJ, Rivellese F, Pitzalis C, Hughes LB, Horowitz D, DiCarlo E, Gravallese EM, Boyce BF, Accelerating Medicines Partnership: RA/SLE Network, Moreland LW, Goodman SM, Perlman H, Holers VM, Liao KP, Filer A, Bykerk VP, Wei K, Rao DA, Donlin LT, Anolik JH, Brenner MB, Raychaudhuri S (2023) Deconstruction of rheumatoid arthritis synovium defines inflammatory subtypes. Nature 623(7987):616–624. https://doi.org/10.1038/s41586-023-06708-y. Epub 2023 Nov 8. PMID: 37938773; PMCID: PMC10651487

Zhu J et al (2024) Mapping cellular interactions from spatially resolved transcriptomics data. Nat Methods 21(10):1830–1842

Integrating QSP and ML to Facilitate Drug Development and Personalized Medicine

Tongli Zhang

Contents

Abstract

In this chapter, the potential integration between quantitative systems pharmacology (QSP) and machine learning (ML) is explored. ML models are in their nature "black boxes", since they make predictions based on data without explicit system definitions, while on the other hand, QSP models are "white boxes" that describe mechanistic biological interactions and investigate the systems properties emerging from such interactions. Despite their differences, both approaches have unique strengths that can be leveraged to form a powerful integrated tool.

T. Zhang (✉)
University of Cincinnati College of Medicine, Cincinnati, OH, USA
e-mail: zhangtl@ucmail.uc.edu

E. L. Bradshaw, M. E. Spilker (eds.), *Quantitative Systems Pharmacology*, Handbook of Experimental Pharmacology 289, https://doi.org/10.1007/164_2024_740

ML's ability to handle large datasets and make predictions is complemented by QSP's detailed mechanistic insights into drug actions and biological systems. The chapter discusses basic ML techniques and their application in drug development, including supervised and unsupervised learning methods. It also illustrates how combining QSP with ML can facilitate the design of combination therapies against cancer resistance to single therapies. The synergy between these two methodologies shows promise to accelerate the drug development process, making it more efficient and tailored to individual patient needs.

Keywords

Cancer resistance · Combination therapies · Drug development · Machine Learning (ML) · Personalized medicine · Quantitative Systems Pharmacology (QSP)

1 Introduction

Machine learning (ML) and quantitative systems pharmacology (QSP) have gained substantial momentum in the pharmaceutical industry (Bai et al. 2021; Liu et al. 2023; Nijsen et al. 2018; Zineh 2019). Despite their common goal of facilitating drug development, QSP and ML analyze data from different perspectives. ML models are particularly good at identifying patterns, discovering relationships, and making predictions based on data without explicit definition of how the data-generating systems operate (a "*black box*" approach), whereas QSP models can extrapolate beyond acquired data sets used in model construction and leverage them as a window to evaluate hypothesized mechanisms of the underlying biological systems. Details of experimental design and hypothesized mechanisms of how the biological system is regulated play important roles in explaining the data and then generating novel predictions. In this way, QSP models describing the mechanistic linkages of drug-target-disease interactions could be thought of as a "*white box*" model, in contrast to a "black box" modeling approach employed by ML algorithms.

Both approaches have pros and cons. A "*white box*" offers a mechanistic description of how a drug works, and how outcomes (e.g., efficacy, lack of response, or toxicity) emerge from the temporal and spatial interactions between drugs, biochemical molecules, cells, and organs. Drug developers, reviewers, doctors, and patients may obtain a picture of how efficacy and/or toxicity emerges through a more comprehensive description of the multi-scale interactions of drugs with cells, organs, etc., within a patient or population through QSP modeling. However, while the "*white box*" approach is based on our understanding of the underlying biological mechanisms and disease pathophysiology, it may be limited in scope or necessitate assumptions. Furthermore, the effort requires both a multidisciplinary team, which may consist of modelers, biologists, clinical pharmacologists, and clinicians, and significant time and resource to develop and evaluate a QSP model of a drug and its targeted disease. Additionally, it is often not a trivial task to update QSP models as data and knowledge increase over time, and the incorporation of new data and improvement of predictivity may require the expertise of a similar team.

On the other hand, a "*black box*" approach can make predictions of interest without the incorporation of physiological or pathological knowledge and may not require understanding a drug's mechanism of action. However, this may limit the understanding of model results and the application of the results within the appropriate contexts; highlighting current concerns of ML modeling given the "*black box*" nature of these models. Additional details regarding the promises and limitations of utilizing artificial intelligence (AI) and ML in drug development can be found in the literature (Liu et al. 2023; Terranova et al. 2024; Ahmadi Daryakenari et al. 2024; Laurie and Lu 2023; Ribba 2023; Yuan et al. 2021).

In this chapter, we will briefly describe basic ML techniques and discuss and illustrate how QSP and ML can be combined to leverage their distinct strengths to provide a powerful tool for drug development. Additional thoughts on the use of machine learning and its encompassing technology of artificial intelligence with QSP modeling are further discussed in "Future Directions for Quantitative Systems Pharmacology".

2 Fundamentals of Machine Learning (ML)

Before discussing the basics of ML, it is useful to illustrate the nested relationship between AI, ML, and neural networks (Fig. 1). According to the AI/ML-Discussion-Paper published by the FDA (FDA 2023), *AI can generally be described as a branch of computer science, statistics, and engineering that uses algorithms or models to*

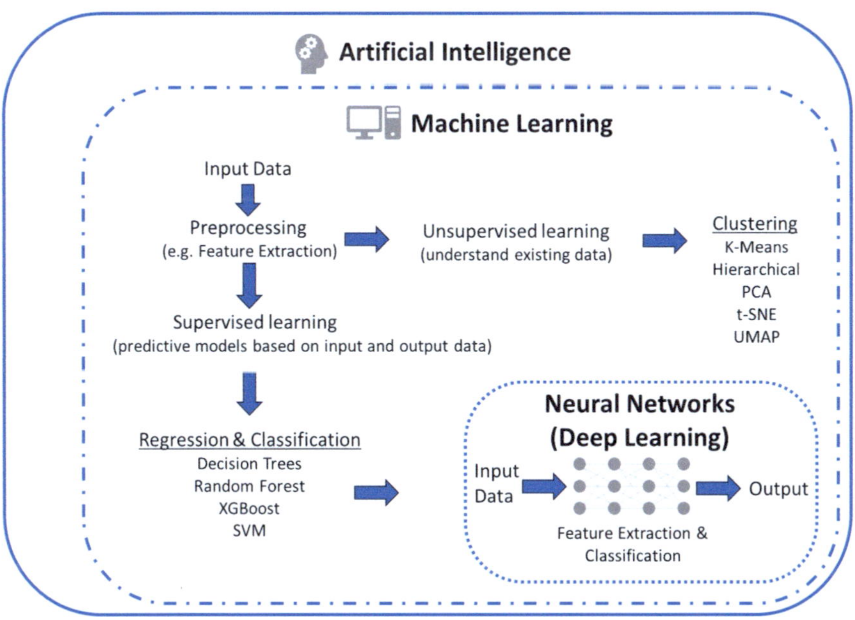

Fig. 1 Nested relationship between AI, ML, and neural networks

perform tasks and exhibit behaviors such as learning, making decisions, and making predictions. ML is considered a subset of AI that allows ML models to be developed by ML training algorithms through analysis of data, without models being explicitly programmed. In other words, AI is what guides a machine's behavior using rules that are developed with the models built by ML and other types of algorithms (including but not limited to neural networks) that extract data features and subsequently identify, classify, and make predictions about relationships within input datasets. In this regard, "big data" in combination with powerful computation facilitates effective ML and neural network models and underpins the advancement of AI.

2.1 ML Basics

Here we provide basic descriptions of common machine learning methods. While new advances in the ML field are continuously improving processes and methodologies, the underlying principles hold.

The first step in ML model development and applications is to preprocess dataset (s), which includes activities such as extracting important features, addressing missing data points, and checking whether the data is consistent with the underlying assumptions of the applied ML methods. It is essential to properly split data into training, test, and validation sets to support model development, evaluation, and validation. This step typically benefits from human intervention and the experience of ML modelers, but the process could be sped up by converting the experience and expertise to automatic or semi-automatic algorithms. Once the data is ready for analysis, the general ML modeling process shown in Fig. 2 is initiated.

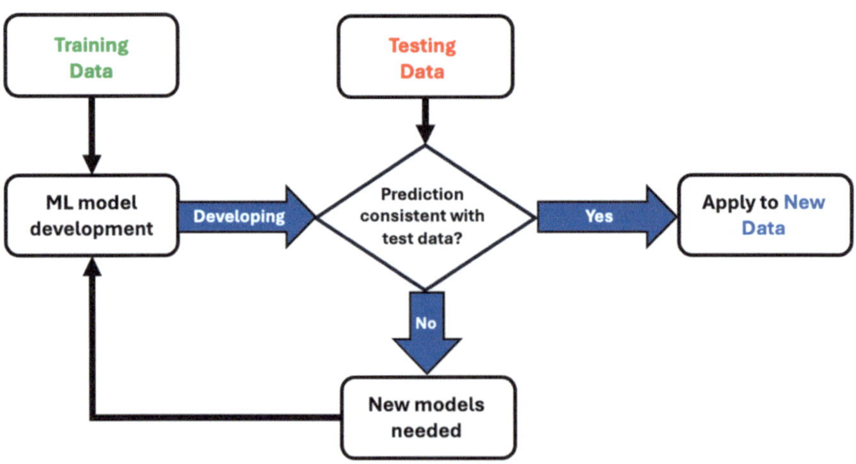

Fig. 2 The general process of developing, testing, and applying ML models

Often, there is no pre-defined starting point for analysis, rather preliminary exploration or experience guides the selection of the initial ML algorithm, and the ML process then starts by building models with training data. Once trained, the models are tested by evaluating their predictions against independent testing data (that were not used for training purposes). If predictions from the models are inconsistent with testing data, ML models may be refined to improve performance. However, if predictions are consistent with test data the model can be applied on new scenarios. Typically, performance of ML models depends on the algorithm, the training data, test data, and the new scenarios.

The experience and expertise of ML modelers and the input from domain experts are important to answer key questions such as the following: How should a "model" be selected? What type of "model" or algorithm to use? Are the testing data sets appropriate for determining the validity selection of model? How rigorous should the criteria be to assess "passing"? All these settings determine the usefulness and the risk of associated ML models.

2.2 Common ML Models

While some common ML models are briefly discussed here, for additional in-depth information on ML models, readers may refer to excellent textbooks and references (Liu et al. 2024; Talevi et al. 2020). ML methods are commonly categorized into supervised learning and unsupervised learning. Supervised learning takes input as heterogeneous populations (e.g., patients responding or resistant to a certain therapy) and identifies features that are associated with different groups. Ultimately, these models use both the input and output (labeled response data) to make future predictions. Whereas unsupervised learning models discover hidden patterns or structures from the input of a homogeneous population. This methodology serves to better understand the existing input data.

2.2.1 Supervised ML Methods

Supervised machine learning is a suitable approach for predicting different types of outcomes (i.e., disease vs healthy, responsive vs resistant, etc.). To accomplish this, supervised machine learning methods such as decision trees and support vector machines can be used.

Decision Trees Decision trees are versatile models that can be used for both classification and regression tasks. They partition the input space into regions and make decisions based on a series of hierarchical rules. Decision trees are intuitive to use, easy to visualize and to communicate with others, and are highly valuable as an exploratory method to better understand data.

For example, a classification and regression tree has been developed to predict the risk of traumatic-induced hemorrhage in patients on oral anticoagulant therapy. With clinical risk factors such as transitory loss of consciousness, the decision tree can separate patients into distinct groups with different prognostic outcomes. In this way, decision trees can be used to guide in the diagnosis and treatment of future patients (Turcato et al. 2022).

Random Forest and XGBoost These two methods are both built upon decision trees. Random forest (RF) is an ensemble method that aggregates the output from multiple decision trees. Since each tree is trained on a subset of the data, followed by aggregation of predictions, random forests can improve performance and reduce overfitting by any single decision tree. Extreme gradient boosting (XGBoost) is a decision tree-based machine learning method characterized by gradient boosting (Moore and Bell 2022).

Random forest algorithms have been used to predict the outcome of stroke patients based on clinical, biochemical, and neuroimaging factors. With the records from over 6,000 patients, the developed random forest model effectively predicted the long-term outcome of stroke patients. (Fernandez-Lozano et al. 2021). In addition to predicting patient outcomes with existing clinical measurements, the Random forest and XGBoost models may also contribute to selection of informative measurements for future clinical practice. Since their results are based on the "collective wisdom" of multiple trees, the conclusions from RF and XGBoost tend to be more robust than single decision trees.

Support Vector Machine (SVM) Support vector machine (SVM) is a powerful machine learning algorithm used for both classification and regression tasks. It works by finding a hyperplane, which is a decision boundary that separates different classes or predicts target values. For classification, SVM seeks the hyperplane that distinctly separates the data points into different classes, while for regression, it aims to predict continuous target values. The key aspect of SVM is the hyperplane's ability to maximize the margin, which is the distance between the hyperplane and the closest data points from either class, known as support vectors.

The support vectors are crucial as they define the margin and determine the position of the hyperplane. By maximizing this margin, SVM ensures that the hyperplane not only separates the classes but does so with the greatest possible distance from the nearest points, enhancing the model's generalization and robustness. With nonlinear kernel functions, SVM is a powerful tool for handling data with complex decision boundaries (Scholkopf and Smola 2001). For application, SVMs have been applied to stratify cancer patients based on continuous biomarkers (Lanza and Parashar 2021).

2.2.2 Unsupervised ML Methods

Unsupervised ML methods aim to identify patterns, relationships, and structures from datasets without any explicit guidance or instruction. Some common unsupervised learning methods include clustering methods, such as K-means clustering and hierarchical clustering, and dimensionality reduction techniques, such as principal component analysis (PCA), t-distributed stochastic neighbor embedding (t-SNE), and uniform manifold approximation and projection (UMAP).

The independence from human knowledge is both a strength and limitation of unsupervised ML methods. On the one hand, these methods might be developed in a "domain-independent" manner and be applied to essentially all types of data.

For example, principal component analysis (PCA) can be applied to various types of data, whether it be house prices or genetic expressions. Although these datasets are inherently different in their nature, the goal of PCA remains consistent: to identify the most informative parameters that capture the underlying patterns and variability within the data. In both cases, PCA helps to reduce dimensionality and highlight key features, making it a versatile tool across different domains.

On the other hand, the lack of strong connections to human knowledge makes it challenging to properly explain the results from these methods. Although the analysis results, such as principal components or clusters, can be clearly defined in statistical and computational terms, it is often less obvious how the results of unsupervised ML models should be used to make appropriate decisions. The proper explanation and application of the ML results require a deep dive into both biological systems and mathematical algorithms.

Principal Component Analysis (PCA)

Principal component analysis (PCA) is a dimensionality reduction technique (Lever et al. 2017; Ringnér 2008). It transforms complex datasets into principal components that explains the variance of the data. In this way, PCA helps in visualizing high-dimensional data by projecting it into lower dimensions and is useful for handling and visualizing complex data.

Its strengths include technical maturity and computational efficiency. However, it is not recommended for handing complex non-linear relationships, and it is sensitive to scaling of input features.

In biomedicine, PCA has been commonly applied as an exploratory tool to reduce the dimensionality of omic data and helps researchers to identify potentially significant patterns or features, but its application can easily go beyond the omics field. For example, O'Dell, et al. recently applied PCA to analyze synaptic density data collected from patients with Alzheimer's disease and then investigated the association between discovered principal components (PC) subject scores with cognitive performance (O'Dell et al. 2023).

Clustering Methods

K-means clustering is a ML method that partitions data into a pre-defined number of clusters (K) based on the similarity between data points. For K-means clustering, each data point is assigned to the cluster with the nearest cluster centroids, and the algorithm iteratively refines the cluster centroids to minimize within-cluster variance. Its simplicity makes it easy to implement and computationally efficient for large datasets. However, its results depend on pre-determined number of clusters (K) and initial centroid placement, and it is challenged by data points of varying shapes and densities.

Hierarchical clustering, on the other hand, builds a hierarchy of clusters by either agglomerative (bottom-up) or divisive (top-down) methods. It does not require pre-specifying the number of clusters (K in K-means) and is able to accommodate additional data complexity with clusters at different levels of granularity. Hence, it is more flexible compared with K-means clustering, though the power comes with a price of computational cost and sensitivity to noise and outliers.

Like other ML methods, clustering algorithms could be used for a broad range of biomedical data. For example, K-means clustering was used for segmentation of cells showing white and gray matters of brain images of Alzheimer's disease by Olle et al. (2024) (Liu et al. 2024), while Luckett et al. applied hierarchical clustering to group biomarkers in autosomal dominant AD (Luckett et al. 2023).

Non-linear Dimensionality Reduction Techniques: T-SNE and UMAP

Non-linear tools such as t-distributed stochastic neighbor embedding (t-SNE) and uniform manifold approximation and projection (UMAP) have been widely applied by biomedical researchers for visualizing complex, high-dimensional datasets, which are commonly encountered in biomedical fields.

Unlike PCA, which focuses on capturing global structure and maximizing variance, t-SNE focuses on preserving the local structure of data. This makes t-SNE particularly effective for visualizing clusters or groups within datasets, and it often reveals patterns that may not be revealed by PCA (Van der Maaten and Hinton 2008). However, t-SNE is challenged by high computational cost for large datasets and difficulties in interpreting the distances in the reduced dimensions.

In the biomedical field, t-SNE has been applied to single-cell RNA sequencing (scRNA-seq) data to visualize cell populations and explore potential cellular heterogeneity. For instance, t-SNE was used in the study by Villani et al. (2017) as a graph-based approach to visualize and cluster cells. Like t-SNE, UMAP is also useful for visualizing high-dimensional data such as single-cell RNA sequencing datasets, and its performance has been benchmarked against other tools. In the study by Becht et al., UMAP was found to generate reproductive and biologically meaningful results in a computationally efficient way (Becht et al. 2019).

2.2.3 Deep Learning Based on Neural Network

Neural networks consist of layers of interconnected nodes which transform input data to output features by adjusting their internal kernel functions (mapping equations) and associated parameters. The "deep learning" often associated with neural network refers to the existence of many hidden layers required to carry out large amounts of computation that translates input to output. Deep learning is a flexible modeling method that can be applied to multiple tasks provided the availability of a sufficiently large quantity of high-quality data. With their efficiency and flexibility, Deep Neural Networks are a good choice of a tool for handling high-dimensional inputs.

The simple form of a neural network operates unidirectionally through multiple layers, which are mathematical functions transforming input values to output values. Recurrent neural network (RNN) is more complex in the sense that each mathematical function in the RNNs can also incorporate feedback in which the output of one function serves as input in the next time step. Such a design enables RNNs to model sequential data, such as natural language and time-series (Liu et al. 2021).

Neural ordinary differential equations (Neural ODEs) are similar to RNN, but they can handle continuous data and hence serve as an attractive method to model time series, such as pharmacokinetics data (Lu et al. 2021).

2.3 Data Requirements for ML

In practice, ML is both science and art: success arises from finding the right pattern from the data with the right computational algorithms, and how to do so is often subject to the experience and intuition of the researchers.

To understand why some ML projects are more successful than the others, the workflow in Fig. 2 is revisited to explore patterns identified from the training data (Fig. 3).

ML methods extract patterns and relationships from the data. As illustrated in Fig. 3, possible ML patterns and relationships extracted from the training data (not data themselves) can be divided into four regions (Fig. 3, left column, labeled as 1–4). For illustration purposes, we'll explore a simple example where red dogs are training data, red owls are testing data, and blue cats are new data.

One possible pattern is "*living creature,*" and this pattern resides in *Region 1*. The pattern also characterizes the testing data and the new data. If our ML method finds this pattern from the training data of red dogs, the pattern will be able to (a) pass the tests with red owls and (b) correctly predict the new data of blue cats. This is the result we would like to achieve with ML analysis.

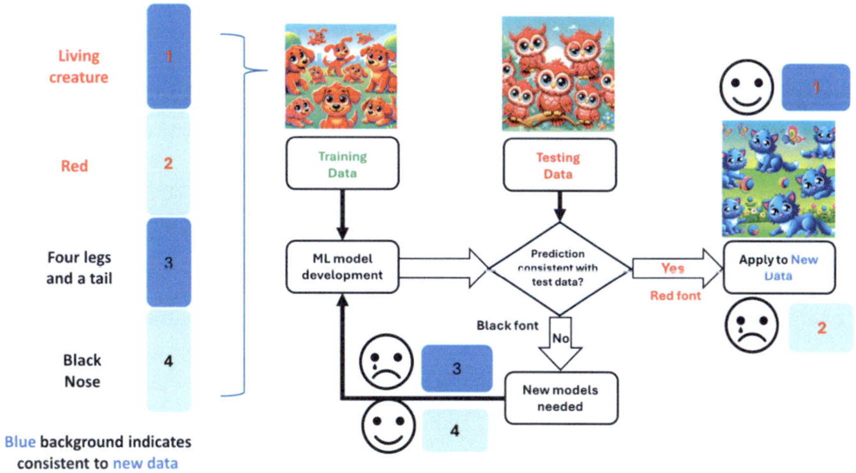

Fig. 3 Rethinking the ML workflow from the patterns extracted from the training data. The cartoon animals are drawn by ChatGPT 4o. See text for additional details

The patterns in *Region 2*, on the other hand, are features of both training and testing data, but not of the new data. In our illustrative example, if our ML methods identified the color red as the pattern from training data (red dogs), the identified pattern will be able to pass the test data (red owls), but they will generate incorrect predictions when applied on new data (blue cats). This would be an error.

Patterns from *Region 3*, on the other hand, would be able to correctly predict new data, but they would have failed the test before they could be pushed into application. In our example, "four legs and a tail" would apply for both the training data and the new data; however, this would be missed since it does not apply to the test data. This would be a missed opportunity.

Patterns from *Region 4* would be rejected by the tests, which will be a desirable result as well. From this view, successful ML projects are those that identify patterns from regions 1 and 4 (since correction rejections are also useful), while the less successful ones would identify patterns from 2 and 3. This exemplifies that the choice of training data, testing data, and new data are as important as the ML algorithms.

Hence, although mathematical algorithms for ML analysis are general and do not change when ML is applied to problems from many fields, the processing and selection of data for the development, testing, interpretation, and application of ML models benefit from an understanding of the biological systems. Similarly, for QSP modelers, the mathematical equations may be general, but the proper development and application of QSP benefit from years of experience, practice, and learning.

3 Enhancing the Potential of ML in Biomedicine with QSP Chaperones

Given that experience and a level of understanding of biological systems can greatly bolster the development, application, and use of ML models, it would be advantageous to have each ML modeler helped by a domain expert. The domain expert should retain all relevant knowledge and data of the biological systems of interest and be able to translate ML results back into the proper biological context with correct biomedical meanings.

We propose that QSP models could serve as "*in silico agents*" for domain experts. With their solid rooting in the underlying biological systems, QSP models could transform the biological data and knowledge into formats that are suited for analysis with ML modeling. In this way, the accumulated biological knowledge and governing biological principles can be utilized to constrain ML models and reduce the risk of biologically implausible results. Furthermore, ML results can be validated and explained with simulations of the corresponding QSP models. The combination between QSP models and ML models bears the best chance to solve the important biological questions of interest.

4 An Example of Combining QSP with ML Methods

To illustrate the process of sequentially combining QSP and ML models, we describe a specific example involving cancer resistance mechanisms and combination therapies, which is based on a previous publication (Zhang et al. 2019).

4.1 The Biological Question

Cancer treatments often fail due to a diverse set of resistance mechanisms occurring in cells (Eduati et al. 2017; Wilson et al. 2009; Holohan et al. 2013; Roux et al. 2015), and effective strategies to fight against such resistance mechanisms are being actively researched. A strategy to minimize resistance is to treat patients with multiple therapies that have different molecular mechanisms of action. Out of many potential reasons for therapy failure, one plausible therapy strategy follows the hypothesis that tumor cells with resistance to one therapy might respond to another therapy and, hence, there is a chance to circumvent resistance by treatment of a differing mechanism of action. The huge number of possible combinations, and the consequential cost, time, and effort, makes it impossible to test all combinations in clinical trials. Therefore, computational modeling and machine learning should be used to pre-examine potential combinations and select only those that are most promising.

4.2 The Input Model

This example begins with the construction of a mechanistic QSP model, which is termed the input model, that incorporates three functional modules that contribute to caspase-3 activation and cellular death in response to pro-apoptosis signals: p53 regulation, BCL family proteins, and caspase proteins (Fig. 4). The current model is the "right size" for the purpose of investigating the fractional killing and p53 dynamics, though it is not comprehensive and additional control elements can be continuously incorporated and validated with knowledge and experimental data. With each model update, the conclusions will update correspondingly.

To begin the computational effort, the wiring diagram of the input model (Fig. 4) is translated into a set of ordinary differential equations (ODE), a standard practice in QSP. Utilizing the same equations, multiple parameter sets, whose resulted model simulations are close to corresponding data, are selected and assembled into a population of tumor cells. In this way, the distributions of the selected parameters explain the observed cell-to-cell variabilities. Temporal simulations of apoptotic biomarkers from a sampling of representative in silico tumor cells demonstrate a varied but physiologically plausible response to a concentration of cisplatin as shown in Fig. 5, where some cells die (red curves), while others survive (blue curves). This behavior can be probed further to extract observations that can be compared to experimental data such as the observation that apoptotic cells activate

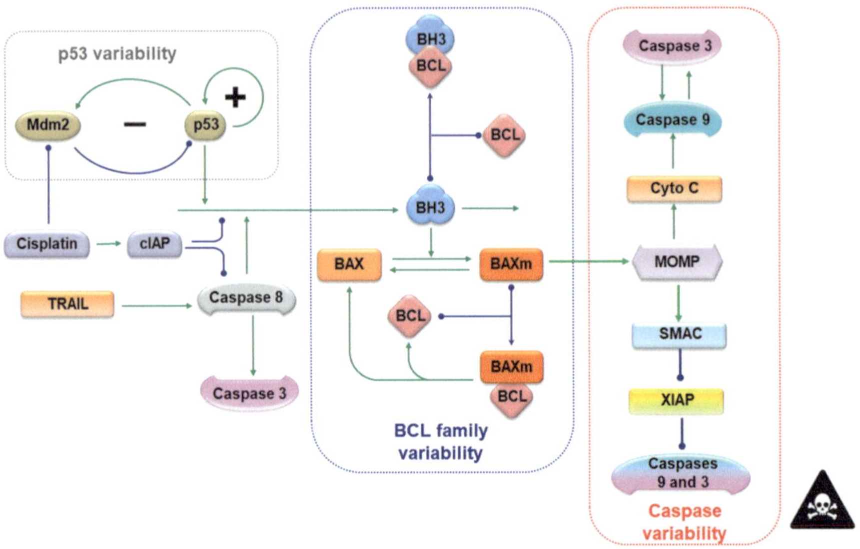

Fig. 4 The regulatory network of the mechanistic model. (Through the molecular control network, cisplatin ultimately leads to activation of caspases-9 and -3, which are then responsible for the death of the cells)

p53 faster than those that survive (Fig. 5a) (Paek et al. 2016). This triggers a sequence of events which can be best understood by evaluating the wiring diagram (Fig. 4) and the simulations together. Through accumulation of BH3 proteins and the rapid accumulation of p53 in apoptotic cells, BAXm proteins are activated. This in turn results in a sequence of events: (1) mitochondria outer membrane permeability (MOMP) (2) the release of CytoC and Smac from mitochondria; (3) the apoptosis inhibitor XIAP are sequestered into Smac:XIAP dimer forms and no longer inhibits apoptosis (Figs. 4 and 5b, c); (4) CytoC results in the activation of caspase-9 (Figs. 4 and 5d) and ultimately the activation of caspase-3 (Figs. 4 and 5e, f) which results in (5) the induction of programmed cell death (Becht et al. 2019; Lu et al. 2021). These simulations demonstrate how the input model describes the sequence of events that occur after tumor cells are treated with cisplatin.

4.3 Experimental Testing of the Model

The success of QSP + ML heavily depends on the quality of the input model; hence, it is advisable to test the input model with additional experiments. In this example, the model was tested with inputs of low doses of tumor necrosis factor (TNF)-related apoptosis-inducing ligand (TRAIL) as monotherapy and in combination with cisplatin. Both the in silico model simulations (Fig. 6a) and experimental data (Fig. 6c) show that TRAIL enhances the killing efficiency of cisplatin with the combination therapy outperforming the monotherapy groups. Furthermore, density plots of the

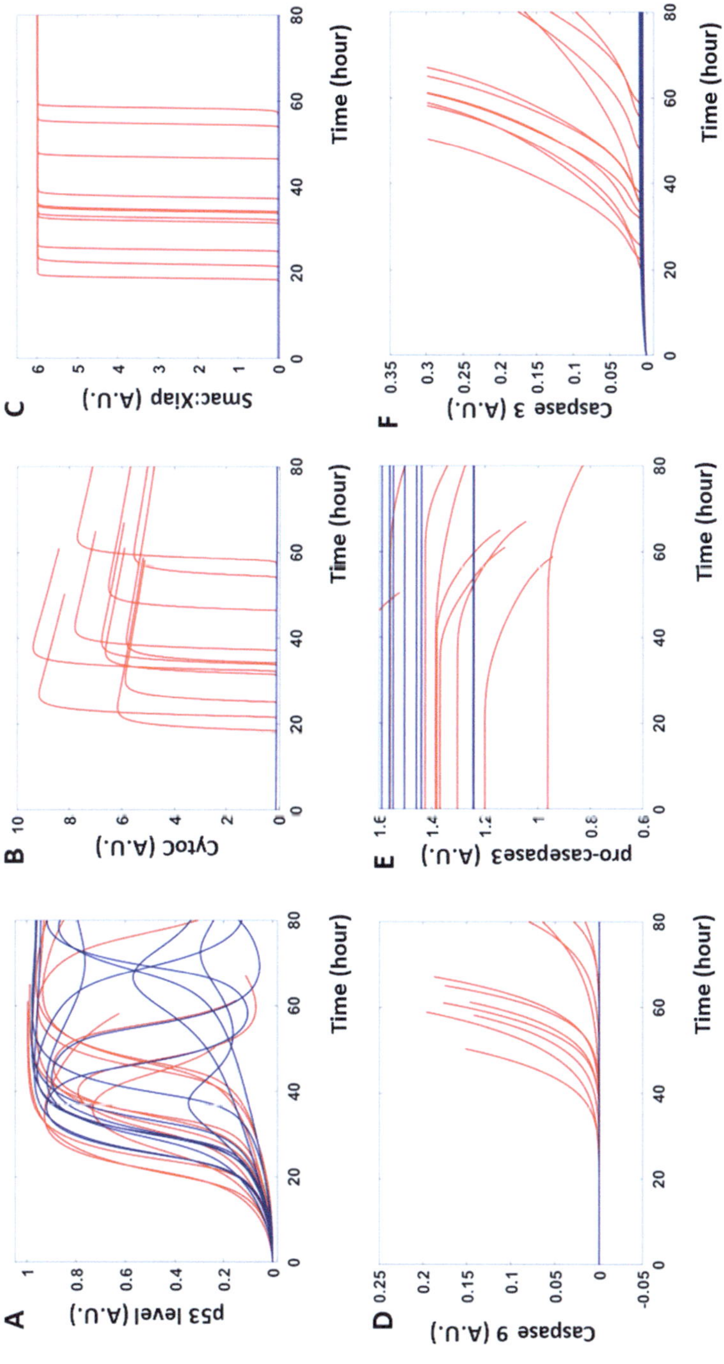

Fig. 5 Time series simulation of the model illustrating the sequential activation of p53, cytoC, Smac, and then caspases. See the text for more details

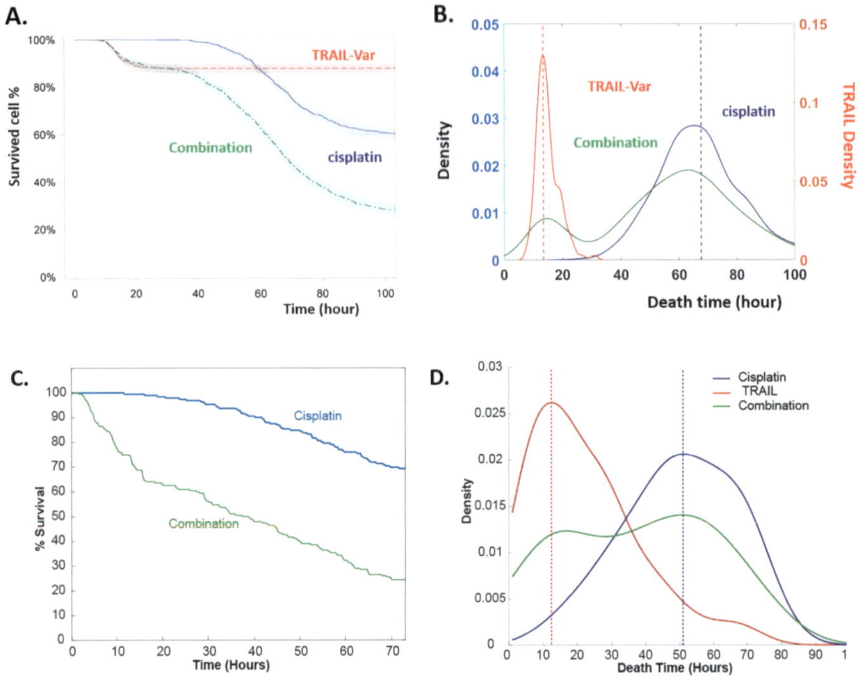

Fig. 6 Simulation of the mechanistic model (**a** and **b**) and the corresponding experimental data (**c** and **d**). (**a**). Kaplan-Meier plot of the simulated data with TRAIL functioning as a dynamical variable. (**b**). Density plots of the distribution of the time of death predicted by the model for TRAIL alone (right y-axis), cisplatin, or cisplatin+TRAIL combination (left y-axis). TRAIL functions as a dynamical variable. (**c**). A Kaplan-Meier plot of the experimental data. 192 single cells were tracked for each dose. (**d**) Density plot of the distribution of the time of death measured in live-cell microscopy time-lapse experiments. Dashed red line represents the peak time for cell death for TRAIL; dashed blue line is the peak time of death for cisplatin

distribution of death times suggest that the cisplatin and TRAIL combination induces "*two waves of death*": the initial peak is around 18 h after treatment and corresponds to the peak time observed with TRAIL alone; and the 2nd peak time is around 65 h, similar to the single peak induced by cisplatin alone (Fig. 6b). This behavior is consistent with the experimental observation (Fig. 6d). The predicted "*two waves of death*" indicates that the level of TRAIL ligand does not remain constant but decays over time, as assumed in the current model. If an alternative assumption is made and the level of TRAIL is held constant, the combined therapy would still result in higher death fraction but with only one large wave of cellular death (simulation in the original publication).

The experimental confirmation of the model's predictions suggests that the model can serve as plausible digital representations of the biological cell, allowing for efficient analysis of cellular fates based on the molecular setting. It is worth emphasizing that such digital representations are virtual reproductions, not the actual biological system, and that the model framework should be continuously refined as new information and biological understanding becomes available.

4.4 Systematic Machine Learning Identifies Chemotherapy Sensitizers

The calibrated QSP model can be used to generate simulated responses that serve as an input for ML algorithms to identify the parameters that contribute most to cell fate (apoptosis). In this example, the input model simulated cellular responses from 6,000 different sets of parameters whose values were randomly sampled from uniform distributions ranging between 70% and 120% of their basal levels. In this way, these simulations explored 6,000 cisplatin-treated tumor cells and are referred to as *digital cells* (DCs) hereafter.

1. *Classifying DCs by Response (data pre-processing)*

Among the population, a DC is labeled as "*dead*" if caspase-3 is activated past a specific threshold and as "*alive*" if caspase-3 activity stays below the threshold. Thus, cell fate, a discrete label, is determined by the peak value of caspase-3, which is a continuous response variable.

2. *Regression Analysis*

As caspase-3 concentration is a systems-level property and controlled by many parameters, partial least squares regression (PLS) was used to reduce the parameter dimension (Hastie et al. 2009). Similar to principal component analysis (PCA), PLS identifies a smaller set of factors that are linear combinations of the original parameters. However, unlike unsupervised PCA, PLS is supervised and the factors identified here do not always include all the original parameters. Rather, only part of the original parameters is included in the PLS-identified factors on the basis of the comparison between the observation and statistical prediction. In the factor space with reduced dimensions, the composed linear regression model bears the smallest distance (least squares) between the measured values and the fitted ones. Two parameter combinations (principal factors) identified by PLS together explain about 60% of the variance that characterizes the change in peak caspase-3 levels (Fig. 7a). Based on their relative contributions to these principal factors, PLS then generated a rank of the contributions by individual parameters (Fig. 7b, top panel) revealing that caspase-3 activity is most sensitive to variabilities in the levels of Bcl-2, Bax, the positive feedback controlling the transcription factor p53, and the stability of BH3.

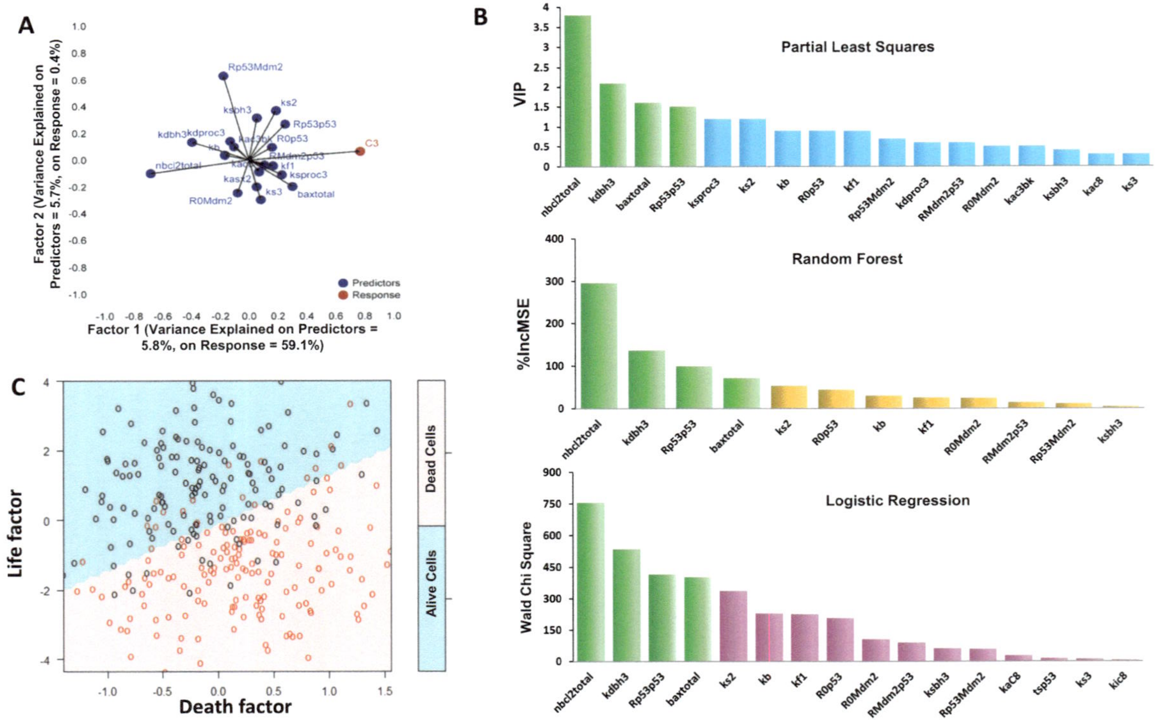

Fig. 7 ML analysis of the simulated data. (**a**) Biplot from PLS analysis, with variances from both parameters and caspase-3 level explained; (**b**) the four components in green are the common ones identified by all these methods. Upper panel: the variable importance in projection (VIP) of parameters achieved by PLS; Middle panel: a hierarchical order of the control parameters based on the percentage increase in the mean squared errors (%IncMSE) in random Forests; Lower panel: the ranking of parameters selected by the logistic regression on the basis of Wald Chi square; (**c**), The SVM plotting carried out with the life-promoting and death-promoting factors. Red circles indicate dead cells, while black circles are surviving cells. nbcl2total = BCL-2 level; kdBH3 = BH3 degradation rate constant; Rp53p53 = self-activation of p53; ksproc3 = background production rate of procaspace 3; see original publication for more details (Zhang et al. 2019)

3. *Cross Validation with Alternative ML Methods*

To avoid bias introduced by any single analysis method, we subjected the discrete life and death responses of DCs and the values of their parameters to two machine learning methods that utilize different algorithms: random forest and logistic regression. Random forest was chosen due to its ability to improve the classification results and generate a rank of how each parameter contributes to the accuracy of predicted cell fates (Hastie et al. 2009). Logistic regression, which models the logistic transformation of probability of the binary response of cell fates (Hastie et al. 2009), was used to rank the contribution of the parameters of the mechanistic model.

For our analysis, the random forest established 2,000 decision trees to calculate the relative contribution of each variable to distinguish living cells from dead ones, while logistic regression identified the dominant parameters using backward elimination (Hastie et al. 2009). Consistent with the results from PLS, these two alternative methods also identified the variability of four components (Bcl-2, Bax, p53 self-activation, and BH3 stability) as the dominant factors contributing to cell death in response to cisplatin (Fig. 7b, middle and bottom panels). In this way, ML methods were used as a computationally efficient way of carrying out global sensitivity analysis.

The ML analysis is much more efficient compared to traditional sensitivity analysis, either local or global. In local sensitivity analysis, the change in some output of a model (say, the steady state of a dynamical variable) is measured in response to a small change in a single parameter. In global sensitivity analysis, multiple parameters are changed simultaneously and the resulting changes in model behavior are analyzed. Although it is straightforward to carry out sensitivity analysis for a single model with a fixed set of parameter values, it is much harder to carry out such analysis for populations of models whose parameters are from data-constrained distributions.

The sensitizers identified by these machine learning methods are consistent with previous experimental observation that cell death in response to cisplatin can be greatly enhanced by inhibiting BCL proteins (Paek et al. 2016). Meanwhile, independent experimental studies using either gain-of-function (Zheng et al. 2005; Pastorino et al. 1998; Li et al. 2001) or loss-of-function (Mizuta et al. 2007; Ke et al. 2015; Ngo et al. 2015) of Bax showed its importance in regulating cell death under diverse conditions. The consistency between the experimental observations and the theoretical results demonstrates the "proof of principle" of such a methodology that combines mechanism-based modeling with machine learning.

Furthermore, the identified controlling components have opposite effects on cellular death. As the elevation of p53 positive feedback strength and Bax level promotes cellular death, the increase of the two other components (BCL level and BH3 stability) promotes cellular survival. Hence, we could further linearly combine apoptosis promoting Bax and p53 control into a "death factor" and the survival promoting BCL-2 level and BH3 degradation rate into a "life factor" with the opposite effect.

4. *Separating Populations of Cells with Different Responses*

With these derived factors, we can also carry out support vector machine (SVM) analysis. The SVM is a classification method that identifies the boundaries to separate data points chosen from different populations, i.e., responding and resistant tumor cells, with maximal margin. The chosen data are referred to as "*support vectors*" since they are used to compute the boundaries.

As a result, SVM plotting identified a two-dimensional plane with two clearly distinguished regions: the top left region is mainly occupied by living cells, while the bottom right region contains mostly dead cells (Fig. 7c). Starting from a high dimensional space of over 30 parameters, this SVM identified a two-dimensional plane that correctly identified a majority of cell fates (84.5%). Thus, the SVM identified that boundary derived from simulated data provides an elegant way to simplify and to understand the complex control of cellular fates.

5 Concluding Remarks

Academic and industrial scientists believe that the interplay between QSP and ML, whether applied sequentially or in an iterative manner, has the potential to improve model-derived results and the scientific impact derived from these results. In 2021, the QSP + ML working group of the International Society of Pharmacometrics QSP Special Interest Group provided an overview on possible ways to integrate QSP and ML (Zhang et al. 2022) and identified four categories for QSP + ML approaches. These included impacting (1) *QSP parameter estimation and extraction* by efficiently comparing model outputs against published data and extracting parameters directly from the literature sources or from large datasets to improve the prediction accuracy of small datasets; (2) *QSP model structure* where ML analysis of large datasets can infer connectivity to help define a QSP model structure and ML methods may subsequently be used to efficiently perform sensitivity analysis of a QSP model; (3) ML *dimension reduction* that focuses the QSP modeling effort on the most informative mechanisms that capture the essential system behavior; and (4) *stochasticity and virtual populations*, where ML methods can aid in the assignment of mechanistic variability to QSP models through analysis of large datasets, including genomic data that may inform on genetic variants.

Much more research has appeared in the public domain since the publication of the white paper, and exciting work is also ongoing without being openly published. With such encouraging momentum of integrating mechanistic modeling and machine learning (Benzekry et al. 2024; Zhang and Tyson 2022), we would expect that further research in these areas will likely contribute to design of novel and effective treatments for a board range of diseases.

Though Modeling Informed Drug Development (MIDD) has been applied with great success in pharmaceutical industry, modeling and simulation activities can still achieve greater impact in the pharmaceutical field. In this sense, let us re-examine how models evolve in other fields. For example, Johannes Kepler proposed the laws of planetary motion:

1. The orbit of a planet is an ellipse with the Sun at one of the two foci
2. A line segment joining a planet and the Sun sweeps out equal areas during equal intervals of time.
3. The square of a planet's orbital period is proportional to the cube of the length of the semi-major axis of its orbit.

The above laws are good examples of data-driven modeling, similar to what may result from the ML algorithms of today, and highlights both the strengths and limitations of ML methods. The strengths are as follows: (1) they can provide accurate results without specifying an underlying mechanism if a large amount of high-quality data is available as input to the ML algorithms; (2) they can summarize a vast quantity of data in simple, user-friendly formats; and (3) they make reliable predictions within the scopes of the input datasets. A major limitation, though, is that these rules are not readily applicable to novel objects that have not been observed before. For example, the Kepler rules cannot be used to predict the trajectories of a new spaceship or inform us how to best design a new spaceship.

While the first three points intrigue us to consider how ML methods can enhance QSP models, the fourth point suggests how QSP models may enhance ML approaches. Just like what Kepler did with data on stars, ML modeling can be used to mine large amount of data accumulated during the drug development process and propose new, potentially unintuitive laws based on the relations between molecular structure of the drug, omics composition of human, and the relevant treatment responses. If ML modeling indeed identified reliable patterns from the underlying data (patterns from regions 1 and 4 in Fig. 3), these models promise to provide satisfactory explanations and actionable predictions.

Similarly, QSP models are in silico platforms that transform our learnings from cells, animals, organoids, and patients into the predicted efficacy and toxicity of new drugs in treating diseases. As ML models can mine the accumulated data and extract summarizing rules, QSP models could serve to investigate the emergent behavior of complex interaction systems with the knowledge of the underlying first principles. By mimicking the outcomes of planned treatments in clinical settings, QSP models help researchers test different hypotheses on how to better treat disease and to do so in a more efficient manner.

References

Ahmadi Daryakenari N, De Florio M, Shukla K, Karniadakis GE (2024) AI-Aristotle: a physics-informed framework for systems biology gray-box identification. PLoS Comput Biol 20(3): e1011916

Bai JPF, Earp JC, Florian J, Madabushi R, Strauss DG, Wang Y et al (2021) Quantitative systems pharmacology: landscape analysis of regulatory submissions to the US Food and Drug Administration. CPT Pharmacometrics Syst Pharmacol 10(12):1479–1484

Becht E, McInnes L, Healy J, Dutertre C-A, Kwok IW, Ng LG et al (2019) Dimensionality reduction for visualizing single-cell data using UMAP. Nat Biotechnol 37(1):38–44
Benzekry S, Mastri M, Nicolo C, Ebos JML (2024) Machine-learning and mechanistic modeling of metastatic breast cancer after neoadjuvant treatment. PLoS Comput Biol 20(5):e1012088
Eduati F, Doldan-Martelli V, Klinger B, Cokelaer T, Sieber A, Kogera F et al (2017) Drug resistance mechanisms in colorectal cancer dissected with cell type-specific dynamic logic models. Cancer Res 77(12):3364–3375
FDA (2023) AIML-Discussion-Paper-05-05-23.pdf
Fernandez-Lozano C, Hervella P, Mato-Abad V, Rodriguez-Yanez M, Suarez-Garaboa S, Lopez-Dequidt I et al (2021) Random forest-based prediction of stroke outcome. Sci Rep 11(1):10071
Hastie T, Tibshirani R, Friedman JH (2009) The elements of statistical learning: data mining, inference, and prediction, 2nd edn. Springer, New York, xxii, 745 p
Holohan C, Van Schaeybroeck S, Longley DB, Johnston PG (2013) Cancer drug resistance: an evolving paradigm. Nat Rev Cancer 13(10):714–726
Ke F, Grabow S, Kelly GL, Lin A, O'Reilly LA, Strasser A (2015) Impact of the combined loss of BOK, BAX and BAK on the hematopoietic system is slightly more severe than compound loss of BAX and BAK. Cell Death Dis 6:e1938
Lanza B, Parashar D (2021) Do support vector machines play a role in stratifying patient population based on cancer biomarkers? Arch Proteom Bioinform 2(1):20–38
Laurie M, Lu J (2023) Explainable deep learning for tumor dynamic modeling and overall survival prediction using Neural-ODE. NPJ Syst Biol Appl 9(1):58
Lever J, Krzywinski M, Altman N (2017) Principal component analysis. Nat Methods 14(7): 641–642
Li X, Marani M, Yu J, Nan B, Roth JA, Kagawa S et al (2001) Adenovirus-mediated Bax overexpression for the induction of therapeutic apoptosis in prostate cancer. Cancer Res 61(1):186–191
Liu X, Liu C, Huang R, Zhu H, Liu Q, Mitra S et al (2021) Long short-term memory recurrent neural network for pharmacokinetic-pharmacodynamic modeling. Int J Clin Pharmacol Ther 59(2):138–146
Liu Q, Huang R, Hsieh J, Zhu H, Tiwari M, Liu G et al (2023) Landscape analysis of the application of artificial intelligence and machine learning in regulatory submissions for drug development from 2016 to 2021. Clin Pharmacol Ther 113(4):771–774
Liu Q, Joshi A, Standing JF, van der Graaf PH (2024) Artificial intelligence/machine learning: the new frontier of clinical pharmacology and precision medicine. Clin Pharmacol Ther 115(4): 637–642
Lu J, Deng K, Zhang X, Liu G, Guan Y (2021) Neural-ODE for pharmacokinetics modeling and its advantage to alternative machine learning models in predicting new dosing regimens. iScience 24(7):102804
Luckett PH, Chen C, Gordon BA, Wisch J, Berman SB, Chhatwal JP et al (2023) Biomarker clustering in autosomal dominant Alzheimer's disease. Alzheimers Dement 19(1):274–284
Mizuta T, Shimizu S, Matsuoka Y, Nakagawa T, Tsujimoto Y (2007) A Bax/Bak-independent mechanism of cytochrome c release. J Biol Chem 282(22):16623–16630
Moore A, Bell M (2022) XGBoost, a novel explainable AI technique, in the prediction of myocardial infarction: a UK biobank cohort study. Clin Med Insights Cardiol 16: 11795468221133611
Ngo J, Matsuyama M, Kim C, Poventud-Fuentes I, Bates A, Siedlak SL et al (2015) Bax deficiency extends the survival of Ku70 knockout mice that develop lung and heart diseases. Cell Death Dis 6:e1706
Nijsen M, Wu F, Bansal L, Bradshaw-Pierce E, Chan JR, Liederer BM et al (2018) Preclinical QSP modeling in the pharmaceutical industry: an IQ consortium survey examining the current landscape. CPT Pharmacometrics Syst Pharmacol 7(3):135–146

O'Dell RS, Higgins-Chen A, Gupta D, Chen MK, Naganawa M, Toyonaga T et al (2023) Principal component analysis of synaptic density measured with [(11)C]UCB-J PET in early Alzheimer's disease. Neuroimage Clin 39:103457

Olle DG, Zoobo Bisse J, Abessolo Alo'o G (2024) Application and comparison of K-means and PCA based segmentation models for Alzheimer disease detection using MRI. Discov Artif Intell 4:11. https://doi.org/10.1007/s44163-024-00106-7

Paek AL, Liu JC, Loewer A, Forrester WC, Lahav G (2016) Cell-to-cell variation in p53 dynamics leads to fractional killing. Cell 165(3):631–642

Pastorino JG, Chen ST, Tafani M, Snyder JW, Farber JL (1998) The overexpression of Bax produces cell death upon induction of the mitochondrial permeability transition. J Biol Chem 273(13):7770–7775

Ribba B (2023) Quantitative systems pharmacology in the age of artificial intelligence. CPT Pharmacometrics Syst Pharmacol 12(12):1823–1826

Ringnér M (2008) What is principal component analysis? Nat Biotechnol 26(3):303–304

Roux J, Hafner M, Bandara S, Sims JJ, Hudson H, Chai D et al (2015) Fractional killing arises from cell-to-cell variability in overcoming a caspase activity threshold. Mol Syst Biol 11(5):803

Scholkopf B, Smola AJ (2001) Learning with kernels: support vector machines, regularization, optimization, and beyond. MIT Press

Talevi A, Morales JF, Hather G, Podichetty JT, Kim S, Bloomingdale PC et al (2020) Machine learning in drug discovery and development part 1: a primer. CPT Pharmacometrics Syst Pharmacol 9(3):129–142

Terranova N, Renard D, Shahin MH, Menon S, Cao Y, Hop C et al (2024) Artificial intelligence for quantitative modeling in drug discovery and development: an innovation and quality consortium perspective on use cases and best practices. Clin Pharmacol Ther 115(4):658–672

Turcato G, Cipriano A, Park N, Zaboli A, Ricci G, Riccardi A et al (2022) Decision tree analysis for assessing the risk of post-traumatic haemorrhage after mild traumatic brain injury in patients on oral anticoagulant therapy. BMC Emerg Med 22(1):47

Van der Maaten L, Hinton G (2008) Visualizing data using t-SNE. J Mach Learn Res 9(11): 2579–2605

Villani A-C, Satija R, Reynolds G, Sarkizova S, Shekhar K, Fletcher J et al (2017) Single-cell RNA-seq reveals new types of human blood dendritic cells, monocytes, and progenitors. Science 356(6335):eaah4573

Wilson TR, Johnston PG, Longley DB (2009) Anti-apoptotic mechanisms of drug resistance in cancer. Curr Cancer Drug Targets 9(3):307–319

Yuan B, Shen C, Luna A, Korkut A, Marks DS, Ingraham J et al (2021) CellBox: interpretable machine learning for perturbation biology with application to the design of cancer combination therapy. Cell Syst 12(2):128–40 e4

Zhang T, Tyson JJ (2022) Understanding virtual patients efficiently and rigorously by combining machine learning with dynamical modelling. J Pharmacokinet Pharmacodyn 49(1):117–131

Zhang Y, Huynh JM, Liu GS, Ballweg R, Aryeh KS, Paek AL et al (2019) Designing combination therapies with modeling chaperoned machine learning. PLoS Comput Biol 15(9):e1007158

Zhang T, Androulakis IP, Bonate P, Cheng L, Helikar T, Parikh J et al (2022) Two heads are better than one: current landscape of integrating QSP and machine learning: an ISoP QSP SIG white paper by the working group on the integration of quantitative systems pharmacology and machine learning. J Pharmacokinet Pharmacodyn 49(1):5–18

Zheng JY, Yang GS, Wang WZ, Li J, Li KZ, Guan WX et al (2005) Overexpression of Bax induces apoptosis and enhances drug sensitivity of hepatocellular cancer-9204 cells. World J Gastroenterol 11(23):3498–3503

Zineh I (2019) Quantitative systems pharmacology: a regulatory perspective on translation. CPT Pharmacometrics Syst Pharmacol 8(6):336–339

Part II

Model Development and Application

Quantitative Systems Pharmacology Development and Application in Neuroscience

Hugo Geerts

Contents

H. Geerts (✉)
Certara Predictive Technologies, Radnor, PA, USA
e-mail: hugo.geerts@certara.com

E. L. Bradshaw, M. E. Spilker (eds.), *Quantitative Systems Pharmacology*,
Handbook of Experimental Pharmacology 289, https://doi.org/10.1007/164_2024_739

Abstract

Successful clinical development of therapeutics in neurology and psychiatry is challenging due to the complexity of the brain, the lack of validated surrogate markers and the nature of clinical assessments. On the other hand, tremendous advances have been made in unraveling the neurophysiology of the human brain thanks to technical developments in noninvasive biomarkers in both healthy and pathological conditions.

Quantitative systems pharmacology (QSP) aims to integrate this increasing knowledge into a mechanistic model of key biological processes that drive clinical phenotypes with the objective to support research and development of successful therapies. This chapter describes both modeling of molecular pathways resulting in measurable biomarker changes, similar to modeling in other indications, as well as extrapolating in a mechanistic way these biomarker outcomes to predict changes in relevant functional clinical scales.

Simulating the effect of therapeutic interventions on clinical scales uses the modeling methodology of computational neurosciences, which is based on the premise that human behavior is driven by firing activity of specific neuronal networks. While driven by pathology, the clinical behavior can also be influenced by various medications and common genotype variants. To address this occurrence, computational neuropharmacology QSP models can be developed and, in principle, applied as virtual twins, which are in silico clones of real patients.

Overall, central nervous system (CNS) QSP is an important additional tool for supporting research and development from the preclinical stage to post-marketing studies and clinical practice. Overall, CNS QSP is an important additional tool for supporting research and development from the preclinical stage to post-marketing studies and clinical practice.

Keywords

Comedications · Computational neuropharmacology · Computational neuroscience · Pharmacodynamic interactions

Abbreviations

ACh	acetylcholine
AChE	acetylcholinesterase
AD	Alzheimer's disease
ADAS-Cog	Alzheimer's Disease Assessment Scale – Cognitive Readout
AI	artificial intelligence
ALS	amyotrophic lateral sclerosis

AMPA	α-amino-3-hydroxy-5-methyl-4-isoxazolepropionic acid
BC	basket cells
BOLD-fMRI	blood-oxygen level dependent functional magnetic resonance imaging
CBF	cerebral blood flow
CBV	cerebral blood volume
CNS	central nervous system
CSF	cerebrospinal fluid
ChAT	choline acetyltransferase
DAT	dopamine transporter
DBS	deep brain stimulation
EEG	electroencephalogram
FFT	fast Fourier transform
GlyT1	glycine transporter 1
GPe	globus pallidus pars externa
GPi	globus pallidus pars interna
ISF	interstitial fluid
mAChR	muscarinic acetylcholine receptors
ML	machine learning
MSN	medium spiny neurons
nAChR	nicotinic acetylcholine receptors
NMDA	N-methyl-D-aspartate
ODE	ordinary differential equations
PANSS	positive and negative symptom scale
PD	Parkinson's disease
PDE	phosphodiesterase
PET	positron emission tomography
Pyr	pyramidal (excitatory) cells
PVS	perivascular system
QSP	quantitative systems pharmacology
STN	subthalamic Nucleus
SOD	superoxide dismutase
SUVR	stand ardized uptake value ratio
TPP	target product profile
TRN	thalamus reticular nucleus
UPDRS	Unified Parkinson's Disease Rating Scale

1 Introduction

Studies assessing the reasons for clinical trial failures across therapeutic areas and indications over the last 30 years have originally identified poor pharmacokinetics, toxicity and lack of efficacy (target engagement) as major drivers (Gribkoff and

Kaczmarek 2017). The first of these issues has now largely been successfully addressed by improved translational assays and advanced physiologically-based pharmacokinetic (PBPK) modeling to the point that most clinical trials now fail because of lack of efficacy at the tolerated doses tested (Cook et al. 2014). This is very much the case for central nervous system (CNS) diseases, where animal models are far from predictive (Geerts 2009); knowledge of the disease and definition of the clinical phenotypes are often still evolving and other translational approaches are lacking. As a consequence, clinical trials in neurology and psychiatry rank at the bottom in terms of success rate (Hay et al. 2014). In particular, psychiatric disorders such as schizophrenia have not seen a major breakthrough in the last 50 years (Brody 2014), and in Alzheimer's Disease, the failure rate is well over 95% (Cummings et al. 2014).

CNS QSP modeling can better integrate the recent developments in human neurophysiology and neuropathology to support successful clinical trial development, especially in cases where biological insights are needed to combine diverse and sparse data sets.

Computational modeling in neurosciences has a long history in academia and has been used to generate hypotheses on clinical phenotypes associated with specific disease indications. Recently this approach has been integrated with pharmacological aspects of therapeutic intervention, with specific attention to drug exposure, target engagement, biomarker and functional clinical scale outcomes, in what is now called quantitative systems pharmacology (QSP). QSP is a discipline aimed at modeling the impact of a drug on the body, with a much greater emphasis on biological mechanisms and the insights they provide rather than on data characterization. This modeling approach applies to many disease indications, but brain disorders stand out because of the sheer complexity and specific challenges that will be highlighted in this chapter.

Note that unlike artificial intelligence and machine learning methods (AI/ML) (discussed in "Integrating QSP and ML to Facilitate Drug Development and Personalized Medicine"), which rely on large databases for developing the predictive network, QSP is ideally suited to situations with limited data that need to be supported and augmented by biological information. Furthermore, QSP explicitly provides a quantitative prediction of time-dependent dynamics and focuses on detail on the most crucial pathways associated with the phenotype under consideration.

QSP is an important additional tool for supporting neuroscience research and development from the preclinical stage to post-marketing studies and clinical practice. However, it is not without its unique challenges and opportunities, as will be outlined in this chapter. To this end, the chapter has been organized to first highlight that while key aspects of the QSP pipeline are generalizable among various disorders, there are specific challenges for CNS QSP models, which will be discussed in Sect. 2. The biological map that reflects the key biological question is usually informed by literature studies of the relevant anatomy, physiology and pathology. While systems biology and bioinformatics can identify the key pathways for a specific phenotype, it is the unique challenge of QSP to bring these dynamic relationships to life, especially with regard to predicting the impact of therapeutic interventions. A pipeline for developing QSP models with a detailed explanation of a particular project in Alzheimer's Disease starting from the biological map is given in Sect. 3, while considerations on developing QSP models for psychiatric diseases are discussed in Sect. 4. Critical to the implementation of QSP models is the software used; therefore, Sect. 5 describes several software platforms for CNS QSP models. Lastly, a few more detailed examples are provided in Sect. 6, before concluding with a reflection about the capabilities and limitations of Neurosciences QSP in Sect. 7.

2 Unique Factors Impacting CNS QSP Modeling

The challenges facing the development of CNS QSP models with regard to biomarker or pathology modeling are similar to the ones encountered in other therapeutic areas in terms of calibration and validation. While a vast amount of preclinical data is available for various neurodegenerative diseases, the translatability to the clinical situation is far from established. In fact, for Alzheimer's Disease, more than 200 different targets, all of them with a certain "validation" in some animal model, have subsequently failed in the clinic (Cummings et al. 2014). Conversely, there are a few robust examples of a successful clinical outcome with acetylcholinesterase inhibitors and amyloid antibodies.

For Parkinson's disease, dopamine modulation has been around for more than 50 years as a standard of care to address symptoms, but so far, no disease-modifying agent has shown any clinical benefit. This makes clinical validation of any QSP model for disease-modifying interventions extremely difficult. The situation is even worse for psychiatric diseases, where a clear pathology is missing, and symptoms are related to dysfunctional interaction between different neuronal circuits. As briefly outlined in Sect. 4, computational neuroscience can model explicitly the interactions between different brain regions and provide support for drug discovery and development in this indication.

A further challenge to CNS QSP modeling is in the nature of the scientific and medical disciplines directly impacting advancement of CNS biology and therapeutics. Apart from peripheral nervous system disorders, CNS disorders are broadly divided into neurological and psychiatric indications. Although focused on the same organ, neurologists and psychiatrists tend to have their own approaches, journals and scientific societies and rarely talk to each other, limiting the dissemination of knowledge.

While translational and organizational challenges exist, useful quantitative clinical data is emerging that can be incorporated into QSP models. This includes imaging endpoints from positron emission tomography (PET) and magnetic resonance imaging (MRI) studies. In addition, the field of computational neuroscience has provided a biological foundation that can bridge therapeutic modulation of mechanistic biological endpoints to functional brain activity with the ultimate goal of informing on clinical scores and meaningful endpoints in neuroscience. Thus, while there are challenges in the application of QSP modeling for CNS applications, progress is being made with the potential to impact human disease. In this section, some unique challenges and promising activities impacting QSP modeling for neurodegenerative disorders are discussed, while additional specific challenges for psychiatric indications will be addressed in Sect. 4.1.

2.1 Nature of the Disease and Complexity of the Brain

Neurological indications include chronic neurodegenerative diseases, such as various forms of dementia (Alzheimer's disease as a prime example), motor dysfunction (i.e., Parkinson's disease) or excitation-inhibition imbalance diseases such as epilepsy. Other indications are the consequence of a well-defined traumatic event, such as stroke and traumatic brain injury, providing a well-defined anchor time point along the clinical trajectory. Furthermore, inflammation-driven pathologies such as multiple sclerosis and virally mediated infections such as Long COVID also affect brain function and are included within neurological indications.

A major recent development is the substitution of the clinical syndrome by a more biologically informed pathology, similar to the evolution in clinical oncology where some cancer types and treatments are defined by genetic mutations or aberrant protein characteristics (e.g., HER2+ tumors). The increasing understanding of the underlying pathology is often based on genetic associations or further refined using specific biomarker signatures. For example, Alzheimer's and Parkinson's diseases are relatively well documented with regard to the multiple proteinopathies driving the clinical progression. This allows mechanistic modeling to start from the observed pathological and molecular pathways and link these to functional clinical readouts. Although developing appropriate QSP models seem daunting, recent efforts have identified common mechanism-based themes useful for modeling across a wide variety of indications (Bloomingdale et al. 2022a), such as misfolding of key proteins, oxidative stress, mitochondrial dysfunction and inflammatory responses. Specific to QSP models that link drug exposure to clinical outcomes, while mechanism-based connections to disease are improving, the proper diagnosis and the many comorbidities in chronic neurodegenerative diseases make calibration to clinical outcomes very challenging.

2.2 (In)Accessibility of the Brain

The brain is a privileged organ with a very effective blood-brain-barrier controlling the permeability of different types of molecules or agents. A lot is known about the optimal properties for good brain penetration for small molecules to the extent that physiologically based pharmacokinetic modeling can very well predict the brain exposure of most new small molecule compounds (Kostewicz et al. 2014). Those small molecules that can pass the blood-brain barrier can achieve substantial brain concentrations, which in some cases can be indirectly confirmed by assessment of the drug exposure at the site of action from PET tracer displacement studies.

However, larger entities, such as biologics and lipid nanoparticles for delivering gene therapies and antisense oligonucleotides, have a much lower penetration. For instance, large molecules will penetrate the brain only by 0.1–0.3%, although new developments in brain shuttling technologies can increase this ratio by a factor of 5–10 (Kurnellas et al. 2023). Well-validated PBPK models exist for central antibody exposure (Chang et al. 2019; Bloomingdale et al. 2021) and even for shuttle-mediated brain penetration (Chang et al. 2022). Validation of these models is restricted to measurements in the CSF, as brain drug concentrations by microdialysis in humans are beyond measurement, with very few exceptions.

Once inside the brain, therapeutic modalities encounter a very compact interstitial fluid (ISF) compartment and densely packed cells (Sykova and Nicholson 2008). In some cases, the target of interest (for example, anti-tau and anti-alpha-synuclein antibodics) rcside inside the synaptic cleft, which is an even more restricted environment with only a few molecules present at any time. In this case, Monte Carlo simulations of individual molecule trajectories within the actual geometry can be used to describe both the drug's presence and the interaction with its target at the site of action (see Sect. 5.3).

Beyond a therapeutic's distribution to the site of action, the drug's mechanism of action also needs to be quantitatively defined and incorporated into a QSP model. To aid in this objective, the effect of pharmacological agents on downstream effectors, such as voltage- and ligand-gated ion channels, can be calibrated using the large dataset of clinical interventions with antipsychotics, benzodiazepines, cholinergic modulators and antidepressants in a variety of indications. To support the incorporation of drug pharmacology (i.e., affinities towards human receptors) into QSP models, the systematic database maintained by the University of North Carolina in the Psychoactive Drug Screening Program (PDSP, PDSP – NIMH Psychoactive Drug Screening Program (unc.edu)) (https://pdsp.unc.edu/databases/kidb.php) is a recommended resource.

2.3 Biomarkers

Absence of Strong Surrogate Biomarkers

In contrast to many other therapeutic areas, where accessible biomarkers have been validated as surrogate markers for functional readouts (for instance, cholesterol levels predicting myocardium infarct), noninvasive markers of brain activity have not been validated as surrogate markers, except for Alzheimer's disease (AD) (Hansson et al. 2023). Plasma and cerebral spinal fluid (CSF) markers of amyloid and tau have been demonstrated to be correlated to central pathologies that are usually measured using PET imaging. However, a major challenge of mechanistic QSP modeling is identifying the molecular processes that link the level of central pathology (i.e., amyloid or tau) to the functional clinical outcome of the patients. Indeed, the predictive value of a QSP model to simulate a clinical outcome is significantly enhanced when a well-validated mechanistic surrogate marker is strongly correlated with clinical outcome. We will present an alternate approach using neuronal network firing characteristics as a close proxy for clinical outcome based on the academic discipline of computational neuroscience (see Sect. 3.4).

Beyond Alzheimer's disease, none of the other CNS indications have achieved the same degree of biomarker validation, although recent decisions by the regulatory agencies like the FDA suggest an openness to forfeit an actual improvement on a clinical scale for relevant biological processes such as stabilization of plasma neurofilament levels, as demonstrated by the decision for accelerated approval of tofersen, an anti-sense superoxide dismutase (SOD) oligonucleotide in amyotrophic lateral sclerosis (ALS).

Emergence of Quantitative Mechanistic and Functional Biomarkers
Imaging technologies such as PET and MRI can provide spatial and temporal information on biological processes within the intact brain. PET imaging is a powerful molecular imaging modality that can inform on the distribution of radiolabeled small molecule therapeutics, receptor occupancy, and select biological processes through its ability to probe the concentration and dynamics of specific proteins inside the living brain. For example, PET imaging can measure concentrations of key pathological proteins in chronic neurodegenerative diseases such as Alzheimer's disease, where amyloid and tau can be measured in the brain tissue using specific radiotracers. Often the resulting images are reported as SUVR (Standardized Uptake Value Ratio (Martínez et al. 2017)) that reflects the retention time of a radioactive marker in one location on the image relative to a reference location.

For functional activity, imaging through blood oxygenation level-dependent functional MRI (BOLD-fMRI) paired with electroencephalogram (EEG) data can be utilized to report on neuronal activity and its modulation by therapeutics. For example, calibration of neuronal firing activity in specific neuronal circuits is becoming more established based on decades of clinical research with BOLD-fMRI and EEG, and recently from intracranial electrodes and noninvasive stimulation paradigms. Together with detailed single-neuron activity in nonhuman primates (Williams and Goldman-Rakic 1995), this provides for a well-calibrated model for healthy human subjects with applications on cognition for instance.

BOLD-fMRI is widely used in CNS clinical studies. Starting from a series of biophysical relationships between excitatory and inhibitory neuronal activity from experimental studies (Smolen et al. 2008; Sotero and Trujillo-Barreto 2007), a relationship between firing activity and predicted voxel-based BOLD-fMRI signal can be derived. It has been demonstrated that the BOLD-fMRI signal is driven by the input and local neuronal processing within a voxel (Logothetis 2002). For the interested reader, a derivation of the BOLD-fMRI signal starting from the temporal neuronal firing properties of the network is provided in the Appendix B.

2.4 Clinical Outcomes

The objective of every drug discovery project is to ultimately improve the lives of patients in a functional way based on clinical scales approved by regulatory agencies. With the absence of surrogate markers, the situation is more challenging in CNS disorders. In Alzheimer's disease, functional improvement can be measured as cognitive performance, which is usually implemented as a structured set of questions. In psychiatry, most clinical scales rely on a subjective evaluation by the patient over a certain period with an ordinal categorical estimation from very good to very bad. It is clear that many confounding factors can influence this outcome and large placebo responses are not uncommon in clinical trials.

The situation is somewhat different for motor diseases where movement dysfunction can be analyzed by an independent observer and can sometimes be quantified. The recent development of powerful wearables can help quantify the impact of therapeutic interventions in Parkinson's Disease (Pagano et al. 2022), although this still needs to be validated to reach regulatory standards. This, however, opens the possibilities of continuous monitoring.

3 Implementation of QSP in Neurology

Like in any disease condition, the general strategy of QSP modeling in Neuroscience is to (1) define a map of the biological processes with an emphasis on causality derived from biological insights, (2) identify quantitative relationships between the different processes, (3) develop a computer model to quantitatively describing the temporal relationship between cause and effect, (4) introduce the pharmacokinetics (PK) and pharmacology of the drugs acting on parts of the network based on existing knowledge as well as experimental preclinical or clinical data, (5) calculate the impact on the key readouts of the model, and (6) calibrate the model using these data (training conditions). Ideally, a QSP model should then be able to prospectively predict the correct outcome of an intervention that was not used for calibration (test condition for validation of the model). Where it differs from QSP approaches in other therapeutic areas is the integration with a comprehensive computational neuroscience model that links functional brain activity resulting from the therapeutic perturbation to the measured clinical scales. While this approach bridges therapeutic administration to clinical outcome, it is acknowledged that for other context-of-use scenarios, where the objective is focused on the response of CNS biomarkers or molecular pathways following pharmacologic perturbation, a similar QSP modeling strategy as in other disease indications can be followed, and the computational neuroscience sub-model would not be included.

Before diving into the technical aspects of CNS QSP modeling, it is worth noting that combining a number of relevant biological processes into a more comprehensive network usually requires input from diverse domain experts. This step is important to define the scope of the model and set expectations with regard to development time and context of use. Crucial in this step and others is communication; scientists, clinicians, and modelers usually do not speak the same language, and people with both a biological knowledge and knowledge of mathematical translation are relatively rare.

Figure 1 illustrates the key elements of a neuroscience QSP model when used to evaluate a therapeutic effect and provides a roadmap for the remaining topics in this

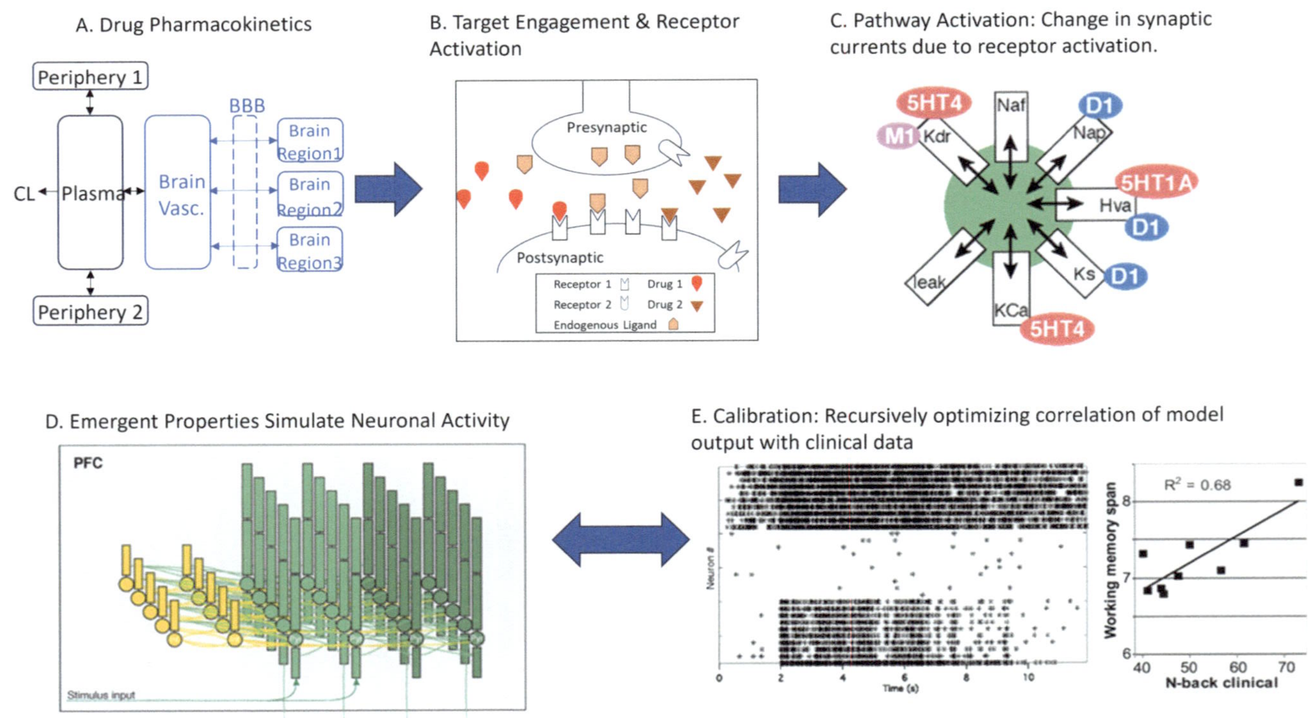

Fig. 1 Overview of CNS QSP modeling. (**a**) Define drug concentrations at the site of action (modified from (Yamamoto et al. 2017)); (**b**) connect drug concentrations to target engagement and subsequent postsynaptic receptor activation; (**c**) the receptor activation is then linked to intracellular pathways and downstream voltage-gated or ligand-gated ion channels; (**d**) which in turn connect to neuronal activity using established computation neuroscience sub-models; (**e**) ultimately linked to measurable biological outcomes such as the neuronal activation, which inform on memory traces (shown here) and can be correlated with clinical scales (in this case, the 2-back working memory test)

section of the chapter. In brief, a QSP model will capture the therapeutic's pharmacokinetics and distribution to the site of action (Fig. 1a) as well as its target engagement (e.g., occupancy or potency model) reflective of the therapeutic's pharmacology (Fig. 1b). In CNS applications, the occupancy or effect on the activation state of membrane-bound receptors can be estimated from the competition between various agents binding to the same site so that the downstream effect on the activation state of appropriate membrane-bound GPCR can be derived. The therapeutic's pharmacology may be further defined by subsequent pathway activation. For example, effects of GPCR activation changes can be coupled through second messenger pathways to modifications in voltage- or ligand-gated receptors of single neurons (Fig. 1c). These changes lead to altered neuronal firing dynamics in specific networks (Fig. 1d) that can be simulated with a computational neuroscience sub-model. Illustrated in the figure is an example of cognitive performance, where the computation neuroscience sub-model simulates neuronal firing traces that align to a memory trace which in turn correlates with a clinical scale (in this case, the 2-back working memory test) (Fig. 1e). Finally, calibration of a CNS QSP model is performed by optimizing the correlation between simulated and observed outcomes in clinical trials on a clinically relevant scale such as Alzheimer Disease Assessment Scale-Cog(nitive) variant (ADAS-Cog). The CNS QSP model elements shown in Fig. 1 may leverage sub-models, such as the inclusion of the computation neuroscience sub-model, which have been previously developed from basic biology or non-therapeutic interventions. In this manner, the wealth of existing knowledge and computation models in the neuroscience space can be utilized.

3.1 Pharmacokinetic Modeling of Brain Exposure

In CNS QSP projects, the general strategy is similar to QSP modeling projects in other diseases where the aim is to seamlessly integrate pharmacokinetics with the subsequent mechanistic pharmacodynamic model. As mentioned in Sect. 2.2, the brain has different barriers and compartments for the distribution of therapeutic modalities; however, usually an existing compartmental or PBPK model can be readily adapted to describe CNS drug exposure, although the more recent modalities such as oligonucleotides and gene therapy need additional work because of the specific injection protocols (often via CSF or intraparenchymal).

While most compartmental and full-body human PBPK models consider the brain as one single compartment, more complex PBPK models explicitly model different CSF compartments: Cisterna Magna (CM), Third and Fourth Ventricle (TFV), Latera ventricle (LV), subarachnoid space (SAS), extracellular fluid (ECF, the channels in between cells in the brain tissue), and intracellular fluid (ICF, cytoplasm of different cell types) as shown for small molecules (Yamamoto et al. 2017) and biologics (Chang et al. 2019; Bloomingdale et al. 2021) given by intravenous infusion or subcutaneous injection. Usually, these compartments are considered well-stirred, and compartmental modeling approaches are used to describe movement of drug between compartments.

More advanced models address the issue of intralumbal or intrathecal CSF administration of larger molecules such as oligonucleotides (Khani et al. 2022) and viral capsids for gene therapy. For therapeutic modalities given into the CSF or directly into the brain parenchyma, an important challenge is the leakage of the oligonucleotides or viral capsids into the plasma, where it can trigger a strong immunogenic response, which may be an important modeling consideration for overall tolerability and performance of the therapeutic. Additionally, the brain parenchyma is a dense tissue with channels between adjacent cells of 40–60 nm wide, requiring additional considerations to properly describe drug distribution into this tissue. To this end, a compound's tortuosity (the ratio of free diffusion over actual diffusion) can be defined and leveraged in models of drug distribution (El-Kareh et al. 1993). A possible approach to model the dynamics of molecules in such small spaces where only a finite number of molecules are present is to use Monte Carlo simulations that describe the trajectory and interaction of individual molecules (Sect. 5.3). This also applies to other scenarios, such as anti-tau and anti-alpha synuclein antibodies that capture misfolded proteins as they move from the presynaptic to the postsynaptic nerve ending. The drug targets are located in and around synapses, where there is an extremely small volume (20 nm wide by 1um by 1 um, corresponding to a 0.04 fL volume). In all of these situations, the number of molecules is very small, and the assumptions of well-stirred volumes break down.

3.2 Calculating Receptor Activation States

While glutamate and GABA are the major neurotransmitters in the brain, G-protein-coupled receptors (GPCRs) extensively modulate neuronal firing; therefore, it is important to estimate their activation state, which can be affected by pathology or medications. Many small molecule CNS therapeutics bind to specific receptors in the brain and decrease (antagonism) or increase (agonism) the activity, depending upon their concentration and relative affinity to the endogenous ligand. From clinical PET tracer displacement studies, where the drug competes with an exogenous radiolabeled small molecule that also binds to the same target, it is possible to estimate the absolute drug concentration at the site of action in the living human brain. The dynamics of this imaging study are governed by a series of differential equations that simulate the competition between different agents to binding sites based on their affinities (Athan Spiros 2012). Dynamic binding and unbinding of the different agents to receptor sites are calculated by the following coupled ODEs ($n = 1 \ldots 4$).

$$\partial[R_n]/\partial t = k_{on}^n \times [A] \times [R_f] - k_{on}^n \times K_d^n \times [R_n] \tag{1}$$

For the nth agent, $[R_n]$ represents the receptor concentrations bound to the nth agent; A is the concentration of the agent at the site of action (A could be a neurotransmitter, imaging tracer, drug, or active metabolite, $n = 1 \ldots 4$); $K^n{}_{\text{on}}$ and $K^n{}_d$ is the association and dissociation rate constants, respectively; $[R_f]$ represents the free receptors concentration. In this modeling framework, dynamic changes in neurotransmitter can be modeled using specific synaptic physiology of presynaptic autoreceptors and vesicle release properties in tonic and phasic firing epochs. Intrasynaptic drug concentration ("drug exposure") can be derived from the dose-response of a PET tracer displacement study as the concentration that results in the observed clinical PET tracer displacement at a certain dose of tracer and drug.

Once the concentration of the drug at a given tissue location is determined from these PET imaging displacement studies to a particular receptor, the same model can be used to derive the change in activation of all the other receptors the drug is engaging with depending upon its specific affinity, receptor concentration and drug tissue concentration.

3.3 Molecular Pathway QSP Modeling

In many applications, the output of the QSP model is focused on a molecular biomarker that is linked to a pathological state (see Appendix A). Examples include amyloid and tau dynamics, neuroinflammation, and glymphatic system dynamics in Alzheimer disease; cell survival, alpha-synuclein aggregation, as well as mitochondrial and oxidative stress in Parkinson's disease.

Here we illustrate the example of amyloid aggregation in Alzheimer disease, specifically the effect of amyloid antibody treatment on different Abeta species, such as monomers, oligomers, protofibrils and plaques (Fig. 2). While these antibodies are provided peripherally by either intravenous or subcutaneous formulation, their targets are only inside the human brain. In this case, there is experimental evidence

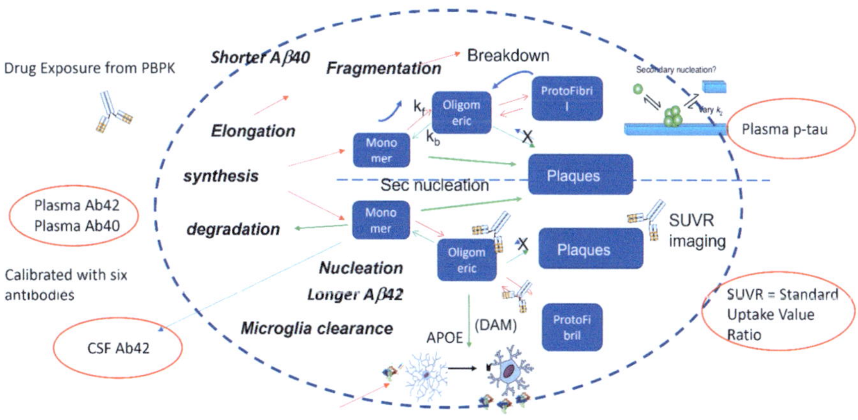

Fig. 2 Schematic of the Amyloid QSP model with aggregation dynamics of short (Ab40) and long (Ab42) forms, based on microscopic forward and backward rate constant with explicit modeling of intermediate monomer, oligomeric, protofibrillar and plaque Abeta peptides. The dashed line represents the brain boundary. Key readouts of the model are plasma Ab42/Ab40, p-tau, CSF Ab40 and Ab42 and SUVR (PET imaging)

linking the dynamics of soluble Abeta forms with neuronal firing through various but different mechanisms.

In general, molecular pathway QSP models follow the rule of mass conservation. For example, the changes in amyloid dimer $a_1(t)$ when interacting with an amyloid antibody in the treatment of Alzheimer's disease (Geerts et al. 2023a) are simulated by Eq. 2.

$$\underbrace{\frac{da_1(t)}{dt}}_{\substack{\text{Rate of change}\\ \text{of unbound dimer}\\ \text{concentration in ISF}}} = \underbrace{k_f^0 * a_0(t) * a_0(t)}_{\substack{\text{Formation of dimer}\\ \text{from 2 mononers}}} - \underbrace{k_f^1 * a_0(t) * a_1(t)}_{\substack{\text{Loss of dimer}\\ \text{to form trimer}}} \; \underbrace{+ k_b^1 * a_2(t)}_{\substack{\text{Formation of dimer}\\ \text{from trimer breakdown}}}$$
$$\underbrace{+ k_f^0 * P_a * a_0(t) * a_0(t) \left(\frac{a_{24}(t)}{a_{24}(t) + \mathrm{SEC}_{50,a}} \right)}_{\substack{\text{Dimer fromation from 2 monomers}\\ \text{with amplification factor}}} \underbrace{- k_f^1 * P_a * a_0(t) * a_1(t) \left(\frac{a_{24}(t)}{a_{24}(t) + \mathrm{SEC}_{50,a}} \right)}_{\substack{\text{Trimer formation from dimer and}\\ \text{monomer with amplification factor}}}$$
$$\underbrace{- f^{Ab-\mathrm{oligo}} * a_1(t) * Ab(t)}_{\substack{\text{Binding of}\\ \text{antibody to dimer}}} - \underbrace{\frac{\left(1 - \sigma_{valu}^{\mathrm{ISF}}\right) * Q_{\mathrm{brain}}^{\mathrm{PVS}} * a_{1u}^{\mathrm{brain}}}{Vol_{\mathrm{brain}}^{\mathrm{Interstitial}}}}_{\substack{\text{Unbound dimer movement from}\\ \text{current location to perivascular space}}} \tag{2}$$

Equations such as these can be developed for the dynamics of all the intermediate Abeta species, which ultimately drives the fluid biomarkers (basically the monomer dynamics in plasma and CSF) and the Standard Uptake Value ratio imaging biomarker (see Sect. 2.3).

A number of other models for amyloid aggregation have been published (Karelina et al. 2017a; Madrasi et al. 2021; Lin et al. 2022; Ramakrishnan et al. 2023) along the same line; however, this model uses the detailed microscopic and biophysically accessible aggregation kinetics for increasingly higher-order oligomers. A challenge is to define the forward and backward rate constants as only the transitions up to a trimer have been measured experimentally (Garai and Frieden 2013). To further define the model, 3 extrapolation parameters have been defined that reproduce the natural history of Abeta accumulation as a function of the age of the patient as well as the absolute concentration of Abeta species as measured in postmortem studies of Alzheimer patients (Geerts et al. 2023a).

3.4 Computational Neuroscience Modeling

What makes CNS QSP modeling unique is its ability to simulate the impact of disease state and therapeutic interventions on actual clinical functional scales. This is based on the premise that human behavior (like cognition, depressive state, aggression and agitation) is driven by the firing activity of relevant neuronal networks. These can be monitored by BOLD-fMRI imaging and therapeutically modulated by active transcranial magnetic stimulation. For instance, complex emotional states such as shame, sadness and fear can be induced by electrical stimulation of the hypothalamus (Parvizi et al. 2022) and analysis of spatio-temporal brain activity patterns on the human connectome linked to cognitive functions (Bertacchini et al. 2022).

While the discipline of computational neuroscience started in the late 1940s with the seminal work of Hodgkin and Huxley, and mostly focuses on understanding the neuronal activity of the normal human brain and to a lesser degree of pathological states, modeling the effect of pharmacological intervention was introduced only recently (Aradi and Erdi 2006), coining the term computational neuropharmacology.

Computational neurosciences (Lytton et al. 2017) provides a template for modeling the opening and closing dynamics of voltage- and ligand-gated ion channels. The idea is to make a biophysically realistic model of the actual human neuronal circuits (see Fig. 3). To support these efforts, modeling platforms such as NEURON and more recently NetPyne – basically a python wrapper for NEURON – allow for coding these interactions in large and complex neuronal networks. A general overview of the biophysics of neuronal networks is provided next, whereas a description of how to expand these models to a QSP computational neuroscience model is presented in Sect. 3.5.

A detailed outline of the Hodgkin and Huxley approach has been previously described (H Geerts, P.R.A.S.R.C 2018). Basically, it provides a general set of equations describing the current through a neuronal membrane. Most of the equations are empirical in nature but were derived from the physics of the movement of a charged particle (ion) in an electric field and describe remarkably well the dynamics of action potential firing. Similar approaches led to the dynamics of ligand-gated ion channels for glutamate, GABA and acetylcholine. The current through these channels is often driven by a sum of exponentials. Subsequently, after cloning of different ion channel subtypes, specific equations for individual subtypes can be derived. Most of these values (and the corresponding code) can be found on the ModelDB website (ModelDB) (https://modeldb.science). There are currently 18 models for various K^+ channels, 6 for Na^+-channels, and 8 for Ca^{++}-channels.

Ion channel properties are affected by GPCR activation state, which can be modulated by psychoactive drugs as outlined in Sect. 3.2. For example, the inward rectifying potassium current, I_{Kir2} in the MSN neuron projecting into the indirect pathway is modified by the dopamine D_1R activation (u) (Kuzhikandathil and Oxford 2002; Falk et al. 2008) so that the total current $I = u \cdot I_{Kir2}$. Other GPCR subtypes are coupled to different voltage-gated ion channels, likely through phosphorylation at the end of a secondary messenger pathway.

Most computational neuroscience models have a stochastic element reflecting the interaction of other brain regional activity on a modeled circuit. In many cases, this is achieved by sampling from a specific Poisson statistic (Destexhe et al. 2001) using a random number generator as the seed. As a consequence, an essential part of model

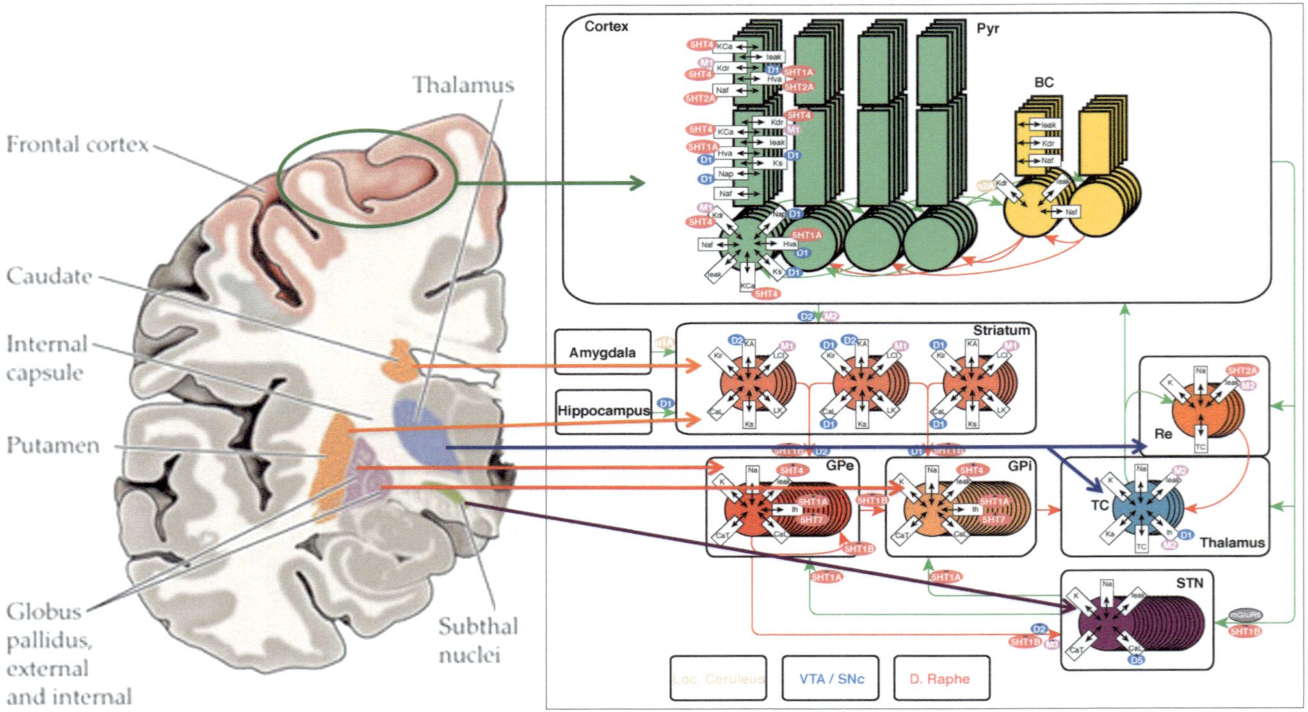

Fig. 3 Computational neuroscience aims to develop biophysically realistic neuronal networks informed by the human brain anatomy, neurophysiology and connectomics. We will elaborate on a model for the frontal cortex, involved in working memory and a cortico-striato-thalamocortical closed loop for motor symptoms in Parkinson's disease

validation is determining the minimal number of runs for a specific condition to achieve convergence. This depends upon the complexity of the model but can range from 30 runs in the 120-neuron Alzheimer ADAS-Cog model to 250 runs in the 225-neuron Parkinson's UPDRS model (Roberts et al. 2016).

In order to derive functional clinical outcomes, as in the case of the amyloid therapy example, readouts of the amyloid aggregation pathway (Geerts et al. 2023a) need to be coupled to changes in neuronal firing through their interaction with voltage- and ligand-gated ion channels. The specific parameters for a baseline disease state are usually informed by both imaging studies in living patients and post-mortem neuropathology studies. For instance, the decrease in acetylcholine levels as a consequence of N basalis pathology in Alzheimer's disease pathology is implemented using a lower synaptic Ach release in the receptor competition model, leading to a proportionally lower activation state of nicotinic and muscarinic receptors. This is accompanied by a gradual loss of synapses and neurons (Roberts et al. 2012). Of interest is the compensatory upregulation of Ach levels in the prodromal Mild Cognitive Impairment (MCI) state as a consequence of increased Choline Acetyl Transferase (ChAT) expression (Ikonomovic et al. 2007).

In our example, preclinical studies have demonstrated that lower levels of the shorter Abeta40 monomer species increase NMDA conductance, but higher levels decrease NMDA conductance, while the longer Ab42 monomer species monotonically decrease the NMDA conductance (Wang et al. 2013). Both Abeta species decrease a7 nAChR in a monotonic way (Geerts et al. 2018a). On the other hand, oligomeric Ab species increase AMPA conductance on pyramidal cells (Fernandez-Perez et al. 2021) and K^+ conductance in inhibitory interneurons (Tamagnini et al. 2015), while they decrease K^+ conductance in pyramidal cells and GABA conductance in interneurons (Ren et al. 2018).

Reduced modeling approaches have been proposed (Chialva et al. 2023), such as the integrate-and-fire model, where the neuron is represented by a single-compartment sphere and simpler differential equations. Other approaches include a neural mass and field model that is used to model functional connectivity between different brain regions (Nakagawa et al. 2013). These approaches can substantially expand the complexity of neuronal network models but are not detailed enough in terms of drug targets to simulate the effects of therapeutic interventions or pathology.

Obviously, capturing the activity of billions of neurons, supported by a myriad of other cells such as microglia and astrocytes, is impossible. While academic computational neurosciences often work with models that encompass millions of cells on large supercomputer clusters (Markram 2006), the development effort involved and the time needed to run different scenarios can be prohibitive for the often time-sensitive questions related to drug discovery and development. Interestingly, recent theoretical investigations have indicated that a reliable prediction of neuronal activity where the connectivity is known is more dependent upon the dimensionality of the task than the size of the network (Clark 2024). This would suggest that modeling the dynamics of smaller networks might be sufficient to predict the outcome of much larger neuronal ensembles.

The granularity and complexity of the QSP model therefore are clearly dependent upon the fine balance between capturing the relevant biological variability and the turn-around time for addressing specific drug-related questions. In the next sections, we provide an example on how to build a computational neuroscience QSP model for the impact of amyloid therapies by incorporating the effect of Abeta species in voltage- and ligand-gated ion channels (Roberts et al. 2012). A similar example is provided in the Appendix C for Parkinson's Disease.

3.5 Setting up a QSP Model of Human Cognition

Predicting a functional clinically relevant outcome on cognition in diseases such as Alzheimer's disease and other dementias requires neuronal cells to be integrated into relevant neuronal network models. Here we describe a computational neuroscience model of a cortical microcolumn with appropriate excitation-inhibition balance that reflects the neurophysiological processes associated with a working memory paradigm.

Working memory is evaluated using a clinical test where a series of letters is shown and the subject needs to react to a test letter only when it is separated by N letters from the target letter, hence the name N-back working memory. The subject needs to keep the memory trace of the target letter active during the intermediate letter presentations. Studies in primates have demonstrated that the circuit continues to fire for several seconds after the target stimulus has disappeared (Williams and Goldman-Rakic 1995). The longer the working memory span, the better performance, which is assumed to be a measure of cognitive strength that is important for the day-to-day functioning of Alzheimer disease patients.

The computational model (Fig. 4a) associated with the working memory paradigm consists of 80 pyramidal cells with 4 compartments, a distal, apical, cell soma and a basal compartment, roughly located to layers I-II to layer V in the cortical microcolumn (Geerts et al. 2013; Roberts et al. 2012). Half of the pyramidal cells are strongly connected and will be stimulated by the target letter, while the remaining half are connected only weakly. The 40 inhibitory basket cells (BC) consisting of a cell soma and one dendritic compartment have 40% of the interneurons synapsing on pyramidal cells and the remaining part forming a recurrent microcircuit. Each pyramidal cell therefore receives (1) synaptic input from other pyramidal cells on its apical and basal dendrites, (2) synaptic input from inhibitory interneurons onto its soma, (3) excitatory and inhibitory synaptic input from a background noise source that represents other pyramidal cells and inhibitory interneurons that are not explicitly represented in the model and is driven by thalamic projections. An mGluR-dependent *delayed afterdepolarization* current (Sidiropoulou et al. 2009) was implemented that can increase the spiking rate of pyramidal cells for several seconds with a current injection shaped as an alpha function initiated at time 2 s.

Fast-spiking, basket cell-type interneurons receive excitatory synaptic input from pyramidal cells onto their dendrites and inhibitory synaptic input from other interneurons onto their soma. The inhibitory interneuron model includes Na and K

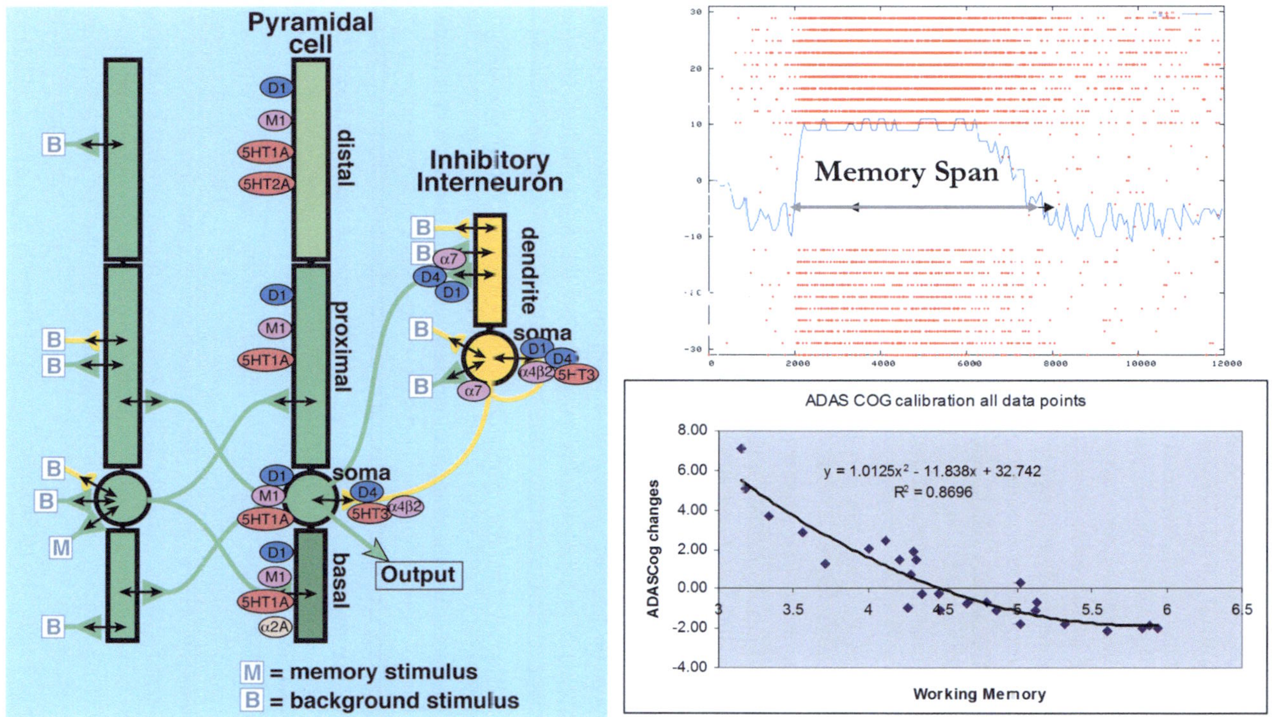

Fig. 4 Schematic of QSP model for human cognition. The network consists of 80 excitatory neurons (green) and 40 inhibitory neurons (yellow) in a cortical microcolumn. Various GPCR affect voltage- and ligand-gated ion channels. The model generates a state diagram (top right, (**a**), essentially a raster diagram of individual neuron firing activity. A memory span is calculated as the duration over which the network tends to fire in synchronization after a short stimulus pulse at 2000 ms (**b**). This in silico readout can then be calibrated to an actual clinical outcome (here change in ADAS-Cog) when simulating a number of historical clinical trial with specific therapeutic interventions (**c**)

voltage-gated channels that control the membrane currents and state of depolarization.

AMPA and NMDA currents have fixed maximal inward depolarizing conductance (g), rise time constant (t_{rise}), decay time constant (t_{decay}) and reversal potential of 0 mV (Traub et al. 2004). Inhibitory chemical synapses represent GABA-A receptor currents using a similar scheme as excitatory synapses, with the GABA-A (chloride) reversal potential and kinetics associated with each cell type. Each model neuron receives fluctuating currents to simulate background synaptic bombardment by excitatory and inhibitory neurons (Destexhe et al. 2001).

The network runs for 2 s before the 50 msec single pulse is provided to half of the pyramidal cells; these then continue to fire in the absence of the stimulus until the synchronized firing slowly disappears. A state diagram is generated showing the action potentials of individual neurons over time, and an obvious in silico biomarker is the length of time the memory trace can be kept active after the stimulus (see Fig. 4b).

The number of neurons and synaptic connections in the network affects the computation time; therefore, a trade-off between granularity and biological relevance needs to be made. Our experience is that a cortical microcolumn network of 80 pyramidal cells and 40 interneurons is sufficient for simulating Alzheimer disease conditions where the pathology leads to deletion of synapses and neurons. To account for the stochasticity of the background noise, we suggest a minimum of 30 runs per condition.

While QSP model readouts for molecular pathways usually focus on biomarker changes that can be detected in biofluids or with imaging, the outcomes in computational neuroscience involve complex properties of the firing pattern of specific neuronal networks. Here we propose the working memory span, which is the time a memory trace is active in the cortical microcolumn after an initial short stimulation. Another endpoint for cognitive readouts is an information content measure (Strong et al. 1998) that describes the complex interspike interval statistics and is related to the total amount of information a spike train (i.e., sequence of action potentials) can carry in units of bits/s, originally proposed by Claude Shannon (Shannon 1948). The concept of entropy has recently been integrated into human cognition theory (Crupi et al. 2018) such that the larger the variability of the interspike interval, the more information that signal contains, as shown in Fig. 5.

As shown in Fig. 5, the spike train is discretized into time bins with size ΔT, where a bin with an action potential is given the value 1. The full spike train is divided in segments with a length T, so that each possible neural response is a word with $T/\Delta T$ symbols, where ΔT is 1, 2, 3 or 4 ms. The maximal word length spans a time segment between 4 and 64 msec, allowing "words" to be defined with maximal length $T/\Delta T$. Using the sliding window approach, a distribution of words generates the normalized count p_i of ith word [$i = 1, 2, 3, \ldots, 2^{16}-1$] so that a naïve estimate of the information bandwidth or Shannon entropy is given by $S_{\text{naïve}}\ (T, \Delta T, \text{size}) = \Sigma\, p_i \log\, (p_i)$. However, true entropy (independent of data set size) is calculated as.

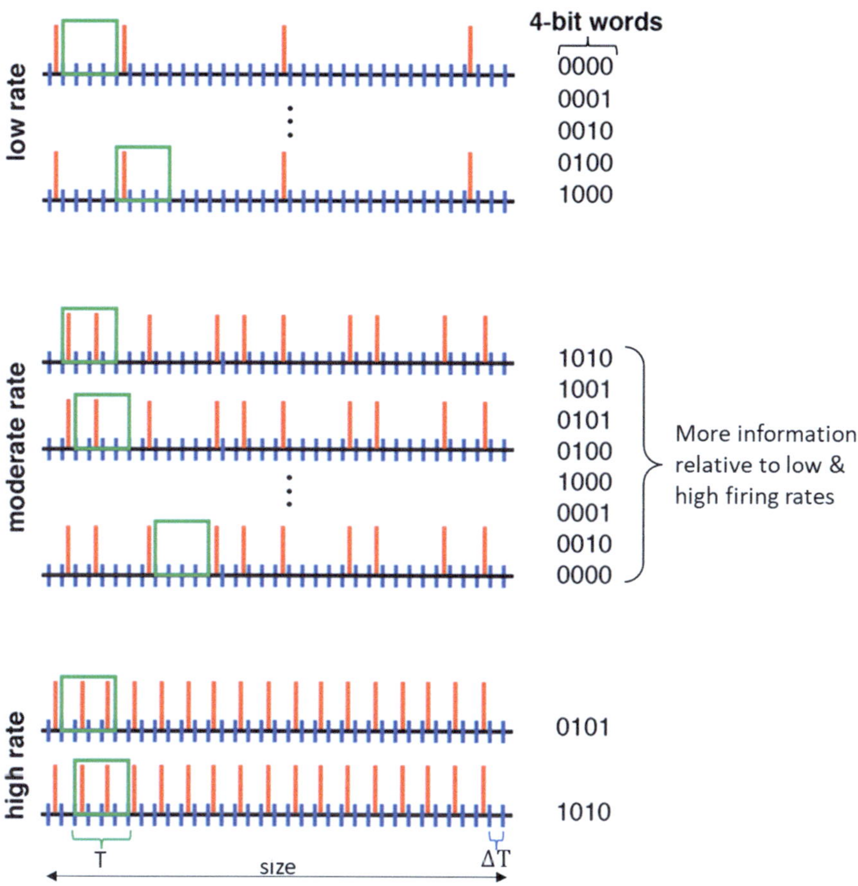

Fig. 5 Illustration of the concept of information content. When using a sliding window (green) over the time course of activation (red indicates action potential), one can form binary words. There is an optimal firing frequency in which the interspike interval can show the greatest variability and therefore carry the most information

$$S(T, \Delta T) = \lim_{\text{size} \to \infty} S_{\text{naïve}}\ (T, \Delta T, \text{size}).$$

A parabolic estimate is obtained by fitting the following equation from a calculated entropy at 3 different action potential train sizes,

$$S_{\text{naïve}}\ (T, \Delta T, \text{size}) = S(T, \Delta T) + S_1(T, \Delta T)/\text{size} + S_2(T, \Delta T)/\text{size}^2 \qquad (3)$$

With $x_i = 1/\text{size}_i$ and with y_i the calculated naive entropy values, the intercept A or $S(T, \Delta T)$ can be calculated. Plotting the readout $S(T,\Delta T)$ as a function of $1/T$ results in the intercept at the x-axis representing the best estimate for the real entropy change. This specific QSP readout is used in the cognitive network for Alzheimer's Disease as well as in the schizophrenia models.

3.6 Calibration and Validation

The final step in QSP model development is calibrating the model, such that the model quantitatively represents the clinical situation, which in our example is a functional scale called ADAS-Cog (Alzheimer Disease Assessment Scale-Cognitive subscale). As mentioned above, some of the parameters linking activation states of GPCR to voltage-gated or ligand-gated ion channels can be optimized for the best correlation between the readout of the QSP model and the functional clinical scale or the changes in the selected biomarker. Care should be taken to not overfit the data, i.e., the number of parameters should always be much lower than the number of data points. If available, objective biomarkers of CNS disease, such as amyloid or tau levels, can be the first level of calibration. However, in drug development projects of symptomatic cognitive enhancement for Alzheimer's disease, motor dysfunction in Parkinson's disease or psychotic symptoms in schizophrenia, the clinical outcomes are more subjective and are affected by a number of intervening factors that are not well understood and therefore beyond the reach of QSP modeling of the individual subject's clinical outcome. One way to mitigate these variabilities is to calibrate the outcome of the QSP model using group average data weighted by the number of subjects in each treatment group rather than individual patient data, assuming that the number of subjects in each treatment arm is sufficiently large.

Calibration of the predominantly cognitive functional scale of ADAS-Cog is achieved in three ways. First, the model parameters for voltage- and ligand-gated ion channels are selected so that the network features realistic firing frequencies of the human neuronal cells. Secondly, the parameters linking dopamine, serotonine, cholinerge and adrenerge neurotransmitter receptor activation are optimized using clinical studies on the N-back working memory task in healthy volunteers with dopaminergic challenges and in schizophrenia patients on different antipsychotics (Geerts et al. 2015a). With these parameters fixed, the remaining parameters of the neuronal network were optimized using 28 drug-dose-timepoints data of treatment in Alzheimer's Disease (Roberts et al. 2012) against clinically reported changes in ADAS-Cog after implementation of the AD pathology. This enabled simulations of the effect of the many comedications such as antidepressants, benzodiazepine, antipsychotics and most importantly cholinerge drugs on ADAS-Cog in the Alzheimer population.

Predictive validation or the prediction of outcome for an intervention not used in the calibration set is the ultimate objective of any QSP model. Examples include the unexpected worsening of a 5-HT4 partial agonist in a human Phase 1 scopolamine challenge trial (Nicholas et al. 2013b) where the model correctly predicted a worsening of the cognitive performance by 25% at the lowest dose where the sponsor expected an improvement. In another example, for an antipsychotic drug with novel and untested pharmacology, an unexpected two-fold higher extrapyramidal side-effect liability compared to the active comparator olanzapine was prospectively and correctly predicted by the model (Geerts et al. 2012). The last example investigates the lack of clinical efficacy of a standalone rather selective 5HT-2C antagonism antipsychotic (Liu et al. 2014), despite showing encouraging effects in preclinical animal models.

3.7 Computational Neuropharmacology Implementation of Drug Pharmacology

To address the challenge of supporting clinical trials, where many subjects are on various comedications in addition to the investigative drug, the pharmacology and downstream impact of more than one CNS active drugs need to be implemented. Many CNS active molecules act on membrane-bound targets such as voltage- and ligand-gated ion channels or G-Protein Coupled Receptors (GPCR). Binding of neurotransmitters to these GPCR initiate second-messenger pathways that for the purposes of computational neuroscience modeling ultimately lead to changes in voltage- or ligand-ion channels, most likely due to phosphorylation, for instance downstream of Protein kinase A.

Modeling second-messenger pathways from Gs (stimulatory GPCRs) or Gi (inhibitory GPCRs) coupled receptors has been developed for many situations, starting from the modeling of slow excitatory changes (Bertrand et al. 2000) to more complex models of calcium dynamics (Wieder et al. 2011) and the integration of spatial component of calcium dynamics in dendritic spines (Ohadi et al. 2019). Of interest to note is that NEURON or Netpyne actually provides a well-characterized tool for modeling diffusion-reaction kinetics. The dynamics of cAMP can be modeled both in terms of microdomains (Oliveira et al. 2010), which allow to simulate the effect of phospho-diesterase (PDE) inhibitors. This opens the possibility to model the impact on Long-term potentiation (LTP), an important contribution to learning and memory (Bouteiller et al. 2008). However, coupling a detailed intracellular pathway model into a complex computational neuroscience model not only leads to a substantial increase in computational burden but also introduces issues around the different time scales of the spiking dynamics (milliseconds) and the intracellular pathways (seconds to minutes).

To disentangle the different time scales, one can make the assumption that the effect of a drug has reached a steady state and the receptor activation changes can be linked to downstream effects on voltage-gated ion channels using a linear transfer function. We assume that CNS active drugs affect membrane receptor states only in a limited way. From preclinical studies, each receptor subtype affects specific voltage- and ligand-gated ion channels in specific cell types (see Table 1).

Starting from the simulations outlined in Sect. 3.2, changes in receptor activation states (target engagement) calculated from the competition between neurotransmitter and drugs (based on their exposure) at the same binding site are then differentially affecting these intracellular pathways. While it is beyond the scope of this chapter to describe these effects in detail, the various downstream effects of D1-R activation

Table 1 List of receptor effects implemented in CNS QSP platform for computational neuropharmacology, their localization and physiological effect

Receptor	Physiological implementation	Location	References
Dopamine D_1R in cortex pyramidal cells	Increases NMDA, decreases AMPA and increases GABA conductance	All compartments of pyramidal cells in cortex	Law-Tho et al. (1994)
Dopamine D_1R in striatum	Affects K_{ir2}	D_1R+ cells in the direct pathway; thalamic neurons	Falk et al. (2008)
Dopamine D_2R	Presynaptic autoreceptor; modulates AMPAR on pyramidal cells	All DA synapses with D_1R, D_2R and D_4R as postsynaptic receptors	Tseng and O'Donnell (2007)
Dopamine D_2R in striatum	Affects A-type K channel & presynaptic Glu release on afferents	D_2R+ cells in the indirect pathway; subthalamic neurons	Bamford et al. (2004), Perez et al. (2006)
Dopamine D_4R	Affects AMPA-R function	All compartments of pyramidal cells in cortex	Yuen and Yan (2009)
Dopamine D_5	Modulates cyclic nucleotide gated K^+ current	Projections on STN neurons	Baufreton et al. (2003)
Serotonin 5-HT_{1A}	Affects Na^+-currents	All compartments of pyramidal cells in cortex; direct effect on GP_e and GP_i neurons; modulates efferent projections from STN to GP_e and GP_i	Foehring (1996) Cardenas et al. (1997)
Serotonin 5-HT_{1B}	Presynaptic autoreceptor in 5HT synapse	Presynaptic receptor for all 5-HTR receptor in circuit; modulates projections of direct & indirect pathway on GP_e and GP_i neurons	Gothert (1990)
Serotonin 5-HT_{2A}	Affects Na^+ and High Threshold Ca^{++}-currents	Distal dendritic compartment of pyramidal cells in cortical model	Carr et al. (2002)
Serotonin 5-HT_{2C} in ventral tegmentum area	Reduces DA release	Localized on projecting DA neurons from VTA; Cholinergic striatal interneurons	Geerts et al. (2013), Abdallah et al. (2009)
Serotonin 5-HT_3	Modulates GABA conductance	GABAergic interneurons in cortex	Puig et al. (2004)
Serotonin 5-HT_3 in ventral striatum	Stimulates DA release	Localized on projecting DA neurons from VTA	De Deurwaerdere et al. (2005)

(continued)

Table 1 (continued)

Receptor	Physiological implementation	Location	References
Serotonin 5-HT_4	Affects serotonergic firing rates; affects K_{dr} and K_s and GABA in pyramidal cells	Direct localization in cortical pyramidal cells; indirect effect on Dorsal raphe, affecting all other 5-HT receptors; direct effect on GP_e and GP_i neurons	Ansanay et al. (1995), Cai et al. (2002)
Serotonin 5-HT_6	Modulates Levels of ACh, and Glu in cortex	On efferent projections from Dorsal Raphe to Basal Forebrain and cortical circuits	Riemer et al. (2003)
Serotonin 5-HT_6 in striatum	Reduces Ventral Tegmentum Area (VTA) firing	Cholinergic striatal interneurons	Minabe et al. (2004), Bonsi et al. (2007)
Serotonin 5-HT_7	Affect cholinergic modulation on MSN neurons	Cholinergic striatal interneurons; direct effect on GP_e and GP_i neurons	Bonsi et al. (2007), Gustafson et al. (1996)
Adrenergic $Alpha_{1A}$ in amygdala	Reduces excitatory afferent input in striatum	Both D_1R+ and D_2R+ medium spiny neurons in striatum	Braga et al. (2004)
Adrenergic $Alpha_{2A}$	Modulates K^+ conductance on inhibitory interneurons	Cortical GABA neurons	Boehm (1999)
Muscarinic M_1R	Modulates Cl leak in MSN & K_{dr} in pyramidal cells	All compartments of pyramidal cells in cortex; efferent projections from cholinergic interneurons on all medium spiny neurons	Perez-Rosello et al. (2005), Shen et al. (2005)
Muscarinic M_2R	Presynaptic autoreceptor for cholinergic synapses	Presynaptic autoreceptor for M_1, M_2 mAChR, a_7 and a_4b_2 nAchR;	Blackwell et al. (2019), Parnas (2005)
Muscarinic M_2R in striatum	Modulates afferent Glu from cortex; affects membrane potential	Efferent projections from pyramidal cells into striatal medium-spiny neurons; on efferent projections from thalamic neurons to reticular inhibitory neurons;	Calabresi and Di Filippo (2008), Nobuhara et al. (2001), Kohlmeier et al. (2012)
Muscarinic M_3R in striatum	Affects DA release	Affects D_2-effect on STN neurons	Zhang et al. (2002)
Nicotinic α_7 AChR	Increases presynaptic Glu release, increases interneuron excitability	Presynaptic projections from pyramidal cells	Parikh et al. (2010), Alkondon and Albuquerque (2001)
Nicotinic $\alpha4\beta2$ AChR	Increases presynaptic GABA release	Presynaptic projections from inhibitory cells	Aracri et al. (2010)

(continued)

Table 1 (continued)

Receptor	Physiological implementation	Location	References
mGluR5	Delayed depolarization on pyramidal excitatory cells	All compartments from pyramidal cells; presynaptic projections on hyperdirect pathway on STN neurons	Sidiropoulou et al. (2009), Awad et al. (2000)
$mGluR_2$	Presynaptic Glu autoreceptor; affects Glu release	Presynaptic projections from pyramidal cells;	Mateo and Porter (2007)

All processes are described for the cortex, except where noted. These relationships can be implemented in models of many different CNS indications, from Alzheimer's and Parkinson's disease to neuropsychiatric indications like schizophrenia, anxiety, and major depressive disorder

changes in different cell types can be further studied in more detailed publications related to Alzheimer's Disease (Roberts et al. 2012), cognitive impairment in schizophrenia (Geerts et al. 2013) and Parkinson's Disease (Hernandez-Lopez et al. 2000).

List of Pharmacological Interactions Implemented in the Quantitative Systems Pharmacology Model

4 Quantitative Systems Pharmacology of Psychiatric Diseases

The term "Computational Psychiatry" was coined by John Krystal (Wang and Krystal 2014) with the aim of modeling neuronal processes involved in behavioral phenotypes in various psychiatric disorders. In this section we document a strategy for developing QSP models in schizophrenia as opposed to neurodegenerative diseases. After describing some of the clinical challenges, we describe the detailed development of a QSP model for psychotic symptoms in schizophrenia followed by other examples of models for different psychiatric indications.

4.1 General Challenges

Psychiatry includes schizophrenia, major depression, anxiety, post-traumatic stress disorder (PTSD) and some behavioral symptoms in dementia, such as apathy and agitation. Other indications at the border of peripheral and central nervous systems include pain, multiple sclerosis and migraine. A major challenge is related to the diagnosis; often patients with the same diagnosis can present with very different symptoms. In addition, many phenotypes such as anhedonia or cognitive deficit are common across many indications. In view of these observations, more biology-focused Research Domains Criteria (RDoc) have been proposed by the NIH (Williams et al. 2024) to account for the cross-diagnostic nature of the clinical phenotypes, as these are defined by activity of well-defined neuronal networks. However, while this makes sense from a biological point of view, this had not generated much traction from a regulatory point of view.

In contrast to many neurodegenerative diseases, psychiatric patients very often are on standard-of-care medications (i.e., antipsychotics in schizophrenia, serotonin-transport inhibitors (SSRI) in depression, benzodiazepines in anxiety). This makes augmentation therapy very challenging as there might be complex pharmacodynamic interactions in relevant neuronal networks.

Many psychiatric diseases such as schizophrenia have either a neurodevelopmental origin or affect multiple neuronal circuits rather than being driven by specific, well-defined molecular pathologies. Clinical symptoms are merely a consequence of altered firing patterns of interconnected brain regions.

The absence of surrogate biomarkers is especially problematic in psychiatry where clear consensus on the molecular pathways giving rise to the clinical phenotype is lacking. Therefore, modeling of key intracellular pathways is very rarely used; rather, QSP modeling focuses on the neuronal activity of interconnected brain regions.

Finally, clinical scales are often structured interviews based on rating scales and are subject to a substantial placebo effect. This makes clinical calibration of QSP models very challenging.

All together and as elaborated in Sect. 3.4, the discipline of computational neurosciences provides the only way to simulate the effect of therapeutic interventions on the neuronal firing activity of these brain circuits, to account for the pharmacodynamic interaction of combination therapy and achieve a level of useful predictability.

4.2 Developing a QSP Model of Psychotic Symptoms

A mechanistic model of the basal ganglia in the association circuit has been applied to schizophrenia-specific reinforcement dysfunction and patient classification (Geana et al. 2022). Because the basal ganglia dorsal motor circuit involved in Parkinson's disease has the same connectomics layout as the limbic-associative circuit for psychotic symptoms, we used the same structure as described in Sect. 3.6 but with completely different implementation of the pathology.

Schizophrenia pathology is introduced as a hyperdopaminergic state, in contrast to Parkinson's disease, where the pathology leads to a dopamine deficit. The size of this dopaminergic increase is calibrated using the receptor competition model (see Sect. 3.2, Eq. 1) to define the presynaptic dopaminergic release dynamics that align with clinically determined PET tracer studies of dopamine metabolism and displacement of specific D_2R tracers in unmedicated schizophrenia patients (Fusar-Poli and Meyer-Lindenberg 2013). Target exposure of pharmaceutical agents, notably approved antipsychotics, at the site of action can be predicted using the receptor competition model from PET tracer displacement studies, usually performed with the D_2R-specific radiotracer ^{11}C-raclopride. Target engagement at the full range of receptors (not limited to dopamine) can be modeled using the competition with endogenous neurotransmitter based on the relative affinities of the different agents. The complex neuropharmacology of antipsychotics, with multi-target affinity of the drugs affecting many downstream voltage-gated ion-channels results in a complex neuronal firing.

The model used a different in silico readout (see Sects. 3.7 and 2.3), the so-called information bandwidth (Spiros et al. 2017). Using this readout in simulations of antipsychotic clinical trials, it was demonstrated that the best correlation with the changes in PANSS (mostly sensitive to changes in positive symptoms) after treatment with antipsychotics was achieved for changes in information content of the neurons in the thalamic reticular nucleus. This region is in a key position to control information flow from the basal ganglia back to the cortex. These results suggest that reduced discrimination of relevant information (corresponding to reduced signal-to-noise detection) could be associated with the onset of symptoms such as hallucinations and delusions.

4.3 Other Examples of Schizophrenia Models

Other clinical phenotypes associated with schizophrenia can be captured with QSP models as demonstrated by an interesting application related to the role of dopamine in the 40 Hz auditory resonance response (Siekmeier and vanMaanen 2014) that links clinical observations of EEG to neurobiological mechanisms. Similarly, a Wilson-Cowan biophysical model was used to link neuronal network activity to the clinical scale of Bush-Francis catatonia rating scale (Roberts and Conour 2022) providing a way to personalize treatment in a cohort of patients.

An important clinical indication is Cognitive Impairment Associated with Schizophrenia (CIAS) (Veselinovic and Neuner 2022) where new therapeutic interventions will be added to antipsychotics leading to multiple complex pharmacodynamic interactions. As a consequence, the QSP model shows that any clinical effect is uniquely dependent upon the nature and dose of the baseline comedication (Geerts and Spiros 2020a). Failure to recognize this important aspect has likely doomed many clinical trials.

Similarly, the introduction of the known interactions and localization of different NMDA-Receptor subtypes with their co-agonist glycine in a QSP model has revealed the underlying biology for the observed inverse U-shape dose-response of a GlyT1 inhibitor, bitopertin (Spiros et al. 2014). More importantly, the modeling suggests that the shape of this dose-response is dependent upon the nature and dose of the baseline antipsychotic medication. This partially explains the challenges to develop successful augmentation therapies in schizophrenia.

5 Technical Aspects of QSP Implementation

5.1 ODE-Based Modeling

Ordinary differential equations (ODEs) are the mainstay of QSP modeling and are covered in "Breadth of Pharmacology Modeling: Fundamentals of Pharmacokinetics, Pharmacodynamics and Mechanistic Modeling" and "A Framework for Quantitative Systems Pharmacology Model Execution". Just like in PBPK modeling, QSP modeling of biomarkers often assumes mass balance conditions. For example, the dynamics of neurotransmitters binding to a receptor needs to be balanced by the corresponding changes in free receptors. These processes can be described adequately in terms of ODE, as they take place in large volumes, such as the brain parenchyma and are related to relatively large numbers of molecules. However, for very small volumes such as synaptic clefts, these assumptions break down and other approaches such as Monte Carlo simulations of single molecule spatial trajectories need to be used.

It is important to consider the solver used for the optimization as well as the solver settings, such as the tolerance criterion. Besides the popular Runge-Kutta methods, backward differentiation formula (BDF), one-step Euler method, two-step Adams method or Generalized Alpha are available. These methods differ in stability, versatility, accuracy and damping of lower frequencies which might lead to numerical instability. The NEURON platform (https://www.neuron.yale.edu/neuron) provides optimized solvers for the differential equations.

5.2 Agent-Based Modeling

Agent-based modeling is a relatively underused approach for modeling spatio-temporal dynamics of molecular processes. Basically, it tracks the evolution of individual agents through space and time based on simple rules for transition. Early studies in the field of Alzheimer's disease (Edelstein-keshet and Spiros 2002) simulated the spatial trajectories of microglia moving in response to an amyloid plaque.

This type of modeling is quite compute intensive, but in contrast to traditional ODE approaches, it can investigate the spatial consequences of, for instance, microglia phenotype conversion as they are affected by the interaction with amyloid plaques (Weathered et al. 2021). It is a reasonable alternative to partial differential equations when space as a dimension is included.

5.3 Monte Carlo Simulation

In some cases, reaction volumes of interest are so small that a finite number of molecules are in play. For instance, antibodies against tau need to access the extremely small synaptic volume (about 4E-17 l). It has been calculated that only one molecule of misfolded tau is being released from the presynaptic ending and can reach its receptor for uptake on the postsynaptic membrane within 10 microseconds. When the normal assumptions of ODEs break down, Monte Carlo simulations model the Brownian-diffusion trajectory of individual tau molecules and the antibodies within the actual geometry, followed by interactions if they happen to collide. A popular tool for this type of modeling is MCell (MCell Home) (https://mcell.org).

Note that this approach is quite compute intensive, especially when considering the interaction between a drug and its (soluble) target and when the geometry of the volume is quite complex. This is the case when considering interaction inside the cytoplasm of a cell or in a neuronal synapse where transsynaptic proteins such as neuroligins and neurexins form a dense crowded space.

5.4 Boolean Network Modeling

A relatively novel approach uses Boolean network modeling to account for the fact that not all parameters are available or known. Boolean network modeling (Kurnellas et al. 2023) aims to connect trajectories in parameter space to experimental data and identifies broad classes such as ordered, asymptotically, chaotic or disordered. This can lead to the identification of emergent neural circuits that can be associated with functional readouts.

5.5 Software Programs

Various software platforms are available for QSP modeling. In principle, programs can be coded in any platform with ODE solving capabilities. A highly specialized platform is NEURON (Hines and Carnevale 1997), which has been used for almost 25 years by the computational neuroscience community with a large repository of well-documented models (ModelDB). Recently, a graphical interface NetPyne (Neuron Python) has been made freely available to the community (Dura-Bernal et al. 2019). It is of interest to note that the current versions also include reaction-diffusion modeling, opening the possibility of modeling molecular pathways and linking the outcomes seamlessly to changes in voltage- and ligand-gated ion channels.

An interesting new modeling toolbox with applications in clinical neurosciences is NeuroBlox (NEUROBLOX), which is a Julia-based set of modules ranging from biophysical models and neural mass models to biomimetic control circuits. It is claimed to be one of the fastest platforms for simulating neuroscience-related projects.

With regard to more traditional QSP modeling of molecular pathways that rely on ODE, many other languages can be used from SimBiology over Matlab to Julia and R. Interestingly, recently a dedicated QSP platform for biological systems, called the QSP Designer, that seamlessly integrates with PBPK modeling has been proposed (2).

6 Examples and Applications of QSP in Neurosciences

The following table lists a number of applications along the spectrum from drug discovery to post-marketing. We will elaborate on some of these applications in the subsequent sections (Table 2).

6.1 QSP Applications in Alzheimer's Disease

The first-generation QSP models focused on the simulation of the dynamics of soluble low-order oligomers after intervention with small molecules, such as BACE-I (van Maanen et al. 2018) and gamma-secretase inhibitors. These models included explicitly data from Stable Isotope Labeling Kinetics (SILK) studies and accompanying kinetic modeling that provided insights on the amyloid dynamics in the living human brain (Potter et al. 2013). Other models studied the aggregation dynamics of amyloid and the impact of therapeutic interventions. These models were based on state transitions between different forms of amyloid species in different brain compartments that made it convenient to integrate with PBPK models (Madrasi et al. 2021; Lin et al. 2022; Ramakrishnan et al. 2022).

However, combining earlier models that incorporated more biophysically realistic microscopic forward and backward rates between different oligomeric forms (Meisl et al. 2016) with explicit interaction of six amyloid antibodies and their effects on microglia-dependent activation (Geerts et al. 2023) allowed the prediction of the dynamics of "invisible" amyloid oligomers starting from clinical observations of PET imaging and fluid biomarkers. Additionally, this model can predict ARIA-E liability directly from the pharmacology of the different antibodies. Another semi-mechanistic model was derived for gantenerumab only (Mazer et al. 2023).

We presented the pipeline for a more complex amyloid aggregation model (Geerts et al. 2023) and its effect on neuronal firing in Sect. 3. One application of this model was the generation of a hypothesis for the very different clinical outcomes of aducanumab's ENGAGE vs EMERGE Phase 3 trial, despite a relatively small difference in biomarker outcome (Geerts and Spiros 2020b). The model outcome suggested a substantial pharmacodynamic interaction with the common genotype variant of the 5-HT transporter gene (5_HTTLPRrs23351) not with the biomarker but with the cognitive outcome to the point that an imbalance of this genotype in the different treatment arms could explain the clinical observations. This is likely due to the fact that soluble amyloid peptides (monomers and oligomers) interact with the same targets (NMDA and GABA) as, for instance, 5-HT3, 5-HT4 and 5-HT6, which are all affected by the different expression of the 5-HT transporter driven by the common genotype variant.

Other applications include predicting changes in fluid biomarkers such as plasma Ab42/Ab40 and phosphorylated tau (p-tau) to (1) determine the duration of treatment to reach central amyloid negativity, (2) explore the consequences of various titration schedules after an ARIA-E incident on the dynamics of amyloid clearance

Table 2 Table listing various applications of Neuroscience QSP along the spectrum of drug discovery and development

Development Stage	Application	Example	References
Drug discovery			
Guide single-target medicinal chemistry programs	Optimize TPP	5-HT4 partial agonist in AD	Nicholas et al. (2013b)
Guide multi-target medicinal chemistry programs	Use target profiles of MedChem output to simulate efficacy in silico	Estimate effect of off-target pharmacology for DA modulators	Spiros and Geerts (2012)
Early clinical development			
Repurposing drugs	Identification for treating tremor	5-HT modulators	Spiros et al. (2013)
Better understanding of dose-response	Phase 2 efficacy	Bitopertin, GlyT1 inhibitor	Spiros et al. (2014)
Identifying biomarkers	PD biomarker for target engagement	Amyloid and Tau biomarkers	Geerts et al. (2023), Bloomingdale et al. (2022b)
Dose-finding	Optimal effect on biomarker	BACE-I in AD	van Maanen et al. (2018)
Clinical development			
Trial design for augmentation therapy	Efficacy studies in Parkinson's disease augmentation therapy	A2A antagonists	Rose et al. (2022)
Virtual Twin	Halted clinical trials	COVID-19	Geerts and van der Graaf (2020)
Power calculations based on PD effects	PD-PD interactions of baseline comedications	AD trials	Geerts and Spiros (2020a)
Post-marketing			
Responder profile	Sensitivity analysis of specific pharmacology	Iloperidone antipsychotic	Geerts et al. (2015b)
Long-term formulations: switching paradigms	Switching scenario	Paliperidone Long-Term-Injectables	Geerts et al. (2018b)
Polypharmacy guidance	Clinical practice	Side-effect in combination of two antipsychotics in schizophrenia	Kadra et al. (2018)

and (3) optimize maintenance dosing after reaching amyloid negativity (Geerts et al. 2023b).

6.2 QSP Extrapolation of Experimental Data from hIPSC

The modest translatability of preclinical animal data has increased interest in the use of human-derived neuronal cell types, or combinations thereof, and/or brain organoids for preclinical efficacy assessments (Cassotta et al. 2022). These models can be derived from patients, which supposedly ensures the biological pathways are properly aligned with the human clinical situation. However, unlike in vivo animal models, these models lack a proper neuronal network to drive emergent properties of brain activity, which is ultimately associated with behavioral readouts. In addition, unless they are coupled with a vascular compartment, they lack the specific PK profile of any therapeutic intervention, especially in dynamics of target exposure associated with clinical dosing. Both of these issues can be mitigated by QSP modeling (Fig. 6). In this approach, an in silico model similar to the actual neuronally differentiated human induced pluripotent stem cell culture is used to generate biological hypotheses underlying the observed experimental phenotypes and are confirmed by appropriate experiments such as genetic (CRISPi-CAS or RNAi) or pharmacology tools. This leads to the identification of the pathway causing the unique phenotype, which can then be integrated with a more complex QSP model that incorporates relevant neuronal networks and may be used to identify possible noninvasive biomarkers. Finally, accounting for the PK profile of the candidate drug allows the effect of the intervention on the relevant clinical scale to be simulated.

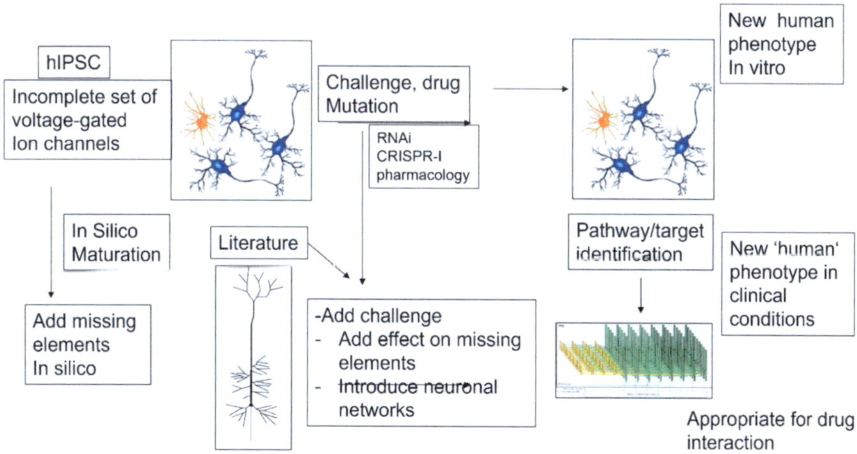

Fig. 6 Incorporation of the QSP modeling strategy into a hIPSC or bran organoids workflow for drug discovery

As an example, the effect of the tau V337M mutation from a Fronto-Temporal Dementia (FTD) patient on the electrophysiological properties of action potential duration and propagation has been documented using multielectrode arrays (MEA) in hIPSC cells (Sohn et al. 2019). In this case, the QSP model suggests that the effect of misfolded tau oligomers on K^+ channels and Na + channels in the axon initial segment can reproduce the experimentally observed changes in axon potential width and propagation speed. This is in line with electrophysiology experiments injecting misfolded tau oligomers in primary neuronal cell cultures (Hill et al. 2021).

7 Epilogue

The combination of traditional QSP modeling of molecular pathways and biomarkers with computational neurosciences opens the way to simulate functional and clinically relevant outcomes. Indeed, neuronal activity patterns in relevant neuronal brain circuits are a proxy for clinical outcome, as human cognitive, executive, motor performance and mental health states are driven by (mal)functioning of brain activity in specific regions.

An interesting position paper (Musuamba et al. 2023) elaborates on the scientific and regulatory evaluation of QSP models. While this applies in general to all disease indications, a few CNS specific issues are worthwhile mentioning. Many computational neuroscience models are stochastic in nature as the impact of other brain regions on the activity of the modeled circuit is represented by sampling from a specific statistical distribution (see Sect. 3.4 for an example on sampling from a Poisson distribution), so an important feature is convergence and consistency. As CNS QSP models often involve the NEURON platform as the central engine, credibility of code verification and software quality relies on the extensive use and applications of this platform in the academic community.

There are two major applications of CNS QSP modeling in the long term. The first is to help identify key therapeutic pathways by reverse engineering many clinical trials. Such trials can be considered as perturbations to the natural progression of the disease, very similar to sound waves used in seismology applications. The objective is to "deconvolute" these responses using mechanistic QSP-computational neuroscience modeling and better understand the key pathways in this "black box." This is quite similar to the concept of brain perturbation mapping (Keller et al. 2014), where single surface or deep brain pulses are triggering complex brain responses that are deconvoluted to better understand dynamics of brain networks. CNS mechanistic modeling of predicted clinical outcomes at the individual patient level that accounts for the effect of comedications, genotypes and disease states can be a powerful tool for hypothesis generation. An example in case is the hypothesis generation about the origin of positive symptoms in schizophrenia as outlined in Sect. 4.2.

The other major application is the development of "virtual twins." With the increasing capability of identifying subpopulations using biomarkers and genetic markers, the variability associated with clinical trajectories becomes better understood. Developing an individual in silico QSP model of the same patient as in the treatment arm with their specific and unique baseline characteristics, such as comedications, genotypes and disease state and simulating the natural "placebo" trajectory allows investigators to ultimately compare the observed functional clinical trial outcome on a per-patient basis. This could support the identification of "responders" based on the alignment of specific biomarkers with the biology of the pathways affected by treatment. In addition, substituting part of the actual placebo patients with the virtual twins of the patients on active medication could reduce the intrinsic variability, reduce the cost of trials and entice patients to enroll in trials as they have a much greater chance of being on active drug.

As QSP modeling started only 10–15 years ago, much later than PBPK, its uptake in the drug development is still evolving. In addition, CNS QSP suffers from the perception that the brain is considered such a complex organ with many knowledge gaps. The examples in this chapter have demonstrated the impact of predictive QSP modeling from target validation and selection of clinical candidates in drug discovery to dosing strategies in the clinic and analysis of failed clinical trials, including rescuing development programs for otherwise safe molecules with excellent target engagement profiles. The hope is that the industry and regulatory agencies will gradually accept this approach to get the right drugs to the right patients faster.

The human brain is complex, yet we might dream 1 day of being able to understand and optimize therapeutic interventions that can really improve patient's lives in functional domains that make the human species really unique.

Appendices

Appendix A: List of Published QSP Models for CNS Disorders

Model	Calibration/ qualification	Application/prediction	References
Amyotrophic lateral sclerosis			
Cell-cell Communication Network	Data from the literature in a mouse model of ALS	Optimization of therapeutic strategies for ALS patients	Shao et al. (2013)
Alzheimer's disease			
Amyloid interventions on Aβ dynamics	Mouse (Tg2576), human, and nonhuman primate data	Retrospective validation of avagacestat and bapineuzumab outcomes	Karelina et al. (2017a), Karelina et al. (2017b)
QSP Model of Tau pathology	Mouse (P301S), human, and in vitro data.	Validation on mouse immunotherapy data	Stepanov et al. (2018)
Calcium Signaling involvement in neuronal cell death	Preclinical studies	Predicted molecular events leading to neuronal death	Ranjan et al. (2018)
Bexarotene PK/PD	APP/PS1 mice	Effects on neuronal viability and Aβ dynamics	Rosenthal et al. (2016)

(continued)

Model	Calibration/ qualification	Application/prediction	References
Periarterial Drainage computational fluid dynamics model	Fluorescent dextran spread in vivo mouse brain	Model for perivascular solute transfer and clearance dynamics of misfolded proteins	Diem et al. (2016)
Impact of b-amyloid on acetylcholine synthesis	Preclinical	Loss of ChAT activity, reduction in choline	Fgaier et al. (2015)
Cell-cell interaction in AD progression	None	Relationship between neuronal death and glial cells/Ab	Puri and Li (2010)
Impact on Abeta on lipid dysregulation	APP/Aβ dynamics from BACEi in NHP and simvastatin clinical data	Modulation byby a S1PR5 receptor agonist	Clausznitzer et al. (2018)
Epidemiological spreading model	Cross-sectional tau pathology imaging	Mechanisms of misfolded tau protein propagation	Fornari et al. (2019)
NMDA-R and downstream postsynaptic [Ca2+]	Experimental (electrophysiology) observations	Aβ effect on NMDA-R and postsynaptic [Ca2+]	Bouteiller et al. (2019)
Neuroinflammation and microglia/neurons associated	Spatial distribution of senile plaques and arrangement of glial cells in human AD brain	Dynamics of plaque formation and neuronal death	Edelstein-keshet and Spiros (2002)
Combined PKPD model of Abeta dynamics	Data for Aβ and antibody therapies in humans	Predicting biomarker response and target engagement	Ramakrishnan et al. (2022)
Multi-factorial causal model (MCM) on imaging dynamics	Clinical study with 561 patients and six different neuroimaging modalities	Identified pathological events and therapeutic strategies	Iturria-Medina et al. (2017)
Multi-scale tractography and amyloid imaging model for EEG readout	Individualized connectomes and amyloid deposition (AV-45 PET)	Reproduces observed slowing of the EEG spectrum	Stefanovski et al. (2019)
Intracellular Tau pathology	Preclinical data	Sensitivity analysis of key pathways for tau aggregation	Yuraszeck et al. (2010)
Cholinergic muscarinic M1-R modulation of pyramidal neuron	Experimental observations on multiple subcellular mechanisms	Effects on pyramidal cell excitability and downstream intracellular calcium dynamics	Mergenthal et al. (2018)
AP5 and memantine on NMDA-R function	Electrophysiological observations	Functional limitations of AP5 versus memantine	Ambert et al. (2010)
PBPK model of anti-tau antibodies	Clinical data	Biomarker response of tau antibodies in CSF	Bloomingdale et al. (2022b)

(continued)

Model	Calibration/ qualification	Application/prediction	References
Amyloid tau pathways combination	Preclinical data	Identification of combination therapies	Karelina et al. (2021)
Parkinson's disease			
Insulin resistance	In vitro experiments and qualitative validation using clinical data.	Identification of potential drug targets and comparing dosing regimens	Braatz and Coleman (2015)
Metabolic model of α-synuclein aggregation	In vitro experiments and qualitative validation using clinical data.	Predict components of disease progression and evaluate different treatment strategies	Sass et al. (2009)
Population-level neuronal model of motor circuit	Clinical EMG data from electrodes in motor thalamus	Generation and propagation of pathological oscillations and DBS effects	Yousif et al. (2020)
Pacemaking dopaminergic neurons and energy consumption	In vitro pacemaking dopaminergic neurons and ion channel blockers	Identification of L-type calcium channel blockers	Francis et al. (2013)
Presynaptic dopamine vesicle release and reuptake	Qualitative comparison with clinical data	Presynaptic mechanisms governing drug response duration	de la Fuente-Fernandez et al. (2004)
Combining levodopa PK, dopamine dynamics and basal ganglia neurotransmission	Parameters from literature, qualitative comparison with clinical trends	Identified mechanisms leading to loss of levodopa effect with denervation	Veronneau-Veilleux et al. (2021)
Parkinson's disease with dementia			
Ab aggregation and reactive microglia	Microglia cluster in	Relationship between reactive microglia and Ab	Kinney et al. (2017)
Bipolar disorder			
Large-scale biophysical type neuronal networks	Clinical data	Understanding ethiogenesis of bipolar disorder	Meyer et al. (2023)
Large-scale Brain Network and electrophysiology after direct brain stimulation	Large database of human imaging	Synchronization explaining changes in spatiotemporal pattern formation after tDCS	Kunze et al. (2016)
Epilepsy			
Large-scale biophysical type neuronal networks	Clinical data	Seizure propagation and personalized therapeutic interventions	Jirsa et al. (2023)
Large-scale biophysical type neuronal networks	Preclinical data	Spreading depression in brain slices	Kelley et al. (2022)

Appendix B: A Primer on BOLD-fMRI

Neuronal cell activity drives change in glucose consumption $g_e'(t) = s_e(t)$ and $g_i'(t) = s_i(t)$ for excitatory (e) cells and inhibitory neurons (i), such that

$$\frac{d}{dt}s(t) = \frac{a}{h}[u(t-d)-1] - \frac{2}{h}s(t) - \frac{[g(t)-1]}{h^2} \tag{B.1}$$

(all variables and parameters with index e or i). $u(t)$ is the neuronal activity, a the efficacy of glucose consumption response, h the time constant of glucose consumption impulse response, $g(t)$ the glucose consumption, and d the delay between neuronal activity and corresponding glucose consumption. The exact derivation of this equation can be found in (Sotero and Trujillo-Barreto 2007).

The oxygen consumption can be described as

$$m_e(t) = \frac{2-x(t)}{2-x_0} g_e(t) \tag{B.2}$$

where $x(t)$, $x(0)$ is the fraction of glucose consumption at time t and at resting conditions (Sotero and Trujillo-Barreto 2007). The sigmoidal relationship is a consequence of glycolysis and glycogenolysis process within the excitatory neurons as outlined in Ref. (Sotero and Trujillo-Barreto 2007). Conversely, the oxygen consumption for inhibitory activity is directly coupled to the glucose consumption, i.e. $m_i(t) = g_i(t)$ leading to a total oxygen consumption of

$$m(t) = \left(\frac{1}{\mathrm{g}}\right)[\mathrm{g}m_e(t) + m_i(t)] \tag{B.3}$$

with m_e and m_i the oxygen consumption of excitatory and inhibitory activity, respectively, and g a proportionality factor.

The cerebral blood flow (CBF), $f(t)$, can be represented by the following extension of the relationship between CBF and neuronal activity (Friston et al. 2000).

$$\frac{d}{dt}f(t) = s_f(t) \tag{B.4}$$

with

$$\frac{d}{dt}s_f(t) = \varepsilon\left[u_e\left(t-\delta_f\right)-1\right] - \frac{1}{\tau_s}s_f(t) - \frac{f(t)-1}{\tau_f} \tag{B.6}$$

with e the efficacy between neuronal activity and increased signal, $u_e(t)$ the excitatory neural activity, d_f the delay between neuronal activity and changes in cerebral blood flow, and t_s the time constant for decay of the signal and t_f time constant for autoregulatory feedback from cerebral blood flow (CBF).

Based on the assumption of an instantaneous link between oxygen consumption and deoxyhemoglobin content (Buxton et al. 2004), the final BOLD-fMRI signal is calculated as

$$y(t) = V_0[a_1(1 - q(t)) - a_2(1 - v(t))] \tag{B.7}$$

where

$$\frac{d}{dt}v(t) = \frac{[f(t) - f_{\text{out}}(v, t)]}{\tau_0} \tag{B.8}$$

and

$$\frac{d}{dt}q(t) = \frac{\left[m(t) - f_{\text{out}}(v, t)\frac{q(t)}{v(t)}\right]}{\tau_0} \tag{B.9}$$

and

$$f_{\text{out}}(v, t) = v(t)^{1/a} + \tau \cdot v(t) \tag{B.10}$$

with t a time constant controlling how fast Cerebral Blood Volume (CBV) adjusts to changes in Cerebral Blood Flow (CBF), $T\text{au}_0$ the transit time through the balloon, a the coefficient determining the steady-state volume relationship, V_0 the baseline blood volume, and a_1 (weight factor for deoxyhemoglobine change) and a_2 (weight factor for blood volume change) are weight constants derived from (Buxton et al. 2004) and $q(t)$ is total deoxyhemoglobin content and $v(t)$ is whole blood volume.

Appendix C: QSP Model for Parkinson's Disease

A more complex example of a cortico-striatal-thalamo-cortical circuit is the motor circuit in Parkinson's disease (Terman et al. 2002) and is informed by a large clinical dataset on functional outcomes both from pharmacological interventions (Hariz and Blomstedt 2022) and recently also from adaptive deep-brain stimulation (Stojsavljevic et al. 2023). Models based on these anatomical connections have been used in outcomes such as exploratory behavior (Maith et al. 2023), perceptual and category learning (Helie et al. 2023) and risky decision making in substance abuse disorder (Guttman et al. 2023).

The computational neuroscience model (see Fig. 7) of the basal ganglia motor circuit has explicit representations of the striatum, Subthalamic Nucleus (STN)-Global Pallidus (GP) circuitry, and thalamo-cortical circuitry, complemented with relevant neuromodulator receptor effects (see Table 2) and subsequently implemented a Parkinson's disease pathology using human imaging Dopamine Transporter (DAT) studies.

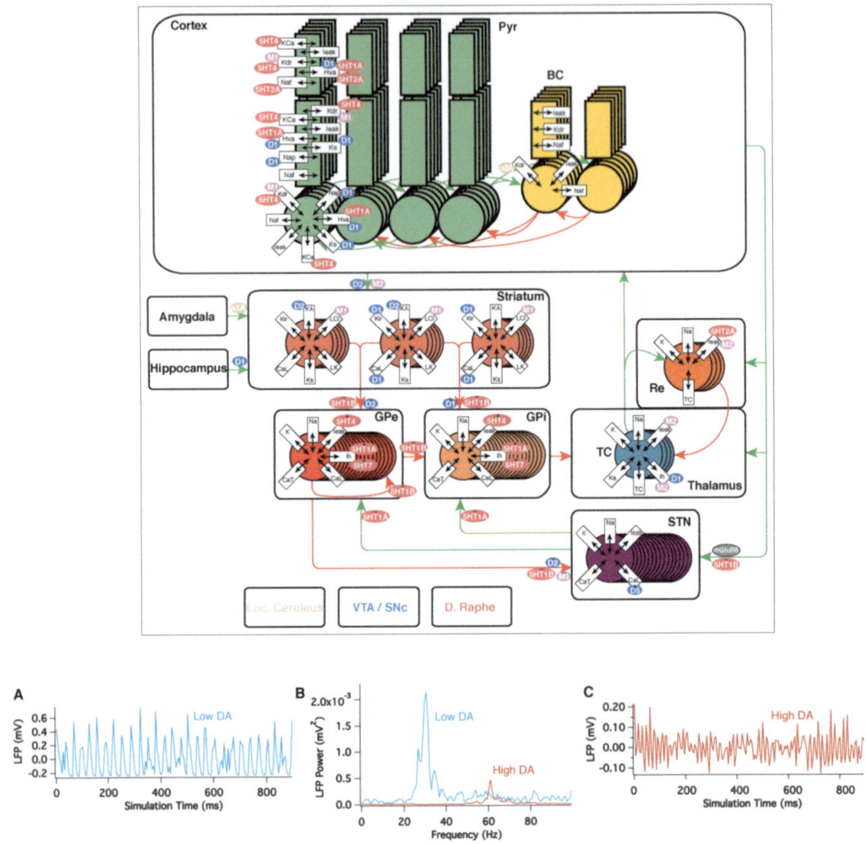

Fig. 7 Overview of QSP model for Parkinson's Disease, consisting of a closed cortico-striata-thalamic-cortical loop with excitatory projections (green) and GABAergic inhibitory projections (Red). Key readout is local field potentials in the subthalamic nucleus (STN). The power spectrum shows two bands, with the ratio of beta power over gamma power an indication of clinical symptoms (bradykinesia and rigidity)

Projections from pyramidal cells with cell soma in layer IV and V drive activity of medium spiny neurons (MSN). The direct pathway for D_1+ MSN neurons (globus pallidus interna) connects to the thalamus; the indirect pathway of D_2+ MSN neurons connects to the globus pallidus externa and the subthalamic nucleus and couples back to the globus pallidus interna. The thalamus projects through the thalamic reticular nucleus back to the cortex, forming a closed loop while a hyperdirect pathway from the cortex drives activity of the thalamus.

Each neuronal cell type is modeled with biophysically realistic membrane conductance as mentioned above; changes in receptor activation as determined by the receptor competition model caused by the pharmacology or the pathological state can change the spiking activity. The striatum model simulates the processing capacity of medium spiny neurons (MSN) in the ventral striatum (Spiros et al. 2012) and calculates the excitability of the D_1 cells in direct pathway and D_2 cells in the indirect pathway, driven by afferent cortical projections. In the direct pathway MSN cells, D_1 receptors affect the K_{ir} channel (Falk et al. 2008) while in the indirect pathway, D_2 receptor activation mainly affects the A-type K^+ current. D_2 receptor activity modulates the presynaptic Glu release on the afferent cortical fibers driving the excitability of all MSN neurons.

Sodium current (Na^+), delayed rectifier potassium current (${K_{dr}}^+$), T-type calcium current (${Ca_T}^{++}$), L-type calcium current (${Ca_L}^{++}$) and a leak current is implemented in STN, GP_i and GP_e (Rubin and Terman 2004; Pirini et al. 2009). The D_2-type and D_1-type MSN cells from the striatum synaptically connects with a GABA current to the GP_e neurons and to the GP_i neurons, respectively. Each GP_e cell receives inhibitory input from three other GP_e cells in addition to excitatory input from three randomly chosen STN cells while each STN cell receives inhibitory input from three GP_e cells. Furthermore, each GP_i cell receives inhibitory input from 1 GP_e neuron and excitatory input from 1 STN neuron. All the TC cells of the thalamus receive inhibitory input from eight GP_i cells.

The thalamus model has four thalamocortical neurons (TC) and four reticular neurons (Re). The TC are excitatory, glutamatergic relay neurons that pass sensory information to the cortex model (Bazhenov et al. 1998). The Re are inhibitory, GABAergic feedback neurons that receive inputs from, and inhibit, TC neurons. The synaptic interaction between these neuronal types, and their intrinsic membrane properties, leads to oscillations, suppression and amplification of multiple input signals. The spiking property of TC neurons is driven by a fast sodium channel, Na^+, a fast potassium channel, K^+, a low-threshold Ca^{++} channel, iTC, a hyperpolarization-activated cation channel, Ih, a potassium A channel, K_a, and a potassium leak channel. Re cells have the same currents, except for the potassium A current (Bazhenov et al. 1998). The cortical part is identical to the "working memory" model described above.

The pyramidal cells project to all MSN neurons in the striatum and to the STN for the hyperdirect pathway. The parameters for the voltage-gated ion-channels and the coupling factors between various GPCR and their effector targets are determined by preclinical data in nonhuman primates (Williams and Goldman-Rakic 1995) and subsequent calibrations (Geerts et al. 2013; Roberts et al. 2012) and predictions of clinical efficacy (Nicholas et al. 2013a; Geerts et al. 2015a, b).

Parkinson's disease state is implemented as a gradual loss of dopaminergic input in the basal ganglia motor circuit (Roberts et al. 2016). This affects the activation levels of D_1 and D_2 receptors on the medium spiny neurons, projecting to different pathways. Simulations using the affinities of dopamine for each of these receptor subtypes suggest a nonlinear relationship with the greatest decline observed for large decreases in dopamine tone, in line with clinical observations that symptoms arise when over 80% of the DA neurons are lost (Wichmann and DeLong 1998).

More complete multiscale models using the Virtual Human Brain have been proposed for deep-brain stimulation (Meier et al. 2022). This is another non-pharmacological application for QSP modeling for treatment of Parkinson's disease patients. Because these patients are often on dopaminergic standard-of-care, combining these approaches with the computational neuropharmacology described above is an exciting future area of development.

References

Abdallah L et al (2009) Impact of serotonin 2C receptor null mutation on physiology and behavior associated with nigrostriatal dopamine pathway function. J Neurosci 29(25):8156–8165

Alkondon M, Albuquerque EX (2001) Nicotinic acetylcholine receptor alpha7 and alpha4beta2 subtypes differentially control GABAergic input to CA1 neurons in rat hippocampus. J Neurophysiol 86(6):3043–3055

Ambert N et al (2010) Computational studies of NMDA receptors: differential effects of neuronal activity on efficacy of competitive and non-competitive antagonists. Open Access Bioinformatics 2:113–125

Ansanay H et al (1995) cAMP-dependent, long-lasting inhibition of a K+ current in mammalian neurons. Proc Natl Acad Sci USA 92(14):6635–6639

Aracri P et al (2010) Tonic modulation of GABA release by nicotinic acetylcholine receptors in layer V of the murine prefrontal cortex. Cereb Cortex 20(7):1539–1555

Aradi I, Erdi P (2006) Computational neuropharmacology: dynamical approaches in drug discovery. Trends Pharmacol Sci 27(5):240–243

Athan Spiros HG (2012) A quantitative way to estimate clinical off-target effects for human membrane brain targets in CNS Research and Development. J Exp Pharmacol 4:53–62

Awad H et al (2000) Activation of metabotropic glutamate receptor 5 has direct excitatory effects and potentiates NMDA receptor currents in neurons of the subthalamic nucleus. J Neurosci 20(21):7871–7879

Bamford NS et al (2004) Dopamine modulates release from corticostriatal terminals. J Neurosci 24(43):9541–9552

Baufreton J et al (2003) D5 (not D1) dopamine receptors potentiate burst-firing in neurons of the subthalamic nucleus by modulating an L-type calcium conductance. J Neurosci 23(3):816–825

Bazhenov M et al (1998) Computational models of thalamocortical augmenting responses. J Neurosci 18(16):6444–6465

Bertacchini F et al (2022) Modelling brain dynamics by Boolean networks. Sci Rep 12(1):16543

Bertrand PP et al (2000) A simple mathematical model of second-messenger mediated slow excitatory postsynaptic potentials. J Comput Neurosci 8(2):127–142

Blackwell KT et al (2019) Molecular mechanisms underlying striatal synaptic plasticity: relevance to chronic alcohol consumption and seeking. Eur J Neurosci 49(6):768–783

Bloomingdale P et al (2021) Minimal brain PBPK model to support the preclinical and clinical development of antibody therapeutics for CNS diseases. J Pharmacokinet Pharmacodyn 48(6): 861–871. https://doi.org/10.1007/s10928-021-09776-7

Bloomingdale P et al (2022a) Hallmarks of neurodegenerative disease: a systems pharmacology perspective. CPT Pharmacometrics Syst Pharmacol. https://doi.org/10.1002/psp4.12852

Bloomingdale P et al (2022b) PBPK-PD modeling for the preclinical development and clinical translation of tau antibodies for Alzheimer's disease. Front Pharmacol 13:867457

Boehm S (1999) Presynaptic alpha2-adrenoceptors control excitatory, but not inhibitory, transmission at rat hippocampal synapses. J Physiol 519(Pt 2):439–449

Bonsi P et al (2007) Endogenous serotonin excites striatal cholinergic interneurons via the activation of 5-HT 2C, 5-HT6, and 5-HT7 serotonin receptors: implications for extrapyramidal side effects of serotonin reuptake inhibitors. Neuropsychopharmacology 32(8):1840–1854

Bouteiller JM et al (2008) Modeling glutamatergic synapses: insights into mechanisms regulating synaptic efficacy. J Integr Neurosci 7(2):185–197

Bouteiller JC et al (2019) Pathogenic processes underlying Alzheimer's disease: modeling the effects of amyloid Beta on synaptic transmission. In: Annual international conference of the IEEE engineering in medicine and biology society. IEEE engineering in medicine and biology society. Annual international conference, 2019, pp 1956–1959

Braatz EM, Coleman RA (2015) A mathematical model of insulin resistance in Parkinson's disease. Comput Biol Chem 56:84–97

Braga MF et al (2004) Stress impairs alpha(1A) adrenoceptor-mediated noradrenergic facilitation of GABAergic transmission in the basolateral amygdala. Neuropsychopharmacology 29(1):45–58

Brody H (2014) Schizophrenia. Nature 508(7494):S1

Buxton RB et al (2004) Modeling the hemodynamic response to brain activation. NeuroImage 23 (Suppl 1):S220–S233

Cai X et al (2002) Activity-dependent bidirectional regulation of GABA(A) receptor channels by the 5-HT(4) receptor-mediated signalling in rat prefrontal cortical pyramidal neurons. J Physiol 540(Pt 3):743–759

Calabresi P, Di Filippo M (2008) ACh/dopamine crosstalk in motor control and reward: a crucial role for alpha 6-containing nicotinic receptors? Neuron 60(1):4–7

Cardenas CG, Del Mar LP, Scroggs RS (1997) Two parallel signaling pathways couple 5HT1A receptors to N- and L-type calcium channels in C-like rat dorsal root ganglion cells. J Neurophysiol 77(6):3284–3296

Carr DB et al (2002) Serotonin receptor activation inhibits sodium current and dendritic excitability in prefrontal cortex via a protein kinase C-dependent mechanism. J Neurosci 22(16):6846–6855

Cassotta M et al (2022) The future of Parkinson's disease research: a new paradigm of human-specific investigation is necessary… and possible. ALTEX 39(4):694–709

Chang HY et al (2019) A translational platform PBPK model for antibody disposition in the brain. J Pharmacokinet Pharmacodyn 46(4):319–338

Chang HY et al (2022) Towards a translational physiologically-based pharmacokinetic (PBPK) model for receptor-mediated transcytosis of anti-transferrin receptor monoclonal antibodies in the central nervous system. J Pharmacokinet Pharmacodyn 49(3):337–362

Chialva U, Gonzalez Bosca V, Rotstein HG (2023) Low-dimensional models of single neurons: a review. Biol Cybern 117(3):163–183

Clark B (2024) Struvtiure of activity in multiregion recurrent neural networks. BioArXiv:2402.12188v2

Clausznitzer D et al (2018) Quantitative systems pharmacology model for Alzheimer disease indicates targeting sphingolipid dysregulation as potential treatment option. CPT Pharmacometrics Syst Pharmacol 7(11):759–770

Cook D et al (2014) Lessons learned from the fate of AstraZeneca's drug pipeline: a five-dimensional framework. Nat Rev Drug Discov 13(6):419–431

Crupi V et al (2018) Generalized information theory meets human cognition: introducing a unified framework to model uncertainty and information search. Cogn Sci. https://doi.org/10.1111/cogs.12613

Cummings JL, Morstorf T, Zhong K (2014) Alzheimer's disease drug-development pipeline: few candidates, frequent failures. Alzheimers Res Ther 6(4):37

De Deurwaerdere P et al (2005) Regionally and functionally distinct serotonin3 receptors control in vivo dopamine outflow in the rat nucleus accumbens. J Neurochem 94(1):140–149

de la Fuente-Fernandez R et al (2004) Presynaptic mechanisms of motor fluctuations in Parkinson's disease: a probabilistic model. Brain J Neurol 127(Pt 4):888–899

Destexhe A et al (2001) Fluctuating synaptic conductances recreate in vivo-like activity in neocortical neurons. Neuroscience 107(1):13–24

Diem AK et al (2016) A simulation model of Periarterial clearance of amyloid-beta from the brain. Front Aging Neurosci 8:18

Dura-Bernal S et al (2019) NetPyNE, a tool for data-driven multiscale modeling of brain circuits. elife 8:e44494
Edelstein-keshet L, Spiros A (2002) Exploring the formation of Alzheimer's disease senile plaques in silico. J Theor Biol 216(3):301–326
El-Kareh AW, Braunstein SL, Secomb TW (1993) Effect of cell arrangement and interstitial volume fraction on the diffusivity of monoclonal antibodies in tissue. Biophys J 64(5): 1638–1646
Falk T et al (2008) Over-expression of the potassium channel Kir2.3 using the dopamine-1 receptor promoter selectively inhibits striatal neurons. Neuroscience 155(1):114–127
Fernandez-Perez EJ et al (2021) Synaptic dysregulation and hyperexcitability induced by intracellular amyloid beta oligomers. Aging Cell 20(9):e13455
Fgaier H et al (2015) Modeling the interaction between beta-amyloid aggregates and choline acetyltransferase activity and its relation with cholinergic dysfunction through two-enzyme/ two-compartment model. Comput Math Methods Med 2015:923762
Foehring RC (1996) Serotonin modulates N- and P-type calcium currents in neocortical pyramidal neurons via a membrane-delimited pathway. J Neurophysiol 75(2):648–659
Fornari S et al (2019) Prion-like spreading of Alzheimer's disease within the brain's connectome. J R Soc Interface 16(159):20190356
Francis F, Garcia MR, Middleton RH (2013) A single compartment model of pacemaking in dissasociated substantia nigra neurons: stability and energy analysis. J Comput Neurosci 35(3):295–316
Friston KJ et al (2000) Nonlinear responses in fMRI: the balloon model, Volterra kernels, and other hemodynamics. NeuroImage 12(4):466–477
Fusar-Poli P, Meyer-Lindenberg A (2013) Striatal presynaptic dopamine in schizophrenia, part I: meta-analysis of dopamine active transporter (DAT) density. Schizophr Bull 39(1):22–32
Garai K, Frieden C (2013) Quantitative analysis of the time course of Abeta oligomerization and subsequent growth steps using tetramethylrhodamine-labeled Abeta. Proc Natl Acad Sci USA 110(9):3321–3326
Geana A et al (2022) Using computational modeling to capture schizophrenia-specific reinforcement learning differences and their implications on patient classification. Biol Psychiatry Cognit Neurosci Neuroimaging 7(10):1035–1046
Geerts H (2009) Of mice and men: bridging the translational disconnect in CNS drug discovery. CNS Drugs 23(11):915–926
Geerts H, Spiros A (2020a) Simulating the effects of common Comedications and genotypes on Alzheimer's cognitive trajectory using a quantitative systems pharmacology approach. J Alzheimer's Dis 78(1):413–424
Geerts H, Spiros A (2020b) Learning from amyloid trials in Alzheimer's disease. A virtual patient analysis using a quantitative systems pharmacology approach. Alzheimers Dement 16(6): 862–872
Geerts H, van der Graaf PH (2020) Salvaging CNS clinical trials halted due to COVID-19. CPT Pharmacometrics Syst Pharmacol 9(7):367–370
Geerts H et al (2012) Blinded prospective evaluation of computer-based mechanistic schizophrenia disease model for predicting drug response. PLoS One 7(12):e49732
Geerts H, Roberts P, Spiros A (2013) A quantitative system pharmacology computer model for cognitive deficits in schizophrenia. CPT Pharmacometrics Syst Pharmacol 2:e36
Geerts H, Spiros A, Roberts P (2015a) Assessing the synergy between cholinomimetics and memantine as augmentation therapy in cognitive impairment in schizophrenia. A virtual human patient trial using quantitative systems pharmacology. Front Pharmacol 6:198
Geerts H et al (2015b) Understanding responder neurobiology in schizophrenia using a quantitative systems pharmacology model: application to iloperidone. J Psychopharmacol 29(4):372–382
Geerts H, Spiros A, Roberts P (2018a) Impact of amyloid-beta changes on cognitive outcomes in Alzheimer's disease: analysis of clinical trials using a quantitative systems pharmacology model. Alzheimers Res Ther 10(1):14

Geerts H et al (2018b) A quantitative systems pharmacology study on optimal scenarios for switching to paliperidone palmitate once-monthly. Schizophr Res. https://doi.org/10.1016/j.schres.2017.11.016

Geerts H et al (2023a) A combined physiologically-based pharmacokinetic and quantitative systems pharmacology model for modeling amyloid aggregation in Alzheimer's disease. CPT Pharmacometrics Syst Pharmacol 12(4):444–461. https://doi.org/10.1002/psp4.12912

Geerts, H., Bergeler, S., Walker, M., Rose, R., & Van der Graaf, P. (2023b). Exploring treatment guidelines for amyloid therapies in Alzheimer's Disease Using Predictive Quantitative System Pharmacology Modeling. Alzheimer's & Dementia, 19(S24). https://doi.org/10.1002/alz.082487

Gothert M (1990) Presynaptic serotonin receptors in the central nervous system. Ann N Y Acad Sci 604:102–112

Gribkoff VK, Kaczmarek LK (2017) The need for new approaches in CNS drug discovery: why drugs have failed, and what can be done to improve outcomes. Neuropharmacology 120:11–19

Gustafson EL et al (1996) A receptor autoradiographic and in situ hybridization analysis of the distribution of the 5-ht7 receptor in rat brain. Br J Pharmacol 117(4):657–666

Guttman Z et al (2023) Decomposing risky decision-making in methamphetamine use disorder: behavioral updating and D2 dopamine receptors. Drug Alcohol Depend 246:109860

H Geerts, P.R.A.S.R.C (2018) Multi-scale modeling of drug acrion in the central nervous system. In: Mager D, Kimpko H (eds) Systems pharmacology and pharmacodynamics, pp 305–329

Hansson O et al (2023) Blood biomarkers for Alzheimer's disease in clinical practice and trials. Nat Aging 3(5):506–519

Hariz M, Blomstedt P (2022) Deep brain stimulation for Parkinson's disease. J Intern Med 292(5): 764–778

Hay M et al (2014) Clinical development success rates for investigational drugs. Nat Biotechnol 32(1):40–51

Helie S et al (2023) A computational model of prefrontal and striatal interactions in perceptual category learning. Brain Cogn 168:105970

Hernandez-Lopez S et al (2000) D2 dopamine receptors in striatal medium spiny neurons reduce L-type Ca2+ currents and excitability via a novel PLC[beta]1-IP3-calcineurin-signaling cascade. J Neurosci 20(24):8987–8995

Hill E et al (2021) Truncating tau reveals different pathophysiological actions of oligomers in single neurons. Commun Biol 4(1):1265

Hines ML, Carnevale NT (1997) The NEURON simulation environment. Neural Comput 9(6): 1179–1209

Ikonomovic MD et al (2007) Superior frontal cortex cholinergic axon density in mild cognitive impairment and early Alzheimer disease. Arch Neurol 64(9):1312–1317

Iturria-Medina Y et al (2017) Multifactorial causal model of brain (dis)organization and therapeutic intervention: application to Alzheimer's disease. NeuroImage 152:60–77

Jirsa V et al (2023) Personalised virtual brain models in epilepsy. Lancet Neurol 22(5):443–454

Kadra G et al (2018) Predicting parkinsonism side-effects of antipsychotic polypharmacy prescribed in secondary mental healthcare. J Psychopharmacol 32:269881118796809

Karelina T et al (2017a) Studying the progression of amyloid pathology and its therapy using translational longitudinal model of accumulation and distribution of amyloid Beta. CPT Pharmacometrics Syst Pharmacol 6(10):676–685

Karelina T et al (2017b) A translational systems pharmacology model for Abeta kinetics in mouse, monkey, and human. CPT Pharmacometrics Syst Pharmacol 6(10):666–675

Karelina T et al (2021) Monoclonal antibody therapy efficacy can be boosted by combinations with other treatments: predictions using an integrated Alzheimer's disease platform. CPT Pharmacometrics Syst Pharmacol 10(6):543–550

Keller CJ et al (2014) Corticocortical evoked potentials reveal projectors and integrators in human brain networks. J Neurosci 34(27):9152–9163

Kelley C et al (2022) Multiscale computer modeling of spreading depolarization in brain slices. eNeuro 9(4). https://doi.org/10.1523/ENEURO.0082-22.2022
Khani M et al (2022) Human in silico trials for parametric computational fluid dynamics investigation of cerebrospinal fluid drug delivery: impact of injection location, injection protocol, and physiology. Fluids Barriers CNS 19(1):8
Kinney et al (2017) Modeling aggregation of proliferating microglia. Spora A J Biomathematics 3:1
Kohlmeier KA et al (2012) Knockouts reveal overlapping functions of M(2) and M(4) muscarinic receptors and evidence for a local glutamatergic circuit within the laterodorsal tegmental nucleus. J Neurophysiol 108(10):2751–2766
Kostewicz ES et al (2014) PBPK models for the prediction of in vivo performance of oral dosage forms. Eur J Pharm Sci 57:300–321
Kunze T et al (2016) Transcranial direct current stimulation changes resting state functional connectivity: a large-scale brain network modeling study. NeuroImage 140:174–187
Kurnellas M et al (2023) Latozinemab, a novel progranulin-elevating therapy for frontotemporal dementia. J Transl Med 21(1):387
Kuzhikandathil EV, Oxford GS (2002) Classic D1 dopamine receptor antagonist R-(+)-7-chloro-8-hydroxy-3-methyl-1-phenyl-2,3,4,5-tetrahydro-1H-3-benzaze pine hydrochloride (SCH23390) directly inhibits G protein-coupled inwardly rectifying potassium channels. Mol Pharmacol 62(1):119–126
Law-Tho D, Hirsch JC, Crepel F (1994) Dopamine modulation of synaptic transmission in rat prefrontal cortex: an in vitro electrophysiological study. Neurosci Res 21(2):151–160
Lin L et al (2022) Quantitative systems pharmacology model for Alzheimer's disease to predict the effect of aducanumab on brain amyloid. CPT Pharmacometrics Syst Pharmacol 11(3):362–372
Liu J et al (2014) Prediction of efficacy of Vabicaserin, a 5-HT2C agonist, for the treatment of schizophrenia using a quantitative systems pharmacology model. CPT Pharmacometrics Syst Pharmacol 3(4):e111
Logothetis NK (2002) The neural basis of the blood-oxygen-level-dependent functional magnetic resonance imaging signal. Philos Trans R Soc Lond Ser B Biol Sci 357(1424):1003–1037
Lytton WW et al (2017) Multiscale modeling in the clinic: diseases of the brain and nervous system. Brain Inf 4(4):219–230
Madrasi K et al (2021) Systematic in silico analysis of clinically tested drugs for reducing amyloid-beta plaque accumulation in Alzheimer's disease. Alzheimers Dement 17(9):1487–1498
Maith O et al (2023) Exploration behavior after reversals is predicted by STN-GPe synaptic plasticity in a basal ganglia model. iScience 26(5):106599
Markram H (2006) The blue brain project. Nat Rev Neurosci 7(2):153–160
Martínez G, Vernooij RW, Fuentes Padilla P, Zamora J, Bonfill Cosp X, Flicker L (2017) 18F PET with florbetapir for the early diagnosis of Alzheimer's disease dementia and other dementias in people with mild cognitive impairment (MCI). Cochrane Database Syst Rev 11(11):CD012216. https://doi.org/10.1002/14651858.CD012216.pub2
Mateo Z, Porter JT (2007) Group II metabotropic glutamate receptors inhibit glutamate release at thalamocortical synapses in the developing somatosensory cortex. Neuroscience 146(3):1062–1072
Mazer NA et al (2023) Development of a quantitative semi-mechanistic model of Alzheimer's disease based on the amyloid/tau/neurodegeneration framework (the Q-ATN model). Alzheimers Dement 19(6):2287–2297
Meier JM et al (2022) Virtual deep brain stimulation: multiscale co-simulation of a spiking basal ganglia model and a whole-brain mean-field model with the virtual brain. Exp Neurol 354:114111
Meisl G et al (2016) Quantitative analysis of intrinsic and extrinsic factors in the aggregation mechanism of Alzheimer-associated Abeta-peptide. Sci Rep 6:18728
Mergenthal AR, Bouteiller JC, Berger TW (2018) Cholinergic modulation of CA1 pyramidal cells via M1 muscarinic receptor activation: a computational study at physiological and Supraphysiological levels. In: Annual international conference of the IEEE engineering in

medicine and biology society. IEEE engineering in medicine and biology society. Annual international conference, 2018, pp 1396–1399

Meyer K et al (2023) Boosting the theory of mind network: specific psychotherapy increases neural correlates of affective theory of mind in euthymic bipolar disorder. Biol Psychiatry Cognit Neurosci Neuroimaging 8(5):572–580

Minabe Y et al (2004) Effect of the acute and chronic administration of the selective 5-HT6 receptor antagonist SB-271046 on the activity of midbrain dopamine neurons in rats: an in vivo electrophysiological study. Synapse 52(1):20–28

Musuamba FT et al (2023) Moving toward a question-centric approach for regulatory decision making in the context of drug assessment. Clin Pharmacol Ther 114(1):41–50

Nakagawa TT et al (2013) Bottom up modeling of the connectome: linking structure and function in the resting brain and their changes in aging. NeuroImage 80:318–329

Nicholas T, Duvvuri S, Leurent C, Raunig D, Rapp T, Iredale P, Rowinski C, Carr R, Roberts P, Spiros A, Geerts H (2013a) Systems pharmacology modeling in neuroscience: prediction and outcome of PF-04995274, a 5HT4 partial agonist, in a Clinical Scopolamine Impairment Trial. Adv Alzheimer's Dis 2(3):83–98

Nicholas T et al (2013b) Systems pharmacology modeling in neuroscience: prediction and outcome of PF-04995274, a 5-HT4 partial agonist, in a clinical scopolamine impairment trial. Adv Alzheimer's Dis 2(3):83–98

Nobuhara K et al (2001) SPET imaging of central muscarinic acetylcholine receptors with iodine-123 labelled E-IQNP and Z-IQNP. Eur J Nucl Med 28(1):13–24

Ohadi D et al (2019) Computational modeling reveals frequency modulation of calcium-cAMP/PKA pathway in dendritic spines. Biophys J 117(10):1963–1980

Oliveira RF et al (2010) The role of type 4 phosphodiesterases in generating microdomains of cAMP: large scale stochastic simulations. PLoS One 5(7):e11725

Pagano G et al (2022) Trial of Prasinezumab in early-stage Parkinson's disease. N Engl J Med 387(5):421–432

Parikh V et al (2010) Prefrontal beta2 subunit-containing and alpha7 nicotinic acetylcholine receptors differentially control glutamatergic and cholinergic signaling. J Neurosci 30(9):3518–3530

Parnas H (2005) Depolarization initiates phasic acetylcholine release by relief of a tonic block imposed by presynaptic M2 muscarinic receptors. J Neurophysiol 93(6):3257–3269

Parvizi J et al (2022) Complex negative emotions induced by electrical stimulation of the human hypothalamus. Brain Stimul 15(3):615–623

Perez MF, White FJ, Hu XT (2006) Dopamine D(2) receptor modulation of K(+) channel activity regulates excitability of nucleus accumbens neurons at different membrane potentials. J Neurophysiol 96(5):2217–2228

Perez-Rosello T et al (2005) Cholinergic control of firing pattern and neurotransmission in rat neostriatal projection neurons: role of CaV2.1 and CaV2.2 Ca2+ channels. J Neurophysiol 93(5):2507–2519

Pirini M et al (2009) A computational modelling approach to investigate different targets in deep brain stimulation for Parkinson's disease. J Comput Neurosci 26(1):91–107

Potter R et al (2013) Increased in vivo amyloid-beta42 production, exchange, and loss in presenilin mutation carriers. Sci Transl Med 5(189):189ra77

Puig MV et al (2004) In vivo excitation of GABA interneurons in the medial prefrontal cortex through 5-HT3 receptors. Cereb Cortex 14(12):1365–1375

Puri IK, Li L (2010) Mathematical modeling for the pathogenesis of Alzheimer's disease. PLoS One 5(12):e15176

Ramakrishnan V et al (2022) Quantitative systems pharmacology model of the amyloid pathway in Alzheimer's disease: insights into the therapeutic mechanisms of clinical candidates. CPT Pharmacometrics Syst Pharmacol. https://doi.org/10.1002/psp4.12876

Ramakrishnan V et al (2023) Quantitative systems pharmacology model of the amyloid pathway in Alzheimer's disease: insights into the therapeutic mechanisms of clinical candidates. CPT Pharmacometrics Syst Pharmacol 12(1):62–73

Ranjan B, Chong KH, Zheng J (2018) Composite mathematical modeling of calcium signaling behind neuronal cell death in Alzheimer's disease. BMC Syst Biol 12(Suppl 1):10

Ren SQ et al (2018) Amyloid beta causes excitation/inhibition imbalance through dopamine receptor 1-dependent disruption of fast-spiking GABAergic input in anterior cingulate cortex. Sci Rep 8(1):302

Riemer C et al (2003) Influence of the 5-HT6 receptor on acetylcholine release in the cortex: pharmacological characterization of 4-(2-bromo-6-pyrrolidin-1-ylpyridine-4-sulfonyl) phenylamine, a potent and selective 5-HT6 receptor antagonist. J Med Chem 46(7):1273–1276

Roberts PD, Conour J (2022) Mechanistic modeling as an explanatory tool for clinical treatment of chronic catatonia. Front Pharmacol 13:1025417

Roberts PD, Spiros A, Geerts H (2012) Simulations of symptomatic treatments for Alzheimer's disease: computational analysis of pathology and mechanisms of drug action. Alzheimers Res Ther 4(6):50

Roberts P, Spiros A, Geerts H (2016) A humanized clinically calibrated quantitative systems pharmacology model for hypokinetic motor symptoms in Parkinson's disease. Front Pharmacol 7:6

Rose R et al (2022) A quantitative systems pharmacology model for simulating OFF-Time in augmentation trials for Parkinson's disease: application to preladenant. J Pharmacokinet Pharmacodyn 49(6):593–606

Rosenthal J, Belfort G, Isaacson D (2016) Early treatment critical: Bexarotene reduces amyloid-Beta burden in silico. PLoS One 11(4):e0153150

Rubin JE, Terman D (2004) High frequency stimulation of the subthalamic nucleus eliminates pathological thalamic rhythmicity in a computational model. J Comput Neurosci 16(3):211–235

Sass MB et al (2009) A pragmatic approach to biochemical systems theory applied to an alpha-synuclein-based model of Parkinson's disease. J Neurosci Methods 178(2):366–377

Shannon CE (1948) A mathematical theory of communication. Bell Syst Tech J 27(3):379–423

Shao H et al (2013) A system mathematical model of a cell-cell communication network in amyotrophic lateral sclerosis. Mol BioSyst 9(3):398–406

Shen W et al (2005) Cholinergic suppression of KCNQ channel currents enhances excitability of striatal medium spiny neurons. J Neurosci 25(32):7449–7458

Sidiropoulou K et al (2009) Dopamine modulates an mGluR5-mediated depolarization underlying prefrontal persistent activity. Nat Neurosci 12(2):190–199

Siekmeier PJ, vanMaanen DP (2014) Dopaminergic contributions to hippocampal pathophysiology in schizophrenia: a computational study. Neuropsychopharmacology 39(7):1713–1721

Smolen P, Baxter DA, Byrne JH (2008) Bistable MAP kinase activity: a plausible mechanism contributing to maintenance of late long-term potentiation. Am J Physiol Cell Physiol 294(2): C503–C515

Sohn PD et al (2019) Pathogenic tau impairs axon initial segment plasticity and excitability homeostasis. Neuron 104(3):458–470 e5

Sotero RC, Trujillo-Barreto NJ (2007) Modelling the role of excitatory and inhibitory neuronal activity in the generation of the BOLD signal. NeuroImage 35(1):149–165

Spiros A, Geerts H (2012) A quantitative way to estimate clinical off-target effects for human membrane brain targets in CNS research and development. J Exp Pharmacol 4:53–61

Spiros A, Roberts P, Geerts H (2012) A quantitative systems pharmacology computer model for schizophrenia efficacy and extrapyramidal side effects. Drug Dev Res 73(4):196–213

Spiros A, Roberts P, Geerts H (2013) Phenotypic screening of the Prestwick library for treatment of Parkinson's tremor symptoms using a humanized quantitative systems pharmacology platform. J Parkinsons Dis 3(4):569–580

Spiros A, Roberts P, Geerts H (2014) A computer-based quantitative systems pharmacology model of negative symptoms in schizophrenia: exploring glycine modulation of excitation-inhibition balance. Front Pharmacol 5:229

Spiros A, Roberts P, Geerts H (2017) Semi-mechanistic computer simulation of psychotic symptoms in schizophrenia with a model of a humanized cortico-striatal-thalamocortical loop. Eur Neuropsychopharmacol 27(2):107–119

Stefanovski L et al (2019) Linking molecular pathways and large-scale computational modeling to assess candidate disease mechanisms and pharmacodynamics in Alzheimer's disease. Front Comput Neurosci 13:54

Stepanov A et al (2018) A mathematical model of multisite phosphorylation of tau protein. PLoS One 13(2):e0192519

Stojsavljevic T, Guo Y, Macaluso D (2023) Adaptive stimulations in a biophysical network model of Parkinson's disease. Int J Mol Sci 24(6):5555

Strong SP et al (1998) On the application of information theory to neural spike trains. Pac Symp Biocomput:621–632

Sykova E, Nicholson C (2008) Diffusion in brain extracellular space. Physiol Rev 88(4):1277–1340

Tamagnini F et al (2015) Intrinsic excitability changes induced by acute treatment of hippocampal CA1 pyramidal neurons with exogenous amyloid beta peptide. Hippocampus 25(7):786–797

Terman D et al (2002) Activity patterns in a model for the subthalamopallidal network of the basal ganglia. J Neurosci 22(7):2963–2976

Traub RD et al (2004) Cellular mechanisms of neuronal population oscillations in the hippocampus in vitro. Annu Rev Neurosci 27:247–278

Tseng KY, O'Donnell P (2007) D2 dopamine receptors recruit a GABA component for their attenuation of excitatory synaptic transmission in the adult rat prefrontal cortex. Synapse 61(10):843–850

van Maanen EMT et al (2018) Extending a systems model of the APP pathway: separation of beta- and gamma-secretase sequential cleavage steps of APP. J Pharmacol Exp Ther 365(3):507–518

Veronneau-Veilleux F et al (2021) An integrative model of Parkinson's disease treatment including levodopa pharmacokinetics, dopamine kinetics, basal ganglia neurotransmission and motor action throughout disease progression. J Pharmacokinet Pharmacodyn 48(1):133–148

Veselinovic T, Neuner I (2022) Progress and pitfalls in developing agents to treat neurocognitive deficits associated with schizophrenia. CNS Drugs 36(8):819–858

Wang XJ, Krystal JH (2014) Computational psychiatry. Neuron 84(3):638–654

Wang Y et al (2013) Multiple effects of beta-amyloid on single excitatory synaptic connections in the PFC. Front Cell Neurosci 7:129

Weathered C et al (2024) Microglial roles in Alzheimer's disease: an agent-based model to elucidate microglial spatiotemporal response to beta-amyloid. CPT Pharmacometrics Syst Pharmacol 13(3):449–463

Wichmann T, DeLong MR (1998) Models of basal ganglia function and pathophysiology of movement disorders. Neurosurg Clin N Am 9(2):223–236

Wieder N, Fink RH, Wegner F (2011) Exact and approximate stochastic simulation of intracellular calcium dynamics. J Biomed Biotechnol 2011:572492

Williams GV, Goldman-Rakic PS (1995) Modulation of memory fields by dopamine D1 receptors in prefrontal cortex. Nature 376(6541):572–575

Williams LM et al (2024) Precision psychiatry and research domain criteria: implications for clinical trials and future practice. CNS Spectr 29(1):26–39

Yamamoto Y et al (2017) Predicting drug concentration-time profiles in multiple CNS compartments using a comprehensive physiologically-based pharmacokinetic model. CPT Pharmacometrics Syst Pharmacol 6(11):765–777

Yousif N et al (2020) A population model of deep brain stimulation in movement disorders from circuits to cells. Front Hum Neurosci 14:55

Yuen EY, Yan Z (2009) Dopamine D4 receptors regulate AMPA receptor trafficking and glutamatergic transmission in GABAergic interneurons of prefrontal cortex. J Neurosci 29(2): 550–562

Yuraszeck TM et al (2010) Vulnerabilities in the tau network and the role of ultrasensitive points in tau pathophysiology. PLoS Comput Biol 6(11):e1000997

Zhang W et al (2002) Multiple muscarinic acetylcholine receptor subtypes modulate striatal dopamine release, as studied with M1-M5 muscarinic receptor knock-out mice. J Neurosci 22(15):6347–6352

Application of Quantitative Systems Pharmacology Approaches to Support Pediatric Labeling in Rare Diseases

Susana Zaph, Randolph J. Leiser, Mengdi Tao, Chanchala Kaddi, and Christine Xu

Contents

Abstract

Quantitative Systems Pharmacology (QSP) models offer a promising approach to extrapolate drug efficacy across different patient populations, particularly in rare diseases. Unlike conventional empirical models, QSP models can provide a

SZ, RJL, MT, CK, and CX are Sanofi employees and may hold shares and/or stock options in the company.

S. Zaph (✉) · R. J. Leiser · M. Tao · C. Kaddi
Translational Disease Modeling – Translational Medicine, Research & Development, Sanofi-US, Bridgewater, NJ, USA
e-mail: Susana.zaph@sanofi.com

C. Xu
Pharmacokinetics, Dynamics, Metabolism – Translational Medicine, Research & Development, Sanofi-US, Bridgewater, NJ, USA

E. L. Bradshaw, M. E. Spilker (eds.), *Quantitative Systems Pharmacology*,
Handbook of Experimental Pharmacology 289, https://doi.org/10.1007/164_2024_734

mechanistic understanding of disease progression and therapeutic response by incorporating current disease knowledge into the descriptions of biomarkers and clinical endpoints. This allows for a holistic representation of the disease and drug response. The mechanistic nature of QSP models is well suited to pediatric extrapolation concepts, providing a quantitative method to assess disease and drug response similarity between adults and pediatric patients. The application of a QSP-based assessment of the disease and drug similarity in adult and pediatric patients in the clinical development program of olipudase alfa, a treatment for Acid Sphingomyelinase Deficiency (ASMD), illustrates the potential of this approach.

Keywords

Efficacy extrapolation · Model-informed drug development (MIDD) · Pediatric drug development · Quantitative systems pharmacology (QSP) · Rare drug development · Virtual twins

1 Drug Development Challenges in Rare Diseases

Efficacious treatments for rare diseases (defined in the United States as diseases affecting less than 200,000 patients) remain a large unmet medical need, with the vast majority of the estimated 7,000 known rare diseases still lacking approved treatments (Congress 2002; FDA 2019a). These diseases tend to have an underlying genetic etiology, with a chronic and often fatal disease progression that is diagnosed in childhood.

Accelerating drug development of efficacious treatments for rare diseases can be life-changing for patients (Miyamoto and Kakkis 2011; Kaufmann et al. 2018; Pearson et al. 2022). However, there are a number of challenges in the development of rare disease treatments, including low patient numbers, a geographically dispersed patient population, and high phenotypic heterogeneity. These factors may hinder the execution of traditionally designed clinical trials and complicate meeting regulatory requirements for drug approval (Augustine et al. 2013; Sun et al. 2017; Wakap et al. 2020; Shah et al. 2021; Adachi et al. 2023). For example, the phenotypic heterogeneity in disease burden and progression can cause large variations in drug response making outcomes of clinical trials difficult to interpret. This can be further compounded by long recruitment periods and low numbers of patients recruited leading to poorly powered clinical studies.

While randomized, double-blind, placebo-controlled studies are considered the gold standard in clinical research, they may be difficult to execute in rare diseases. Practical and ethical considerations often make rare disease clinical trial designs with placebo-controlled arms unfeasible (Bell and Smith 2014). Patients may be reluctant to participate in placebo-controlled trials due to the possibility of non-treatment. In indications where efficacious treatment options are already available, enrollment in a placebo-controlled trial may not be in the patients' best interest, as clinical trial inclusion criteria may require patients assigned to the placebo arm to discontinue other treatments for the duration of the trial or in some instances include a washout period before the initiation of the trial. Furthermore, ethical concerns arise in severe or rapidly progressive diseases, due to greater risk of harm for patients who are assigned to the non-treatment arm. All these challenges are magnified in pediatric drug development in rare indications, where additional ethical concerns need to be considered given the vulnerable nature of the pediatric population (Welch et al. 2015; Rose 2019). One of the hardest ethical challenges in pediatric research is balancing the potential clinical benefit from an investigational drug against the harm and risks, related not only to drug-induced safety concerns, but also procedures during the clinical trial that could cause pain or suffering to pediatric patients. In addition, enrollment of patients across the pediatric age continuum is often required, and such patient recruitment criteria in rare indications are at times unfeasible due to the rarity and geographic dispersion of the population.

Regulatory agencies have provided some guidelines and workshops to formalize their positions and assist sponsors in drug development for rare indications (FDA 2019b). Some of the recommendations include the use of model-informed drug development (MIDD) to optimize drug development (EMA 2006; FDA 2014; FDA 2017; FDA 2019b; FDA and M-CERSI 2023). MIDD can play a critical role in supporting various stages of drug development, aiding in robust decision-making, optimizing clinical trial design, de-risking uncertainty, and shortening clinical trial duration (Milligan et al. 2013; Marshall et al. 2019; Madabushi et al. 2022). MIDD approaches can even contribute significantly to the totality of evidence provided to regulatory agencies for drug approvals (Bi et al. 2019; Wang et al. 2019; Fan et al. 2020; Xiong et al. 2022).

2 Pediatric Extrapolation of Efficacy in Rare Diseases

Given the high incidence of pediatric onset and challenges in executing traditional trial designs in rare indications, there is great interest in extrapolating treatment efficacy from well-controlled trials in adult patients to pediatric patients. Extrapolation, defined as inferring or predicting unknown information from trends in the known data, plays a crucial role in pediatric drug development by leveraging efficacy data from adult trials to support the pediatric efficacy of new drugs (Dunne et al. 2011). Pediatric extrapolations of efficacy from adult trials or other pediatric studies (such as from adolescents to younger children) can bridge knowledge gaps with data to optimize pediatric drug development to ensure the use of safe and efficacious drugs in children while minimizing the need for extensive pediatric clinical trials, leading to an increase the number of drug approvals for pediatric use (Dunne et al. 2011).

The International Council for Harmonization (ICH) E11 guidance provides an outline of critical issues in pediatric drug development to facilitate the development of medicinal products for pediatric use (FDA 2018) and promotes the harmonization of pediatric extrapolation methodology to support the development of pediatric drugs. The pediatric extrapolation framework encompasses an iterative process for understanding the existing information available on the pediatric population and the knowledge gaps needed to inform pediatric drug development.

Pediatric extrapolation requires the evaluation of two key aspects when comparing adult and pediatric patients: 1) similarity of disease; and 2) similarity of response to treatment (FDA 2018; EMA 2018; FDA 2018). Evaluating disease similarity includes comparing the disease pathophysiology, the clinical presentation, and identifying the commonalities of disease progression. This also includes an examination of whether there is a shared etiology of the disease, diagnostic criteria, or any age of onset differences. Other factors to consider include whether there are subtypes (based on severity, genetics, etc.) occurring in either population or how comparable the distributions of these disease subtypes are in both populations.

The evaluation of treatment response similarity includes assessing the similarity of the mechanism of action and pharmacology of the drug, and treatment guidelines in the two populations. Potential differences in pharmacokinetic (PK) properties (absorption, distribution, metabolism, and excretion) are to be considered, with body size, maturation and ontogeny of pathways, and other relevant factors that could impact pharmacokinetic properties. Evaluating the similarity of exposure-response relationships is key for extrapolation. Frequently pediatric dose selection is informed by selecting dosing regimens that achieve pediatric exposures "matched" to those observed in adults. The extrapolation assumption of a similar exposure-response relationship in pediatric and adult patients infers that the same exposure in adult and pediatric patients leads to the same efficacy response. However, there are factors other than drug exposure that impact efficacy, may differ between the adult and pediatric patient population (e.g., differences in drug target due to age or disease burden).

A pediatric extrapolation plan needs to address the similarity of the two key aspects of the extrapolation concept, outlining the pre-existing clinical evidence, and describing assumptions, knowledge gaps, and uncertainties. The plan should also list what studies will be conducted and how the data collected will contribute to inform the identified knowledge gaps, specifying how much data will need to be generated from the reference (adult) population and the target (pediatric) population. In the latest ICH guidelines, a nuanced approach is suggested for assessing the assumptions of similarity between adult and pediatric populations in terms of disease and drug response. The guidelines acknowledge that a binary "yes or no" answer may not suffice. The degree of similarity can vary, and while there may not be a complete overlap in disease and treatment response between pediatric and adult patients, the objective is to ascertain if the differences are significant enough to rule out pediatric extrapolation. The extrapolation plan should accommodate uncertainties in the supporting data for these similarity assumptions. This can be achieved by incorporating the input of clinicians and assessing the acceptable levels of uncertainty and risk for pediatric patients (ICH 2024).

Modeling and simulation (M&S) can provide a systematic method to quantify the degree of similarity in disease, drug response, and exposure-response relationship between adults and pediatrics, maximizing the use of pre-existing information and data collected (EMA 2018). Several M&S techniques that integrate available data have been widely used to inform decision-making in pediatric drug development. These M&S techniques include but are not limited to population PK and pharmacodynamic (PD) models (popPK or popPK-PD), physiologically based pharmacokinetic (PBPK) models, and quantitative systems pharmacology (QSP) models. These have gained recognition and acceptance in MIDD by both the pharmaceutical industry and regulatory agencies and have been utilized to guide study design, dose selection, and authorization of pediatric drugs (Barrett et al. 2008; Barrett et al. 2012; Vinks and Barrett 2021; Azer and Barrett 2022; Chan et al. 2022; Madabushi et al. 2022). FDA Advisory Committee for Pharmaceutical Science and Clinical Pharmacology stated that M&S should be considered for all pediatric drug development programs, supporting the value these approaches can bring (FDA 2012). When taking into account factors such as physiological maturation, organ function, and body weight in adult and pediatric populations, the model-based extrapolation approach often involves scaling adult PK and/or PD parameters to pediatric populations based on age, weight, and/or other relevant factors. The scaled models are used to predict PK, PD, and efficacy of drugs in pediatric patients as well as identify important gaps in knowledge to inform what additional data are required (Azer and Barrett 2022; Barrett et al. 2008, 2012; Madabushi et al. 2022; Vinks and Barrett 2021).

3 QSP Approaches as Tools to Quantify Disease and Drug Response Similarity Can Support Extrapolation of Efficacy in Rare Indications

QSP models hold great promise to support the extrapolation of efficacy to different patient populations, especially in rare diseases. Empirical-based M&S approaches mathematically describe drug exposure and the trajectories of biomarkers and clinical endpoints from a defined dataset without considering a biological mechanism. This is unlike QSP models, which provide a basis for understanding and predicting disease progression and therapeutic response by incorporating the current knowledge of the disease into the descriptions of biomarkers and clinical endpoint trajectories. To mechanistically describe the disease, QSP approaches contain mathematical descriptions of key pathophysiological processes, often spanning multiple scales of biology from signaling pathway dysregulation to organ-level dysfunction. QSP approaches also link a drug's mechanism of action (MoA) to the appropriate target reactions at the molecular scale, allowing the propagation of the drug efficacy to pathophysiological processes to simulate therapeutic responses of biomarkers and clinical endpoints. The value of this approach is its ability to connect molecular processes to biomarkers and clinical endpoints at the organ or organismal level to provide a holistic representation of the disease and its response. Thus, the bottom-up nature of QSP models permits the simulation of untreated and treated states of the disease in the target patient population to support various aspects of drug development such as drug target predictions, drug candidate prioritization, biomarker justification, dose optimization, exploration of different dosing regimens, and efficacy predictions in patients subpopulations considering age, severity, disease heterogeneity and other factors that may affect the exposure-response relationship (Bradshaw et al. 2019). QSP approaches have been even applied to predict and understand the risks of adverse events from therapeutics (Varshneya et al. 2021).

QSP models are informed by diverse data sets, beyond what is traditionally collected during a clinical drug development program (such as PK, PD, and clinical endpoint time courses for each patient). QSP models often consider prior knowledge already available in the public domain and combine it with clinical data from the drug candidate (Musante et al. 2017; Bradshaw et al. 2019). This includes integrating knowledge on anatomy, physiology, biochemistry, and pharmacology with preclinical and clinical datasets relevant to the drug candidate to inform key features of the disease and drug response. For instance, preclinical datasets (ranging from in vitro assays to animal disease models) can be used to inform the different biological scales of the disease representation, such as receptor activation that may be inhibited by a drug as measured in a cell-based assay. In fact, preclinical data are at times essential to establish the underlying contribution of pathways (Ekerot et al. 2013; Gotta et al. 2016; Snelder et al. 2016), especially if detailed biochemical descriptions of the disease biology are in scope for the model to address the modeling question of interest.

Natural history datasets and clinical datasets from other drugs in the same indication can be incorporated into the modeling effort due to the mechanistic nature of QSP (Peterson and Riggs 2010; Kaddi et al. 2018; Riggs and Cremers 2019; Lin et al. 2022). All these datasets are combined within the context of the current understanding of the disease and the pharmacology/mode of action of the drug to develop a cohesive, integrative view of the disease and possible response in virtual patients. The level of granularity required to describe the dynamics of biological processes that give rise to the model's biomarker and/or endpoint temporal changes is defined by the type of question intended for the modeling effort.

The QSP approach is compatible with the pediatric extrapolation concept as it provides a quantitative approach to quantify disease and drug response similarity between adults and pediatric patients. The mechanistic feature inherent to QSP models enables the extrapolation of efficacy predictions beyond the data used to construct the model. In some instances, this capability can support the virtual replacement of clinical trials in vulnerable populations where traditional trials may not be feasible. QSP models used for extrapolation of efficacy leverage the concept that similarity in clinical outputs results from similarity in underlying pathophysiology processes (i.e., disease similarity), and drug effects on these processes. The significance of PK/PD modeling in drug development is widely acknowledged, with its primary objective being to describe the relationship between PK and selected PD endpoints while considering some of the factors that could affect the relationship.

However, due to their empirical nature, PK/PD models may have limited capacity to extrapolate beyond the data sets used during the modeling effort.

QSP models contain quantitative descriptions of biological processes and interpretable parameters, such as the descriptions of receptor binding dynamics, or enzyme activation and their associated binding affinities or kinetic parameters. If a QSP model intended to describe adult clinical data can describe with similar accuracy pediatric clinical data, then it can be inferred that the disease presentation and drug response is mechanistically similar in both populations. This is achieved by calibrating the QSP model to the biomarker and clinical endpoint profiles of individual adult reference patients to establish the expected degree of variability in key pathophysiological processes and their associated parameters to accurately capture the observed adult data. The variation observed for the same processes and parameters in pediatric patients can then be evaluated against this benchmark, by calibrating the model against the pediatric dataset and comparing the adult model descriptions to the pediatric model descriptions. Thus, the QSP model can be used to mechanistically compare the exposure and response relationships in these populations and support pediatric extrapolation. The following section describes an example where a QSP-based assessment of the similarity of disease and treatment response provided a valuable tool in pediatric drug development in a rare disease.

4 Olipudase Alfa: The First Disease-Specific Treatment for ASMD

Acid sphingomyelinase deficiency (ASMD, historically known as Niemann-Pick disease, type A, B, and A/B), is a rare, serious, life-threatening lysosomal storage disease stemming from insufficient activity of the lysosomal hydrolase acid sphingomyelinase (ASM). ASMD is caused by recessively inherited mutations in the sphingomyelin phosphodiesterase 1 (*SMPD1*) gene, encoding the enzyme ASM, which catabolizes sphingomyelin into ceramide.

Deficiency of ASM results in sphingomyelin accumulation in macrophages residing in various organs and hepatocytes. The progressive accumulation of sphingomyelin, and other lipids, gives cells a foamy appearance histologically (Thurberg et al. 2012). Infiltration of the resident tissue foam cells is responsible for tissue damage and resultant organ dysfunction in patients with ASMD. In the liver, abnormal sphingomyelin accumulation can lead to liver dysfunction and eventual liver failure. Because macrophages actively internalize cellular debris, such as sphingomyelin-rich membranes of damaged red blood cells in circulation, the spleen is involved, resulting in splenomegaly, which may be massive in some patients (McGovern et al. 2017). The lung is vulnerable as alveolar macrophages fill with sphingomyelin, leading to disruption of the transfer of oxygen from blood (Guillemot et al. 2007; Martins et al. 2022). This impairment results in decreased measures of pulmonary function, such as the diffusing capacity of the lung for carbon monoxide (reported as % predicted DL_{CO}).

ASMD is a progressive genetic disease that, depending on the degree of enzyme deficiency, may be severe and fatal early in life or result in other manifestations as the patients age. Thus, the ASMD phenotypes constitute a continuous clinical spectrum involving variable ages of onset, clinical manifestations, disease severity, and rates of progression. Different phenotypic subtypes have been observed in the ASMD population, ranging from severe, infantile, acute rapidly progressive neuronopathic phenotypes resulting in death in early childhood (McGovern et al. 2006) to more heterogeneous phenotypes of a chronic visceral non-neuronopathic form with later symptom onset and accompanied by a longer lifespan (Wasserstein et al. 2004).

Olipudase alfa (Xenpozyme®), a recombinant form of human acid sphingomyelinase (rhASM), was developed as an enzyme replacement therapy (ERT). ERT is an etiology-specific approach for treating lysosomal storage disorders like ASMD by supplementing the deficient endogenous ASM enzyme to prevent the toxic accumulation of sphingomyelin in tissues. This is expected to reverse many of the non-neurological disease manifestations, reduce disease morbidity, and increase survival. The clinical development program for olipudase alfa in ASMD included five olipudase alfa clinical trials and five non-interventional natural history studies. Among the five trials, one was a phase I, single-dose escalation study. The efficacy of olipudase alfa for the treatment of non–central nervous system manifestations of (ASMD) was established with remaining four clinical trials, which were multiple dose studies: 1) an open-label, multicenter ascending dose study (DFI13412); 2) a single-arm open-label study (DFI13803) in pediatric patients; 3) a long-term extension trial (including participants from DFI13412 and DFI13803); and 4) a multicenter, randomized, double-blinded, placebo-controlled, repeat-dose phase II/III trial (DFI12712) in adult patients with ASMD. The two primary efficacy endpoints were % predicted diffusion capacity for carbon monoxide (% predicted DL_{CO}) and spleen volume in combination with the splenomegaly-related score. The adequate and well-controlled pivotal study (DFI12712) in 31 adults with ASMD showed a clinically meaningful and nominally statistically significant improvement in lung function and spleen size reduction for patients randomized to olipudase alfa compared to those treated with placebo at week 52. In pediatric patients with ASMD from a single-arm open-label study, treatment with olipudase alfa resulted in similar improvements in lung function, and spleen volume at week 52 as compared to baseline. The effectiveness of olipudase alfa in pediatric patients was based on this dataset and complemented with the framework of partial extrapolation.

5 QSP-Based Quantification of Disease and Olipudase Alfa Response Similarity in Pediatric and Adult ASMD Patients

A QSP-based analysis was part of the M&S package that included complementary empirical approaches such as popPK-PD and exposure-response analysis to support the partial pediatric extrapolation of olipudase alfa efficacy from adult to pediatric

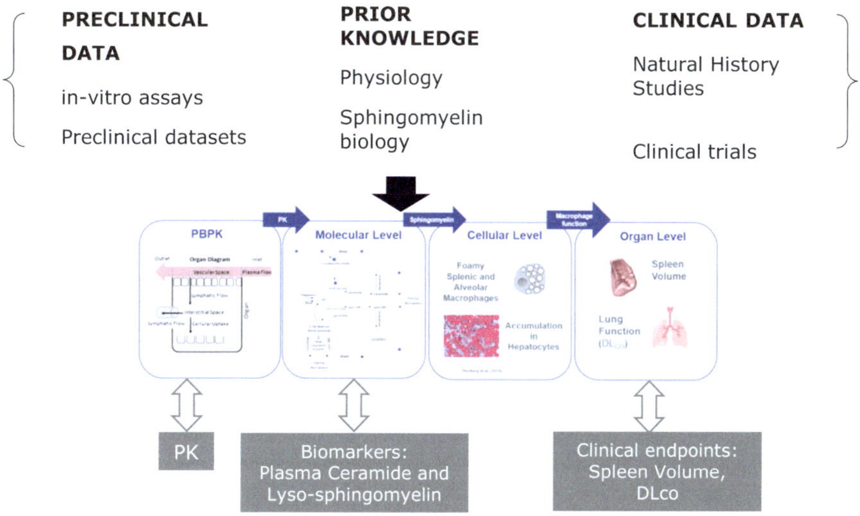

Fig. 1 Overview of the structure of the QSP model for ASMD and the response to olipudase alfa. Model was informed by diverse datasets, along with the PK, biomarker, and clinical endpoint datasets from the olipudase alfa clinical program. DLco, diffusing capacity of the lung for carbon monoxide; PBPK, physiologically based pharmacokinetics

ASMD populations. The QSP-based analysis was used to quantify the mechanistic similarity of disease and response in adult and pediatric ASMD patients. The ASMD QSP model was previously developed (Kaddi et al. 2018) as an ordinary differential equation-based model comprised of four sub-models that cover the multiple biological scales of ASMD and olipudase alfa action (Fig. 1). This includes a PK sub-model, a molecular-level sub-model, a cellular-level sub-model, and an organ-level sub-model. The model describes two biomarkers and two clinical endpoints that are utilized for the clinical assessment of ASMD disease severity and the response to olipudase alfa. Plasma ceramide and plasma lyso-sphingomyelin (lyso-sphingomyelin) are biomarkers described as outputs of the molecular level sub-model. Spleen volume and % predicted DLco are clinical endpoints described as outputs of the organ-level sub-model. The scope of the model was parsimonious and fit for purpose, including descriptions of those key biological processes required to represent the enzymatic deficiency that drives the accumulation of sphingomyelin in tissue, and the observed plasma biomarkers and clinical endpoints in ASMD patients.

A reduced-order PBPK model based on (Shah and Betts 2012) was used to describe the PK for olipudase alfa in this framework. Nonclinical biodistribution data as well as clinical exposure data were used to calibrate this sub-model. The lysosomal concentration of olipudase alfa predicted by the PK sub-model served as input to the molecular portion of the QSP model. In addition, age-appropriate physiological descriptions were also included such as blood flows, interstitial

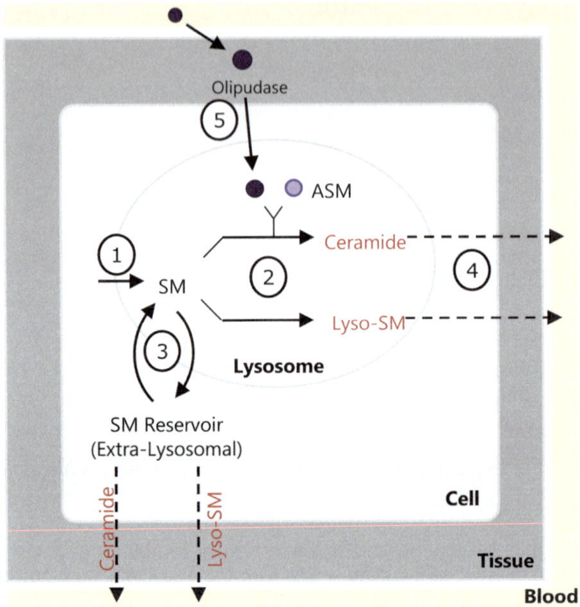

Fig. 2 Molecular sub-model diagram of ASMD QSP model. The sub-model includes key steps in the dynamics of intracellular sphingomyelin (SM) in ASMD, such as (1) the uptake of SM into the lysosome; (2) a lysosomal sphingomyelin fraction is converted into ceramide by olipudase alfa and residual endogenous ASM. The endogenous ASM activity is dependent on the parameter REA. The lysosomal production of lyso-sphyngomyelin from sphingomyelin is also included; (3). A "reservoir" fraction (extra-lysosomal) is metabolized by other enzymes besides ASM in the intracellular compartment. The "reservoir" sphingomyelin is not accessible to olipudase alfa due to lysosomal targeting. Sphingomyelin can be exchanged between the two fractions. (4) Efflux of ceramide and lyso-sphyngomyelin into plasma; (5) Olipudase alfa is targeted to the lysosome, where it supplements the endogenous ASM activity. SM = sphingomyelin, lyso-SM lyso-sphingomyelin

space, and vascular space along with weight-based scaling of organs, plasma, interstitial, and vascular space volumes (Leggett and Williams 1995; Valentin 2002; Edginton et al. 2006).

The molecular sub-model contains the representation of the enzymatic conversion by ASM of sphingomyelin to ceramide within the lysosome (Fig. 2). This is the key reaction that is disrupted in ASMD and restored through the supplementation of olipudase alfa, allowing the accurate representation of its mode of action. Plasma ceramide, the product of the breakdown of tissue sphingomyelin by olipudase alfa, was also used to refine the predicted lysosomal concentration of olipudase alfa. The degree of enzyme deficiency in ASM is encoded in the model by the parameter REA (residual enzymatic activity), with a value of 100% representing full healthy activity, and a value of 30% or less representing the ASMD state (Vanier et al. 1993; Pavlů-Pereira et al. 2005). The molecular sub-model contains a representation of the conversion of sphingomyelin to lyso-sphingomyelin, which is a biomarker of

ASMD (Chuang et al. 2014), and the export of the generated lyso-sphingomyelin into the plasma compartment. The model output describes the decrease in plasma lyso-sphingomyelin throughout treatment with olipudase alfa. The predicted accumulated sphingomyelin in the cells of each organ (splenic macrophages and alveolar macrophages) serves as the input into the cellular and organ sub-models. The spleen sub-model was updated as described in Leiser et al. (2023).

The QSP analysis entailed evaluating the similarity of baseline and treatment response of biomarkers (plasma ceramide and lyso-sphingomyelin) and clinical endpoints (spleen volume and % predicted DLco), between adult and pediatric patients by:

1. Developing virtual twins for adult and pediatric ASMD patients enrolled in the olipudase alfa clinical program. Virtual twins are fits or calibrations of the model to each patient's set of data using a small subset of the model parameters. Each patient's PK, biomarker, and endpoint profile is represented using the same QSP model structure employing a subset of parameters that are personalized or patient-specific. This allows for the accounting of each patient's intrinsic variability in their disease presentation.
2. Comparing pediatric vs adult virtual twins against observed data. This evaluation scrutinizes the ability of virtual twins to accurately represent the observed PK, PD, and clinical endpoints of their corresponding pediatric or adult patients. Ideally, the accuracy remains consistent across all age groups, indicating that the model does not favor pediatric data over adult data, or vice versa.
3. Quantifying how similar parameter distributions from adult virtual twins are to those from pediatric virtual twins. The similarity in these parameters between adult and pediatric patient distributions supports the mechanistic similarity of the biological processes impacted by those parameters.
4. Performing parameter sensitivity analysis for virtual twins to identify and compare the impact of each parameter on the baseline and olipudase alfa treatment response in adult and pediatric ASMD patients.

A summary of the steps in the analysis is shown in Fig. 3 and is mentioned briefly in the subsequent section. Further details can be found in Leiser et al. (2023).

6 Generation of QSP-Based Virtual Twins

Twelve model parameters spanning all sub-models were used to generate each virtual twin along with their demographic characteristics. The parameters to construct each virtual twin were identified based on several criteria that included: 1) parameter sensitivity analysis ranking, 2) biological knowledge as a source of variability, and 3) availability of literature reference values for the base parameter value. Parameters with high sensitivity with known sources of biological variability, (such as the REA parameter controlling the degree of ASM activity) or those with empirically derived values, (such as some of the parameters that describe spleen

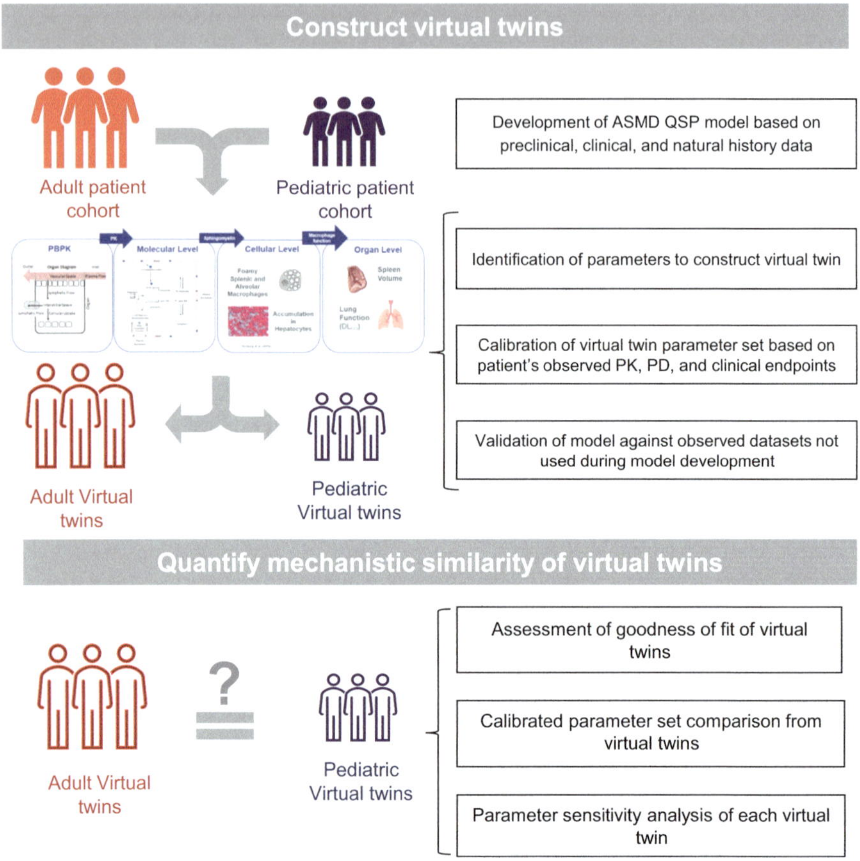

Fig. 3 Overview of QSP-based assessment of mechanistic similarity between adult and pediatric ASMD patients

volume dynamics) were included to capture the individual-level biomarker and endpoint data. All virtual twins' calibrations (or estimations) were performed in a similar manner so that cost functions and initial parameter bounds (confined to appropriate biologically relevant ranges) were kept constant for all pediatric and adult patients.

The calibrations of each of the virtual twins were performed in a sequential manner, beginning in the most mechanistically rich sub-models (i.e., PBPK and molecular sub-models). The PBPK sub-model parameter was calibrated based on the observed plasma exposure. The molecular sub-model was calibrated using the observed plasma biomarkers, defining the REA parameter and parameters controlling ceramide and lyso-sphingomyelin dynamics. The organ-level parameters were calibrated last based on the observed clinical endpoint trajectories. This type of sequential sub-model calibration allowed more empirical representations of the organ sub-models to be grounded by the mechanistic features of the molecular sub-model, increasing the identifiability of the organ sub-model parameters and reducing the size of the parameter space searched during the optimization procedure, resulting in more accurate fits and identifiable parameters (Cheng et al. 2017).

7 QSP Model Validation

An essential step in preparation for using the model to quantify similarity of disease and therapeutic response in these two populations was model validation. Since ASMD is a monogenic disease, with its pathology driven by the single event of ASM enzyme deficiency, datasets from healthy individuals could be used for validation purposes by comparing them to a virtual population without enzyme deficiency. A virtual population of healthy subjects was constructed based on the adult virtual twins, by sampling the covariance matrix of the virtual twin parameter sets. The resulting virtual population (n = 1,000) was then modified to represent healthy individuals by increasing the REA parameter to 100%, normalizing the enzymatic activity, and the predicted untreated plasma ceramide and lyso-sphingomyelin levels were compared against reported values observed in healthy individuals (der Vlugt et al. 2008; Kuchar et al. 2017), which demonstrated adequate agreement. Additional validation exercises included:

1. Olipudase alfa-mediated liver sphingomyelin reductions. The sphingomyelin dynamics in the liver from virtual twins were simulated and compared to their corresponding observed sphingomyelin measurements from liver biopsies.
2. Biomarker and clinical endpoint validations in the adult virtual patient population. Virtual populations for adult patients were constructed based on the adult virtual twins-derived parameter sets, and then compared to the clinical data from the long-term extension period not used during model development.
3. Biomarker and clinical endpoint validations in a pediatric virtual patient population. A bootstrapping approach was utilized to generate a small (n = 10) "personalized" virtual population for each pediatric patient due to a limited number of pediatric patients. The initial virtual population for each pediatric patient was generated without their respective virtual twin's parameter set contributing to the covariance matrix that parameters were sampled from. The resulting "personalized" virtual population was then compared to the observed data of the patient left out.

The validations conducted provided confidence in the model's performance representing each biological scale with adequate quality. Furthermore, these validations supported the suitability of the QSP-based analysis as a tool for assessing mechanistic similarity.

8 QSP-Based Quantification of Disease Similarity and Response to Olipudase Alfa

All virtual twins showed reasonable fits to their corresponding observed clinical data, suggesting that the model structure and parameter values were adequate to represent the disease presentation and olipudase alfa response in pediatric and adult patients. Example virtual twins for an adult and pediatric patient are shown in Fig. 4. The QSP model described the PK profiles of olipudase alfa, PD profiles of plasma levels of ceramide and lyso-sphingomyelin, spleen volume, and percentage predicted DLco after olipudase alfa treatment.

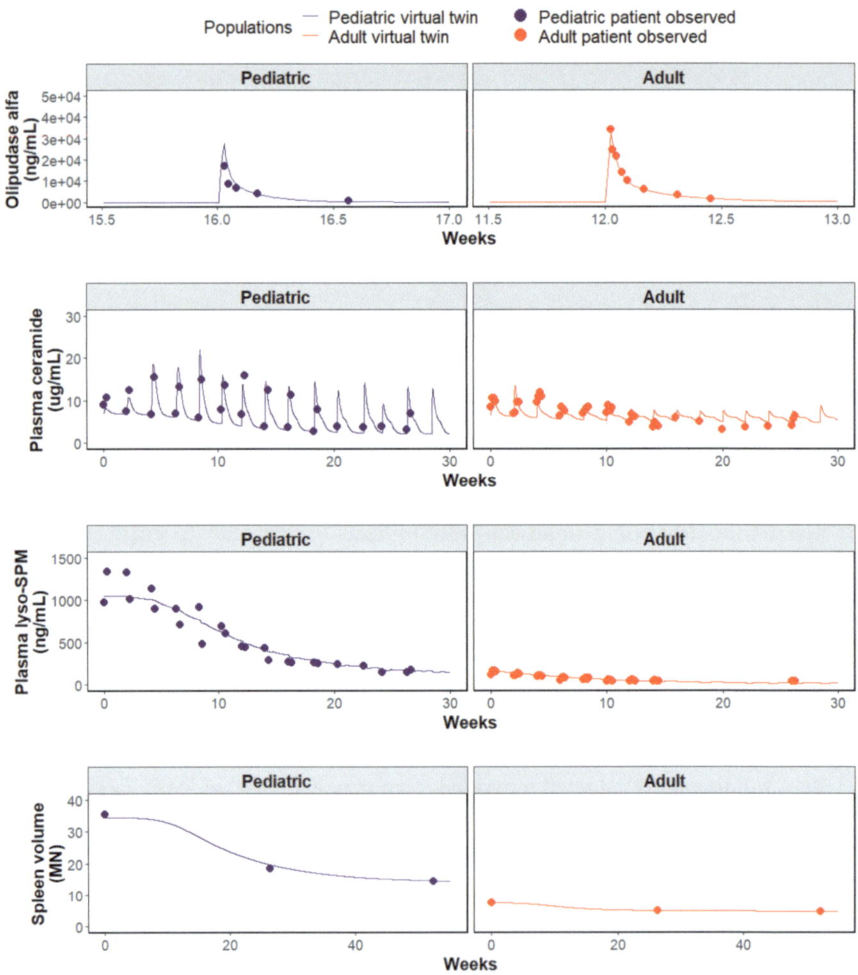

Fig. 4 Virtual twins of PK, PD, and clinical endpoints in a representative patient of each age cohort

Table 1 Criteria and metrics for determining similarity of disease and response between pediatric and adult patients

Criteria	Reasoning	Metrics
Virtual twin quality	*Same model structure* The same mathematical representation of key pathophysiology is used to construct both pediatric and adult virtual twins	• Visual checks of individual patient fits • Visual checks goodness of fit plots • Visual checks of bias plots
	Similar accuracy Biology represented in the QSP model describes observed pediatric and adult patient data in a similar manner	• Pearson correlation coefficient • Absolute average fold error • Normalized root mean squared error • Percent predicted within two standard deviations of the observed mean
Virtual twin parameter values	Parameters that describe biologically interpretable processes are similar in magnitude in pediatric and adult virtual twins	• Two sample Kolmogorov-Smirnov test • 95% confidence interval for group-wise medians
Virtual twin parameters sensitivity	The influence of each key parameter on an endpoint or biomarker is similar in pediatric or adult virtual twins	• Local sensitivity index

A list of the criteria and related metrics to assess the similarity of disease and treatment response to olipudase alfa between pediatric and adult patients was developed and listed in Table 1. Further details can be found in Leiser et al. (2023).

Each virtual twin was compared to their respective observed data using goodness-of-fit plots and visual predictive checks, supporting that the model structure used for all virtual twins was able to represent the observed biomarker and endpoint profiles of all patients adequately. The quality of the virtual twins was assessed using the metrics listed in Table 1, and it was concluded that the virtual twins recapitulated observed clinical patient data with similar accuracy and error, irrespective of age or severity.

The parameter value sets from pediatric and adult virtual twins were compared based on a series of criteria metrics applied (see Table 1). Based on these criteria, most parameter values were similar between the two populations, supporting mechanistic similarity across age cohorts. The only clinically meaningful parameter value distribution that showed a difference between adult and pediatric populations was the parameter used to represent the degree of enzyme deficiency in each virtual twin, the residual enzyme activity (REA) parameter. This discrepancy suggests differences in the degree of disease severity in the two populations, with pediatric virtual twins exhibiting parameter values that tended to be lower (less residual enzyme activity or more enzymatic deficiency), indicating more severe disease presentation. Parameter sensitivities were also qualitatively similar in both populations, suggesting that the influence of each parameter on the biomarkers or clinical endpoint behavior was similar irrespective of age. The totality of this analysis provided supportive evidence that the pathophysiology of ASMD and the response to olipudase alfa are mechanistically similar between adult and pediatric patients.

9 Impact of QSP-Based Analysis on Olipudase Alfa Clinical Development Program

Biologics License Application (BLA) for olipudase alfa was approved on August 31, 2022, for the treatment of non-central nervous system manifestations of ASMD in adult patients and pediatric patients. Substantial evidence for olipudase alfa effectiveness in ASMD patients was established with one adequate and well-controlled trial (DFI12712) with confirmatory evidence. Confirmatory evidence providing strong mechanistic support includes the well-established etiology of the disease, the mechanism of action of olipudase alfa, and pharmacodynamic biomarker data showing reductions in plasma ceramide and lyso-sphingomyelin, metabolites of sphingomyelin. Evidence of effectiveness for pediatric patients was based on the comparable efficacy results (change from baseline) between adult and pediatric patients observed in an open-label, single-arm study of pediatric patients (type B, type A/B). The QSP analysis supported partial extrapolation from the adequate and well-controlled trial in adults, given the similar pathogenesis of ASMD between pediatric and adult patients and the shared mechanism of action of olipudase alfa.

The QSP model of ASMD, as part of an overall M&S strategy, enabled partial pediatric extrapolation: QSP-based virtual twins captured each patient's observed biomarker and endpoint profile, their demographic data, and the intrinsic variability in disease severity, all connected to a unified representation of ASMD. The adult virtual twins were established to benchmark the expected dynamics of these parameters and processes and then compared to those derived from the pediatric virtual twins. Applying the same model structure, individual model fits to each patient showed comparable value ranges for parameters that control key disease processes across age and disease severity. Consistent parameter sensitivities were also identified in both pediatric and adult patients.

In summary, the QSP analysis results provided mechanistic insight into ASMD and suggested that there are no distinct patient sub-populations defined by age, but a continuum of disease presentations due to variability in disease severity. Applying the QSP model of ASMD in the olipudase alfa pediatric investigation plan informed drug development decision-making, supported disease and treatment response similarity assessment, validated extrapolation assumptions as well as facilitated the regulatory assessments and pediatric approval.

10 Conclusion and Perspectives

The QSP approach can leverage the current mechanistic understanding of rare diseases in adult and pediatric patients to quantify disease similarity by benchmarking the biomarker and endpoint profile from a reference population against the target population. These mechanistic M&S approaches could play an important role in rare diseases and pediatric drug development where traditional randomized controlled trial designs have been considered unfeasible or unethical to enroll a placebo group for demonstration of efficacy.

The quantification of the similarity of disease and drug response is a crucial component of the pediatric extrapolation concept. A QSP-based analysis can be applied to facilitate the comparison of the similarity of disease and drug response in adult and pediatric patients and provides the basis for leveraging the well-controlled studies in adult patients to support pediatric drug development, making similarity assessments between these two populations more data-driven. However, this approach has requirements that may preclude its broad application in pediatric drug development. Application of QSP-based approaches for pediatric extrapolation purposes requires a clear understanding of the etiology of the disease, connecting the known pathophysiology and a biologically meaningful description of the drug's MoA to observed biomarkers and clinical endpoints to create virtual twins. Large ambiguity or knowledge gaps that preclude the mechanistic description of the disease or the observed response would limit this type of QSP-based assessment. A QSP-model-based assessment of similarity of disease could be challenging to perform for multifactorial diseases, where the interplay of multiple genes and environmental factors influence the disease phenotype and progression and these factors may not be documented for each patient. In addition, the lack of quantitative data that can inform all the biological scales needed to represent the disease can preclude a robust QSP-based analysis or limit predictive performance. Overfitting of noise in the data is a constant risk given the complexity of QSP models, thus a fit-for-purpose and parsimonious modeling approach to describing the pathophysiology of interest is needed. Striking the right balance between including the features of disease pathophysiology for the desired model application while maintaining a focused scope is key. The model must also be able to appropriately capture the observed heterogeneity in the patient populations and develop a quantitative understanding of the relationship between disease severity and this heterogeneity (Bai et al. 2019; Cucurull-Sanchez 2024).

Despite these challenges, QSP approaches hold promise to impact pediatric drug development and expedite approvals for rare disease treatments. The potential for these models to be ubiquitously integrated into the drug development pipeline is considerable. Yet, there is a pressing need for stringent criteria and measurable metrics, comparable to those in established modeling approaches, for evaluating the performance and quality of QSP models (Androulakis 2022). The comprehensive implementation of such metrics could enhance trust in QSP-centric analyses, thereby fostering wider acceptance among stakeholders and regulatory bodies.

Acknowledgments The authors thank Thomas Klabunde, Monica Kuman, Kelly George, and the Sanofi teams working on ASMD for their support. The authors thank the editors and the reviewers for their thoughtful input.

References

Adachi T, El-Hattab AW, Jain R, Crespo KAN, Lazo CIQ, Scarpa M, Summar M, Wattanasirichaigoon D (2023) Enhancing equitable access to rare disease diagnosis and treatment around the world: a review of evidence, policies, and challenges. Int J Environ Res Public Health 20(6):4732. https://doi.org/10.3390/ijerph20064732

Androulakis IP (2022) Towards a comprehensive assessment of QSP models: what would it take? J Pharmacokinet Pharmacodyn:1–11. https://doi.org/10.1007/s10928-022-09820-0

Augustine EF, Adams HR, Mink JW (2013) Clinical trials in rare disease. J Child Neurol 28(9): 1142–1150. https://doi.org/10.1177/0883073813495959

Azer K, Barrett JS (2022) Quantitative system pharmacology as a legitimate approach to examine extrapolation strategies used to support pediatric drug development. CPT Pharmacometrics Syst Pharmacol 11(7):797–804. https://doi.org/10.1002/psp4.12801

Bai JPF, Earp JC, Pillai VC (2019) Translational quantitative systems pharmacology in drug development: from current landscape to good practices. AAPS J 21(4):72. https://doi.org/10.1208/s12248-019-0339-5

Barrett JS, Fossler MJ, Cadieu KD, Gastonguay MR (2008) Pharmacometrics: a multidisciplinary field to facilitate critical thinking in drug development and translational research settings. J Clin Pharmacol 48(5):632–649. https://doi.org/10.1177/0091270008315318

Barrett JS, Alberighi ODC, Läer S, Meibohm B (2012) Physiologically based pharmacokinetic (PBPK) modeling in children. Clin Pharmacol Ther 92(1):40–49. https://doi.org/10.1038/clpt.2012.64

Bell SA, Smith CT (2014) A comparison of interventional clinical trials in rare versus non-rare diseases: an analysis of ClinicalTrials.gov. Orphanet J Rare Dis 9(1):170. https://doi.org/10.1186/s13023-014-0170-0

Bi Y, Liu J, Wang J, Epps RE, Kettl D, Marcus K, Seo S, Zhu H, Wang Y (2019) Model-informed drug development approach supporting approval of adalimumab (HUMIRA) in adolescent patients with Hidradenitis Suppurativa: a regulatory perspective. AAPS J 21(5):91. https://doi.org/10.1208/s12248-019-0363-5

Bradshaw EL, Spilker ME, Zang R, Bansal L, He H, Jones RDO, Le K, Penney M, Schuck E, Topp B, Tsai A, Xu C, Nijsen MJMA, Chan JR (2019) Applications of quantitative systems pharmacology in model-informed drug discovery: perspective on impact and opportunities. CPT Pharmacometrics Syst Pharmacol 8(11):777–791. https://doi.org/10.1002/psp4.12463

Chan JR, Allen R, Boras B, Cabal A, Damian V, Gibbons FD, Gulati A, Hosseini I, Kearns JD, Saito R, Cucurull-Sanchez L, Selimkhanov J, Stein AM, Umehara K, Wang G, Wang W, Neves-Zaph S (2022) Current practices for QSP model assessment: an IQ consortium survey. J Pharmacokinet Pharmacodyn:1–13. https://doi.org/10.1007/s10928-022-09811-1

Cheng Y, Thalhauser CJ, Smithline S, Pagidala J, Miladinov M, Vezina HE, Gupta M, Leil TA, Schmidt BJ (2017) QSP toolbox: computational implementation of integrated workflow components for deploying multi-scale mechanistic models. AAPS J 19(4):1002–1016. https://doi.org/10.1208/s12248-017-0100-x

Chuang W-L, Pacheco J, Cooper S, McGovern MM, Cox GF, Keutzer J, Zhang XK (2014) Lyso-sphingomyelin is elevated in dried blood spots of Niemann–Pick B patients. Mol Genet Metab 111(2):209–211. https://doi.org/10.1016/j.ymgme.2013.11.012

Congress 107th (2002) Rare diseases act of 2002 (2002; 107th congress H.R. 4013) – GovTrack.us. https://www.govtrack.us/congress/bills/107/hr4013#:~:text=The%20Rare%20Disease%20Act%20of,provide%20educational%20activities%20for%20researchers. Accessed on 24 Aug 2023

Cucurull-Sanchez L (2024) An industry perspective on current QSP trends in drug development. J Pharmacokinet Pharmacodyn:1–10. https://doi.org/10.1007/s10928-024-09905-y

der Vlugt KG, Langeveld M, Poppema A, Kuiper S, Hollak CEM, Aerts JM, Groener JEM (2008) Prominent increase in plasma ganglioside GM3 is associated with clinical manifestations of type I Gaucher disease. Clin Chim Acta 389(1–2):109–113. https://doi.org/10.1016/j.cca.2007.12.001

Dunne J, Rodriguez WJ, Murphy MD, Beasley BN, Burckart GJ, Filie JD, Lewis LL, Sachs HC, Sheridan PH, Starke P, Yao LP (2011) Extrapolation of adult data and other data in pediatric drug-development programs. Pediatrics 128(5):e1242–e1249. https://doi.org/10.1542/peds.2010-3487

Edginton AN, Schmitt W, Willmann S (2006) Development and evaluation of a generic physiologically based pharmacokinetic model for children. Clin Pharmacokinet 45(10):1013–1034. https://doi.org/10.2165/00003088-200645100-00005

Ekerot P, Ferguson D, Glämsta E-L, Nilsson LB, Andersson H, Rosqvist S, Visser SAG (2013) Systems pharmacology modeling of drug-induced modulation of thyroid hormones in dogs and translation to human. Pharm Res 30(6):1513–1524. https://doi.org/10.1007/s11095-013-0989-4

EMA (2006) Guideline on CT in small populations – Adopted. https://www.ema.europa.eu/en/documents/scientific-guideline/guideline-clinical-trials-small-populations_en.pdf. Accessed on 22 Aug 2023

EMA (2018) Reflection paper on the use of extrapolation in the development of medicines for paediatrics – Final. https://www.ema.europa.eu/en/documents/scientific-guideline/adopted-reflection-paper-use-extrapolation-development-medicines-paediatrics-revision-1_en.pdf. Accessed on 22 Aug 2023

Fan J, Zhang X, Liu J, Yang Y, Zheng N, Liu Q, Bergman K, Reynolds K, Huang S-M, Zhu H, Wang Y (2020) Connecting hydroxychloroquine in vitro antiviral activity to in vivo concentration for prediction of antiviral effect: a critical step in treating COVID-19 patients. Clin Infect Dis 71(12):ciaa623. https://doi.org/10.1093/cid/ciaa623

FDA (2012) Advisory Committee for Pharmaceutical Science and Clinical Pharmacology

FDA (2014) Report: complex issues in developing drugs and biological products for rare diseases and accelerating the development of therapies for pediatric rare diseases. https://cacmap.fda.gov/media/89051/download. Accessed on 22 Aug 2023

FDA (2017) Pediatric rare diseases — A collaborative approach for drug development using Gaucher disease as a model guidance for industry. https://www.fda.gov/media/109465/download. Accessed on 22 Aug 2023

FDA (2018) E11(R1) Addendum: clinical investigation of medicinal products in the pediatric population guidance for industry. https://www.fda.gov/media/101398/download. Accessed on 22 Aug 2023

FDA (2019a) Rare diseases: natural history studies for drug development guidance for industry. https://www.fda.gov/media/122425/download. Accessed on 22 Aug 2023

FDA (2019b) Rare diseases: common issues in drug development: guidance for industry. https://www.fda.gov/media/119757/download. Accessed on 22 Aug 2023

FDA, M-CERSI (2023) Creating a roadmap to quantitative systems pharmacology-informed rare disease drug development

Gotta V, Yu Z, Cools F, Ammel K, Gallacher DJ, Visser SAG, Sannajust F, Morissette P, Danhof M, Graaf PH (2016) Application of a systems pharmacology model for translational prediction of hERG-mediated QTc prolongation. Pharmacol Res Perspect 4(6):e00270. https://doi.org/10.1002/prp2.270

Guillemot N, Troadec C, de Villemeur TB, Clément A, Fauroux B (2007) Lung disease in niemann–pick disease. Pediatr Pulmonol 42(12):1207–1214. https://doi.org/10.1002/ppul.20725

ICH (2024). https://database.ich.org/sites/default/files/ICH_E11A_Guideline_Step4_2024_0821.pdf

Kaddi CD, Niesner B, Baek R, Jasper P, Pappas J, Tolsma J, Li J, van Rijn Z, Tao M, Ortemann-Renon C, Easton R, Tan S, Puga AC, Schuchman EH, Barrett JS, Azer K (2018) Quantitative systems pharmacology modeling of acid sphingomyelinase deficiency and the enzyme replacement therapy Olipudase alfa is an innovative tool for linking pathophysiology and pharmacology. CPT Pharmacometrics Syst Pharmacol 7(7):442–452. https://doi.org/10.1002/psp4.12304

Kaufmann P, Pariser AR, Austin C (2018) From scientific discovery to treatments for rare diseases – the view from the National Center for Advancing Translational Sciences – Office of Rare Diseases Research. Orphanet J Rare Dis 13(1):196. https://doi.org/10.1186/s13023-018-0936-x

Kuchar L, Sikora J, Gulinello ME, Poupetova H, Lugowska A, Malinova V, Jahnova H, Asfaw B, Ledvinova J (2017) Quantitation of plasmatic lysosphingomyelin and lysosphingomyelin-509 for differential screening of Niemann-Pick A/B and C diseases. Anal Biochem 525:73–77. https://doi.org/10.1016/j.ab.2017.02.019

Leggett RW, Williams LR (1995) A proposed blood circulation model for reference man. Health Phys 69(2):187–201. https://doi.org/10.1097/00004032-199508000-00003

Leiser R et al (2023) Pediatric extrapolation: application of a quantitative systems pharmacology (QSP) model to quantify degree of mechanistic similarity of disease and therapeutic response between pediatric and adult patient populations. Presented at: ACOP 14 (2023) https://www.go-acop.org/?abstract=598. November 8, 2023. National Harbor, MD, USA

Lin L, Hua F, Salinas C, Young C, Bussiere T, Apgar JF, Burke JM, Muralidharan KK, Rajagovindan R, Nestorov I (2022) Quantitative systems pharmacology model for Alzheimer's disease to predict the effect of aducanumab on brain amyloid. CPT Pharmacometrics Syst Pharmacol 11(3):362–372. https://doi.org/10.1002/psp4.12759

Madabushi R, Seo P, Zhao L, Tegenge M, Zhu H (2022) Review: role of model-informed drug development approaches in the lifecycle of drug development and regulatory decision-making. Pharm Res 39(8):1669–1680. https://doi.org/10.1007/s11095-022-03288-w

Marshall S, Madabushi R, Manolis E, Krudys K, Staab A, Dykstra K, Visser SAG (2019) Model-informed drug discovery and development: current industry good practice and regulatory expectations and future perspectives. CPT Pharmacometrics Syst Pharmacol 8(2):87–96. https://doi.org/10.1002/psp4.12372

Martins RS, Rocha S, Guimas A, Ribeiro R (2022) Niemann-pick Type B: a rare cause of interstitial lung disease. Cureus 14(1):e21230. https://doi.org/10.7759/cureus.21230

McGovern MM, Aron A, Brodie SE, Desnick RJ, Wasserstein MP (2006) Natural history of Type A Niemann-Pick disease. Neurology 66(2):228–232. https://doi.org/10.1212/01.wnl.0000194208.08904.0c

McGovern MM, Avetisyan R, Sanson B-J, Lidove O (2017) Disease manifestations and burden of illness in patients with acid sphingomyelinase deficiency (ASMD). Orphanet J Rare Dis 12(1):41. https://doi.org/10.1186/s13023-017-0572-x

Milligan PA, Brown MJ, Marchant B, Martin SW, van der Graaf PH, Benson N, Nucci G, Nichols DJ, Boyd RA, Mandema JW, Krishnaswami S, Zwillich S, Gruben D, Anziano RJ, Stock TC, Lalonde RL (2013) Model-based drug development: a rational approach to efficiently accelerate drug development. Clin Pharmacol Ther 93(6):502–514. https://doi.org/10.1038/clpt.2013.54

Miyamoto BE, Kakkis ED (2011) The potential investment impact of improved access to accelerated approval on the development of treatments for low prevalence rare diseases. Orphanet J Rare Dis 6(1):49. https://doi.org/10.1186/1750-1172-6-49

Musante C, Ramanujan S, Schmidt B, Ghobrial O, Lu J, Heatherington A (2017) Quantitative systems pharmacology: a case for disease models. Clin Pharmacol Ther 101(1):24–27. https://doi.org/10.1002/cpt.528

Pavlů-Pereira H, Asfaw B, Poupčtová H, Ledvinová J, Sikora J, Vanier MT, Sandhoff K, Zeman J, Novotná Z, Chudoba D, Elleder M (2005) Acid sphingomyelinase deficiency. Phenotype variability with prevalence of intermediate phenotype in a series of twenty-five Czech and Slovak patients. A multi-approach study. J Inherit Metab Dis 28(2):203–227. https://doi.org/10.1007/s10545-005-5671-5

Pearson C, Schapiro L, Pearson SD (2022) The next generation of rare disease drug policy: ensuring both innovation and affordability. J Comp Eff Res 11(14):999–1010. https://doi.org/10.2217/cer-2022-0120

Peterson MC, Riggs MM (2010) A physiologically based mathematical model of integrated calcium homeostasis and bone remodeling. Bone 46(1):49–63. https://doi.org/10.1016/j.bone.2009.08.053

Riggs MM, Cremers S (2019) Pharmacometrics and systems pharmacology for metabolic bone diseases. Br J Clin Pharmacol 85(6):1136–1146. https://doi.org/10.1111/bcp.13881

Rose K (2019) The challenges of pediatric drug development. Curr Ther Res 90:128–134. https://doi.org/10.1016/j.curtheres.2019.01.007

Shah DK, Betts AM (2012) Towards a platform PBPK model to characterize the plasma and tissue disposition of monoclonal antibodies in preclinical species and human. J Pharmacokinet Pharmacodyn 39(1):67–86. https://doi.org/10.1007/s10928-011-9232-2

Shah KK, Kogut S, Slitt A (2021) Challenges in evaluating safety and efficacy in drug development for rare diseases: a review for pharmacists. J Pharm Pract 34(3):472–479. https://doi.org/10.1177/0897190020930972

Snelder N, Ploeger BA, Luttringer O, Rigel DF, Webb RL, Feldman D, Fu F, Beil M, Jin L, Stanski DR, Danhof M (2016) Characterization and prediction of cardiovascular effects of fingolimod and siponimod using a systems pharmacology modelling approach. J Pharmacol Exp Ther 360(2):jpet. 116.236208. https://doi.org/10.1124/jpet.116.236208

Sun W, Zheng W, Simeonov A (2017) Drug discovery and development for rare genetic disorders. Am J Méd Genet Part A 173(9):2307–2322. https://doi.org/10.1002/ajmg.a.38326

Thurberg BL, Wasserstein MP, Schiano T, O'Brien F, Richards S, Cox GF, McGovern MM (2012) Liver and skin histopathology in adults with acid sphingomyelinase deficiency (Niemann-Pick disease Type B). Am J Surg Pathol 36(8):1234–1246. https://doi.org/10.1097/pas.0b013e31825793ff

Valentin J (2002) Basic anatomical and physiological data for use in radiological protection: reference values. Ann ICRP 32(3–4):1–277. https://doi.org/10.1016/s0146-6453(03)00002-2

Vanier MT, Ferlinz K, Rousson R, Duthel S, Louisot P, Sandhoff K, Suzuki K (1993) Deletion of arginine (608) in acid sphingomyelinase is the prevalent mutation among Niemann-Pick disease type B patients from northern Africa. Hum Genet 92(4):325–330. https://doi.org/10.1007/bf01247328

Varshneya M, Irurzun-Arana I, Campana C, Dariolli R, Gutierrez A, Pullinger TK, Sobie EA (2021) Investigational treatments for COVID-19 may increase ventricular arrhythmia risk through drug interactions. CPT Pharmacometrics Syst Pharmacol 10(2):100–107. https://doi.org/10.1002/psp4.12573

Vinks AA, Barrett JS (2021) Model-informed pediatric drug development: application of pharmacometrics to define the right dose for children. J Clin Pharmacol 61(S1):S52–S59. https://doi.org/10.1002/jcph.1841

Wakap SN, Lambert DM, Olry A, Rodwell C, Gueydan C, Lanneau V, Murphy D, Cam YL, Rath A (2020) Estimating cumulative point prevalence of rare diseases: analysis of the Orphanet database. Eur J Hum Genet 28(2):165–173. https://doi.org/10.1038/s41431-019-0508-0

Wang Y, Zhu H, Madabushi R, Liu Q, Huang S, Zineh I (2019) Model-informed drug development: current US regulatory practice and future considerations. Clin Pharmacol Ther 105(4):899–911. https://doi.org/10.1002/cpt.1363

Wasserstein MP, Desnick RJ, Schuchman EH, Hossain S, Wallenstein S, Lamm C, McGovern MM (2004) The natural history of Type B Niemann-Pick disease: results from a 10-year longitudinal study. Pediatrics 114(6):e672–e677. https://doi.org/10.1542/peds.2004-0887

Welch MJ, Lally R, Miller JE, Pittman S, Brodsky L, Caplan AL, Uhlenbrauck G, Louzao DM, Fischer JH, Wilfond B (2015) The ethics and regulatory landscape of including vulnerable populations in pragmatic clinical trials. Clin Trials 12(5):503–510. https://doi.org/10.1177/1740774515597701

Xiong Y, Fan J, Kitabi E, Zhang X, Bi Y, Grimstein M, Yang Y, Earp JC, Zheng N, Liu J, Wang Y, Zhu H (2022) Model-informed drug development approaches to assist new drug development in the COVID-19 pandemic. Clin Pharmacol Ther 111(3):572–578. https://doi.org/10.1002/cpt.2491

Quantitative Systems Pharmacology Modeling in Immuno-Oncology: Hypothesis Testing, Dose Optimization, and Efficacy Prediction

Hanwen Wang, Theinmozhi Arulraj, Alberto Ippolito, and Aleksander S. Popel

Contents

H. Wang · T. Arulraj · A. Ippolito
Department of Biomedical Engineering, Johns Hopkins University School of Medicine, Baltimore, MD, USA

A. S. Popel (✉)
Department of Biomedical Engineering, Johns Hopkins University School of Medicine, Baltimore, MD, USA

Department of Oncology, and the Sidney Kimmel Comprehensive Cancer Center, Johns Hopkins University School of Medicine, Baltimore, MD, USA
e-mail: apopel@jhu.edu

E. L. Bradshaw, M. E. Spilker (eds.), *Quantitative Systems Pharmacology*, Handbook of Experimental Pharmacology 289, https://doi.org/10.1007/164_2024_735

Abstract

Despite an increasing number of clinical trials, cancer is one of the leading causes of death worldwide in the past decade. Among all complex diseases, clinical trials in oncology have among the lowest success rates, in part due to the high intra- and inter-tumoral heterogeneity. There are more than a thousand cancer drugs and treatment combinations being investigated in ongoing clinical trials for various cancer subtypes, germline mutations, metastasis, etc. Particularly, treatments relying on the (re)activation of the immune system have become increasingly present in the clinical trial pipeline. However, the complexities of the immune response and cancer-immune interactions pose a challenge to the development of these therapies. Quantitative systems pharmacology (QSP), as a computational approach to predict tumor response to treatments of interest, can be used to conduct in silico clinical trials with virtual patients (and emergent use of digital twins) in place of real patients, thus lowering the time and cost of clinical trials. In line with improved mechanistic understanding of the human immune system and promising results from recent cancer immunotherapy, QSP models can play critical roles in model-informed drug development in immuno-oncology. In this chapter, we discuss how QSP models were designed to serve different study objectives, including hypothesis testing, dose optimization, and efficacy prediction, via case studies in immuno-oncology.

Keywords

Cancer immunotherapy · Mechanistic modeling · T cell engager · Cytokine release syndrome · Clinical trial simulation

1 Introduction

Cancer is a leading cause of death worldwide with over 19 million estimated new cases and nearly ten million deaths in 2022 (Bray et al. 2024). Although the number of clinical trials has increased over the past decade, the overall success rate of trials in oncology is among the lowest in the major therapeutic areas (Wong et al. 2019). Between 2016 and 2021, 207 cancer drugs were approved by the U.S. FDA for a variety of patient subgroups that differ in cancer type, disease stage, germline mutation status, etc., with only 28 drugs displacing standard-of-care treatments (Benjamin et al. 2022). Further, the median time between drug discovery/patent application and marketing approval is 12 years (Stewart et al. 2018). From target identification to dose optimization, many factors, such as drug specificity, selectivity, efficacy, and toxicity, have to be investigated, which contributes to the prolonged drug development process and the low success rate of clinical trials (Sun et al. 2022). More importantly, as immunotherapy and other targeted therapies have become the standard-of-care treatments for multiple patient subgroups, it is increasingly difficult for clinical trials using newly discovered drugs to recruit enough patients from these subgroups to achieve statistical validity (Thoma et al. 2010).

Since 2013, quantitative systems pharmacology (QSP) models, along with other types of mechanistic models, have begun to play important roles in the drug approval process in the U.S. to accelerate this process. The number of regulatory submissions to the FDA that contain QSP models has continuously increased from 2013 to 2020, with 22% of the submissions for oncology (Bai et al. 2021). Although QSP models were used for dose optimization in the vast majority of the new drug applications, they have also been developed to waive required clinical studies or address safety concerns in a few cases (Bai et al. 2021). In addition, QSP models have shown the capability to identify potential drug targets, investigate drug-drug/drug-target interactions, and make prospective efficacy predictions in academic immuno-oncology research (Butner et al. 2022).

Starting from model formulation, QSP model development in oncology faces similar challenges to that in all complex diseases. In most cases, model scope (e.g., mechanisms to be incorporated, time scale, stochasticity, and the necessity of spatial heterogeneity characterization) is first determined based on the study objective, data availability, desired computational efficiency, and accuracy (Gadkar et al. 2016). For example, whole-body models of the immune system would help investigate interactions among therapeutic agents, immune cells, and cancer cells throughout the body for better prediction of tumor response to treatments; and localized models of signaling pathways involving the targets of drugs of interest would help answer specific questions, such as drug-drug interaction and target identification (Chelliah and van der Graaf 2022; Zhang et al. 2022). Chelliah et al. analyzed mechanistic models that describe different aspects of the cancer-immunity cycle (Chen and Mellman 2013) and summarized the cell types and molecular components involved in 141 models (Chelliah et al. 2021).

Chapter 5, *Data to Inform Models*, provides an overview of the data and data sources used to develop and validate QSP models. Data used to support QSP model development may come directly from in vitro and in vivo experimental data generated by a pharmaceutical company, academic lab, or contract research organization to support either a drug development project or specifically to support model parameterization. Additionally, data found in published in vitro and in vivo experimental studies on different cancer types may be leveraged for QSP model parameterization. Leonov et al. developed a database that stores publicly available in vivo human data on cellular and molecular concentrations measured in different tissues in 280 diseases including various cancers (Leonov et al. 2023). In addition, Lazarou et al. proposed informing mechanistic models with multi-omics data from public databases, such as the iAtlas and TSNAdb databases (Lazarou et al. 2020). Notably, data that do not directly correspond to model parameters/species need to be converted into proper units. For example, Zhang et al. analyzed scRNA and whole-exome sequencing data with tools like netMHCpan 4.0 to calculate neoantigen expression in each patient, which is an input parameter in their QSP model (Zhang et al. 2021). Once a QSP model is parameterized, model applications vary according to study objectives, and methods for model evaluation are similar across all therapeutic areas (Braakman et al. 2022). In this chapter, we analyze 3 cases of QSP modeling in immuno-oncology with a focus on how models were formulated given different study objectives.

2 Case Study 1: Inferring Optimal Dosage and Mechanistic Insights of GITR Agonist by QSP Modeling

In this case study, we discuss the applicability of immuno-oncology QSP models for generating mechanistic hypotheses and dose optimization for a class of agonist antibodies with multiple mechanisms of action. We consider a representative murine QSP model of anti-GITR antibodies (Ji et al. 2023) and discuss the motivation, steps involved in model development, and applicability of the corresponding QSP model.

2.1 Study Objective

Glucocorticoid-induced tumor necrosis factor receptor-related protein (GITR) is a co-stimulatory molecule expressed on regulatory T cells (Treg), effector T cells (Teff), and other immune cells, such as macrophages and natural killer cells. Agonist antibodies are capable of binding to GITR and mimicking the action of GITR ligands. In preclinical studies (Turk et al. 2004; Ko et al. 2005), GITR agonist antibodies have been shown to promote tumor clearance and induce long-lasting anti-tumor activity. Anti-GITR agonist antibodies display a dual mechanism of action by 1) stimulating Teff proliferation and effector functions and 2) promoting FcγR-mediated depletion of Treg, which together result in a favorable immune microenvironment. Nonetheless, the effectiveness of GITR agonism depends on the interplay among Teff stimulation, Treg depletion, and potential Teff depletion at high doses. Therefore, a quantitative understanding of tumor response is critical to identify the optimal dose of anti-GITR antibodies. To this end, Ji et al. developed the QSP model to recapitulate the action of an anti-GITR antibody DTA-1.mIgG2a on two different murine tumor models derived from: 1) Colon26, a colorectal carcinoma cell line and 2) A20, a murine B lymphoma cell line.

2.2 Model Design

The model includes ordinary differential equations (ODEs) consisting of about 130 reactions and 135 parameters to consider the pharmacokinetics (PK) of the anti-GITR antibody, dynamics of soluble GITR, a simplified GITR signaling cascade, IL-2, IL-10, and cellular components, including CD8+ Teff, Treg, and macrophage expressing FcγIV receptors. The model was developed in a step-by-step manner using a combination of in vivo and in vitro data. This includes data generated by Ji et al. from experiments specifically designed to support model development, such as measurements of serum antibody/soluble GITR concentrations, Teff/Treg

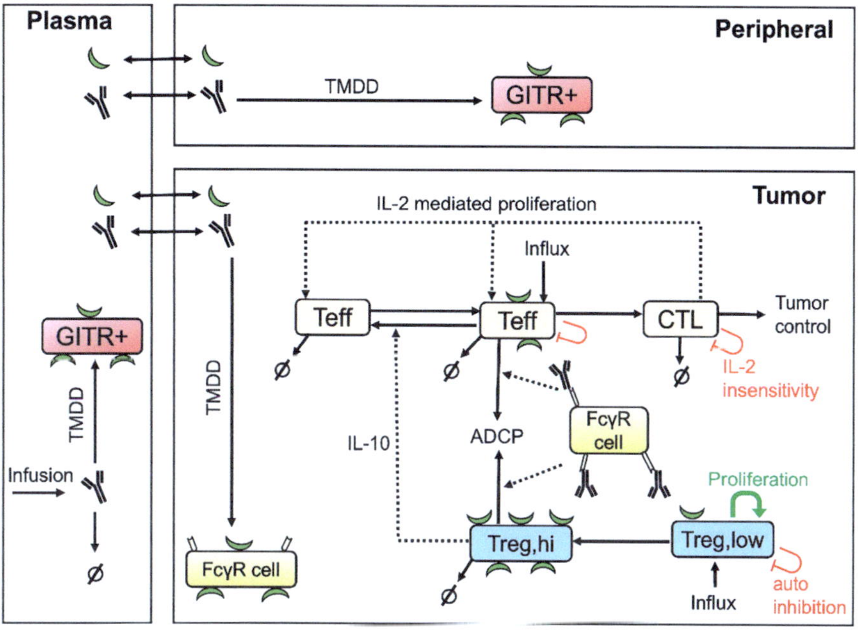

Fig. 1 Schematic representation of the QSP model by Ji et al. ADCP: antibody-dependent cellular phagocytosis; TMDD: target-mediated drug disposition; Teff: CD8+ effector T cells; CTL: cytotoxic T lymphocyte. Green crescents represent GITR receptors (membrane-bound and soluble)

profiles, tumor size from Colon26 and A20 mice models treated with anti-GITR antibody, and cell proliferation estimates from in-vitro T cell stimulation assays, in addition to previously generated data from the literature. A simplified schematic of the model is shown in Fig. 1.

The PK module consists of two compartments, central and peripheral compartments that account for the absorption, distribution, metabolism, and elimination (ADME) of the drug. PK parameters were estimated by fitting the model to serum anti-GITR antibody concentration profiles from Colon26 and A20 tumor models. The antibody is assumed to bind to cells expressing GITR and soluble GITR, which leads to target-mediated drug disposition (TMDD), thus recapitulating a fast drug clearance at low doses.

Cell types explicitly considered include Treg, CD8+ Teff, and macrophage expressing FcγIV receptors. The dynamics of Treg and Teff upon drug administration were substantially different between Colon26 and A20 tumor models. To account for this behavior, the authors considered two subpopulations of Tregs: 1) Treg with high GITR (Treg-hi) and 2) Treg with low GITR expression (Treg-low). They hypothesized that the Colon26 tumor had a higher level of Treg-hi subset than the A20 tumor. In addition, the Treg-hi subset was assumed to produce IL-10, which was a surrogate for all Treg-mediated suppressive effects in the model, and Teff was assumed to be inactive in the tumor due to suppression by Treg. These hypotheses were sufficient to recapitulate the differences in Treg dynamics and anti-tumor responses between the two mouse models, which demonstrates the applicability of QSP models in hypothesis testing to explain observed behavior. Subsequently, subsets of Treg and the prevalence of Treg subsets could be experimentally examined in Colon26 and A20 models to verify these hypotheses.

The PK module was coupled with the cellular module of the tumor microenvironment by considering the drug distribution into the tumor. In the tumor compartment, the antibody binds to GITR receptors on Treg. GITR-bound drugs can also bind to FcγIV receptors on macrophages, which leads to the depletion of Treg-hi cells via antibody-dependent cellular phagocytosis (ADCP) by macrophage through a Michaelis-Menten mathematical function. In contrast, Treg-low cells are assumed to be insensitive to ADCP and rather lead to GITR activation due to the formation of trans-complexes by receptor clustering. Depletion of Treg-hi cells leads to a reduction in IL-10 levels that promotes the transition of Teff to an active state. During this transition, Teff is assumed to upregulate GITR expression. Engagement of GITR on Teff in the activated state is assumed to trigger proliferation and differentiation into cytotoxic T lymphocytes (CTL). A positive feedback loop is considered in the model where the activated Teff and CTL secrete IL-2 that triggers Teff activation, proliferation, and differentiation via a stimulatory rate constant and active cell ratio. In addition, activated Teff and CTL tend to lose their sensitivity to IL-2 over time, which leads to a negative feedback loop. It is also assumed that CTL has a limited lifespan, so the number of CTL declines over time in the model.

Soluble GITR, a potential biomarker of CTL activation, was also incorporated into the QSP model. Experimentally measured soluble GITR levels showed an initial increase after drug administration followed by a steep decline. As the differentiation of CD8+ Teff to CTL was assumed to be associated with the shedding of GITR, the model reproduced the increase in soluble GITR levels. The authors added an empirical mechanism of anti-drug antibody (ADA)-mediated clearance of soluble GITR to recapitulate the decline observed in soluble GITR profiles. In addition to the negative feedback by reduced IL-2 sensitivity and limited lifespan of CTLs, ADA-mediated drug clearance contributes to the decrease in the anti-tumor activity over time. Figure 2 shows the model-generated time profiles of Treg and Teff in the Colon26 mouse model.

2.3 Model Application

Given the competing mechanisms of anti-GITR agonists, the QSP model developed was then used to predict the optimal dose. Different doses of anti-GITR antibodies were simulated using the QSP model to predict the peak number of CTL that was activated under these conditions. As tumor clearance by CTL was not explicitly considered in the model, the minimum number of CTL required to promote tumor clearance in each of the mouse tumor models was estimated using information, such as the tumor size, density of cells in the tumor, and the number of tumor cells that can be typically killed by a CTL. The lowest dose in the model simulations that was

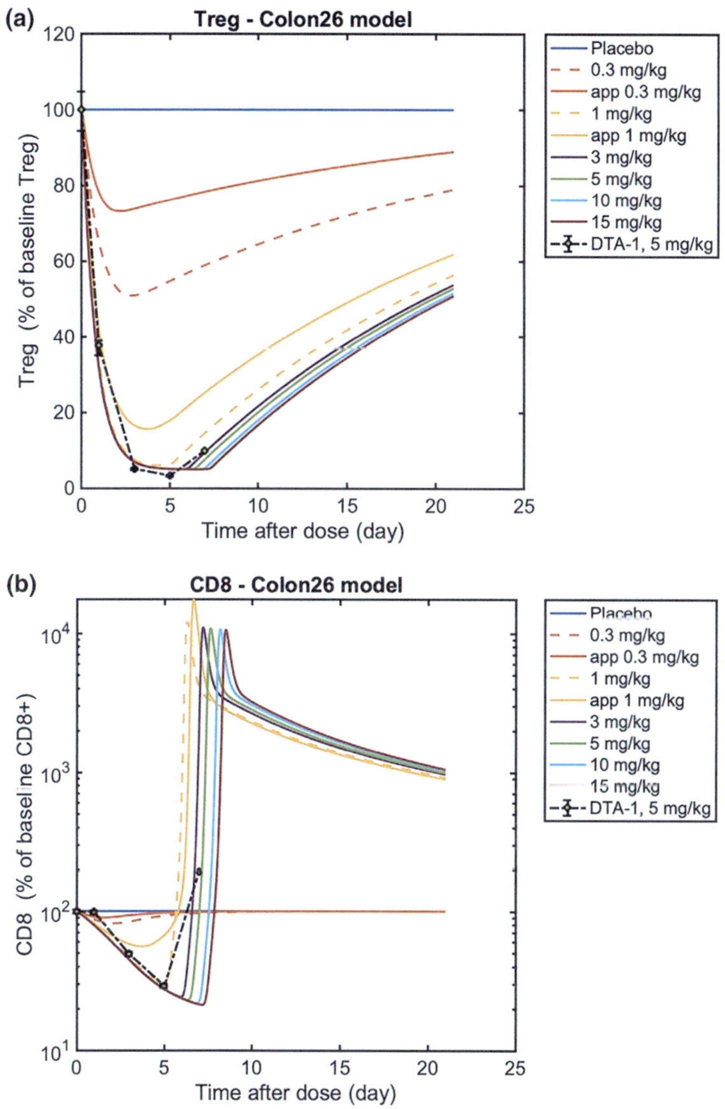

Fig. 2 Model fitting of the time profiles of Treg (a) and Teff (b) in the Colon26 model. The dots and error bars in black represent experimental measurements and 95% confidence intervals. Solid lines/curves represent model-generated profiles for different doses of anti-GITR antibody. Due to deviations in model-predicted pharmacokinetics from experimental measurements at certain doses, estimated doses that better matched the observed pharmacokinetic data for nominal doses were also considered (dashed lines/curves). Figure reused from Ji et al. with permission

predicted to attain the minimum number of CTL estimated to clear the tumor was considered the optimal dose. The model-predicted doses to promote tumor clearance were compared to the in vivo data for model validation.

A local sensitivity analysis was also performed to identify the most influential individual parameters in both tumor models. Parameters were perturbed around their reference parameter values corresponding to Colon26 and A20 tumor models in this analysis. In both tumor models, the most sensitive parameters were those influencing the formation of GITR-drug-FcγIVR complex, tumor drug disposition, drug clearance, and macrophage-related parameters, such as the number of macrophages, GITR expression, and FcγRIV receptor level. These results also suggested that FcγRIV-expressing macrophages and GITR expression may potentially act as predictive biomarkers for anti-GITR therapy.

Overall, this case study demonstrated how a QSP model was built to analyze mechanistic hypotheses and predict optimal doses without considering the full complexity of the tumor microenvironment. The preclinical model by Ji et al. provides an opportunity to study the translatability of preclinical findings to clinical studies. Using human-specific parameters and mechanisms, the murine QSP model could be modified to develop a corresponding human QSP model and used to predict the efficacy of GITR agonists in combination with other drugs. Moreover, the dual mechanism of action of anti-GITR antibodies highlights the importance of QSP models in predicting the starting dose for first-in-human studies.

3 Case Study 2: A Parsimonious QSP Model of Cell Therapy to Mitigate Cytokine Release Syndrome

3.1 Study Objective and Model Design

Chimeric antigen receptor (CAR)-T cell therapy is another type of immunotherapy in oncology. Through transduction of the patient's or a donor's T cells with genes encoding the CAR molecule, CAR-T therapy aims to redirect the specificity and functions of T cells to cancer (Sadelain et al. 2013). CARs are recombinant receptors that bind to target molecules on the cancer cell surface and activate CAR-T cells through activation and costimulatory domain(s). One of the most common targets of CAR-T therapies is CD19, which is a protein ubiquitously and exclusively expressed on B lymphocytes and thus an ideal target in B-cell malignancies (Sermer et al. 2023). In 2017, tisagenlecleucel became the first CAR-T therapy approved by the FDA for relapsed and refractory B-cell acute lymphoblastic leukemia (ALL) (O'Leary et al. 2019). When administered with a single weight-based or flat dose of up to 2.5e8 cells, tisagenlecleucel achieved an overall remission rate of 83% among 63 patients with ALL (O'Leary et al. 2019).

Although CAR-T therapies have achieved high efficacy in many blood cancers, one of the limitations is cytokine release syndrome (CRS) caused by the rapid release of cytokines into the blood from activated CAR-T cells (Sterner and Sterner 2021). In fact, grade 3 or higher CRS was observed in 49% of the patients in the clinical trial of tisagenlecleucel (O'Leary et al. 2019). To investigate associations among CRS, disease burden, and CAR-T cell doses, Hardiansyah and Ng designed a QSP model to describe the dynamics of activated and memory CAR-T cells in two compartments (peripheral blood and tissues), malignant B cells, and three cytokines

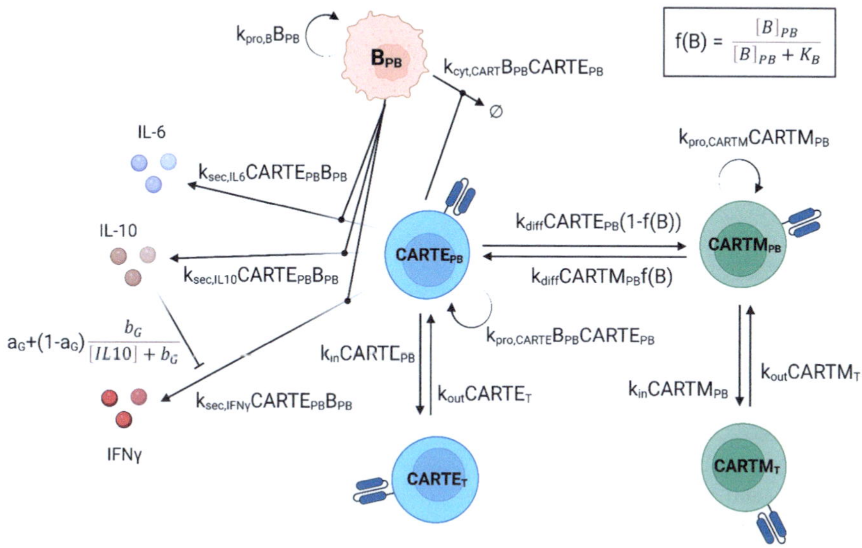

Fig. 3 Structure of a QSP model of CAR-T therapy. Endogenous production and degradation of IL-6, IL-10, and IFNγ, and natural death of CAR-T cells in peripheral blood are omitted from the diagram. PB, peripheral blood. T, tissues. CARTE, effector CAR-T cells. CARTM, memory CAR-T cells. Created with BioRender.com

contributing to the severity of CRS (IL-6, IL-10, and IFN γ) in peripheral blood (Hardiansyah and Ng 2019).

The choice of cells and molecules to be incorporated into the model was based not only on the known mechanisms of CAR-T cells and cytokines, but also on the availability of relevant clinical data to reduce model uncertainty. Specifically, kinetic profiles of IL-6, IL-10, IFN γ, and total number of CAR-T cells were available from two patients with advanced chemotherapy-resistant chronic lymphocytic leukemia (CLL) from a pilot clinical study of tisagenlecleucel (Kalos et al. 2011). Although there are other sources of cytokines related to CRS, such as macrophages that secrete IL-6 and IL-1 in response to CAR-T therapy, they are not incorporated due to the lack of clinical measurements. For the same reason, the kinetics of CAR-T cells and malignant B cells in bone marrow, which are also important components in CLL, are not explicitly described by the model.

Figure 3 demonstrates the incorporated model mechanisms. In addition to cytokine secretion by endogenous immune cells and activated CAR-T cells upon recognition of CD19-expressing tumor B cells, IL-10 inhibits IFN γ production directly by targeting activated CAR-T cells, and also indirectly by reprogramming macrophages and preventing APC maturation, thereby limiting pro-inflammatory cytokine secretion (Iyer and Cheng 2012). Here, the direct effect of IL-10 was described by a multiplier empirically determined in a model of helper T cell trans-differentiation (Kogan et al. 2013). Particularly, $\mathbf{1} - \boldsymbol{a_G}$ is the maximum inhibitory effect of IL-10 on IFN γ production, and $\boldsymbol{b_G}$ is the IL-10 concentration that would result in its half-maximal inhibitory effect. Since macrophages and other immune cells were not explicitly incorporated into the model, the indirect effect of IL-10 on IFN γ production was reflected via parameterization, which will be discussed below.

Kinetics of CAR-T cells in peripheral blood (PB) involves proliferation, death, and two reversible processes: 1) transport between peripheral blood and tissues (T) and 2) differentiation between effector and memory CAR-T cells. Differentiation between effector (CARTE) and memory (CARTM) CAR-T cells was assumed to follow a Michaelis-Menten model of B cells, f(B). Further, the kinetics of B cells (B) in peripheral blood involves proliferation, death, and effector CAR-T cell-mediated killing.

3.2 Model Parameterization

Among the 23 model parameters, 13 parameters were obtained from the literature. Degradation rates of the cytokines were estimated by experimentally measured half-lives; $\boldsymbol{a_G}$ and $\boldsymbol{b_G}$ were estimated based on the inhibitory effect of IL-10 on IFN γ production by type 1 helper T cells, which can be potentially improved by data from personalized CAR-T cells if available as they contain different proportions of $CD8^+$ and $CD4^+$ cells. The proliferation rate of memory CAR-T cells was assumed to be the same as that of regular $CD8^+$ $CD45RO^+$ T cells whose doubling times were measured by in vivo experiments (Macallan et al. 2003). With the same assumption, rates of transport between peripheral blood and tissues, $\boldsymbol{k_{in}}$ *and* $\boldsymbol{k_{out}}$, were estimated by clinically measured rates of intravascular and extravascular CD8 T cells (Storek et al. 2002); and the rate of memory CAR-T cell activation, $\boldsymbol{k_{diff}}$, was assumed to be 1 day^{-1} (De Boer and Perelson 2013). The death rate of memory CAR-T cells was estimated by the rate of exponential decline after effector CAR-T cell expansion observed in patients treated with CAR-T therapy, which was much higher than that of regular memory T cells (Kalos et al. 2011). For malignant B cells, the rate of proliferation and natural death were measured in 15 patients with CLL (Messmer et al. 2005); and the B-CLL cell level that leads to half-maximal CAR-T cell activation, $\boldsymbol{K_B}$, was originally fitted to an experiment of macrophage-mediated killing of tumor cells (De Boer et al. 1985). Notably, saturation parameters in Michaelis-Menten kinetics are often hard to be directly derived from published experimental data and thus could be estimated by indirect experimental results, a dedicated experiment designed for parameter estimation, or an educated guess.

The remaining 10 model parameters were fitted to data from each patient in the pilot clinical study, including cytokine secretion by endogenous immune cells and CAR-T cells, rates of effector CAR-T cell proliferation, death, differentiation to memory CAR-T cells, and CAR-T cell-mediated killing. The clinical data contain measurements of cytokine concentrations (i.e., IL-6, IL-10, and IFN γ) and total CAR-T cell level (i.e., $CARTE_{PB} + CARTM_{PB}$) at multiple time points from day 0 to day 176 upon CAR-T cell injection (Kalos et al. 2011). Importantly, since endogenous immune cells were not explicitly incorporated into this model, the impact of CAR-T therapy on endogenous cytokine secretion was reflected through parameterization. Specifically, the rate of endogenous cytokine secretion before CAR-T cells were administered was fitted to the patient's baseline cytokine concentration, while after the therapy began, it was fitted to the steady-state cytokine concentration at the end of the therapy.

3.3 Model Applications

As introduced at the beginning of this case study, the model was designed specifically to investigate the associations between CRS and potential risk factors, including disease burden and CAR-T dose amount. To this end, 8 ODEs were used to describe 5 cellular and 3 molecular species/cytokines (Fig. 3). The initial condition of all model species, by which model simulation was started, was personalized according to the data from two adult patients UPN1 and UPN3 in the clinical study (Kalos et al. 2011). Cytokine concentrations were initialized by their measured baseline levels, and the pre-treatment number of B-CLL cells was estimated by the measured number of white blood cells in each patient with the assumption that B cells account for about 40% of the total white blood cells. The initial number of effector CAR-T cells corresponds to the number of cells administered for each patient, while memory CAR-T cells and all CAR-T cells in tissues were initialized with 0 cells. CAR-T cells were administered via 3 doses of intravenous infusion, which was simulated by SimBiology.

For each of the two patients, Hardiansyah and Ng conducted 1,000 model simulations with different CAR-T doses and the initial number of B-CLL cells in the peripheral blood (i.e., disease burden) varied within the clinically observed ranges. For visual validation, the model-predicted pharmacokinetics of CAR-T cells, including maximal CAR-T level in blood and the area under the plasma drug concentration-time curve from day 0 to day 28, were compared with clinical measurements in a later study of tisagenlecleucel (Hardiansyah and Ng 2019). In addition, the model-predicted relationships between pre-treatment disease burden and maximal CAR-T/IFN γ level in blood were consistent with clinical observation in ALL treated with anti-CD19 CAR-T therapy (Hardiansyah and Ng 2019).

Once validated, the QSP model was used to investigate the correlation between CAR-T cell dose/disease burden and concentration of cytokines that contribute to CRS in CLL. To this end, the CAR-T cell dose and disease burden were varied by 0.5, 1, 1.5, and two-fold from baseline for each patient. As shown in Fig. 4, the model predicted that the concentration of all cytokines would increase faster and reach a higher maximal concentration as the pre-treatment disease burden increased, while the dose of CAR-T cells had little impact on the cytokine kinetics. Interestingly, in a recent clinical study that analyzed data from 15 pediatric oncology centers, the dose of tisagenlecleucel was not found to be associated with CRS when administered with 1.3–5.1e6 cells/kg (Stefanski et al. 2023). In summary, this case study demonstrated that a relatively simple QSP model trained by data from early-stage clinical trials can be used to address questions relating to pharmacokinetics and certain toxic effects of the drug(s). By generating random deviations from the two patient models, QSP can inform clinicians about the risk factors for CRS without extensive clinical data collection. Based on the model prediction, future

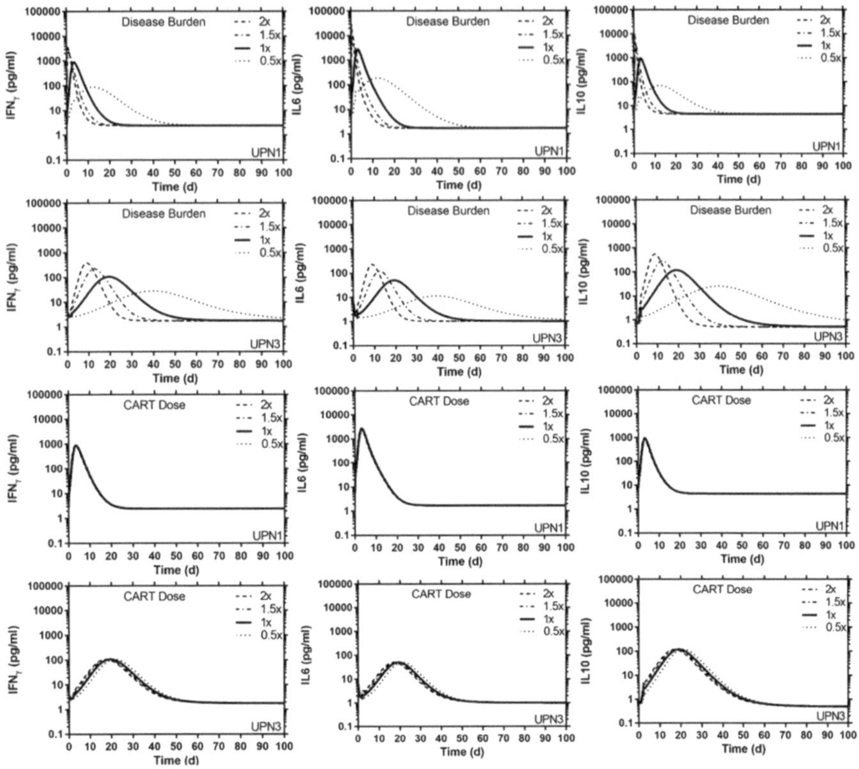

Fig. 4 Model-predicted cytokine profiles with different disease burdens and anti-CD19 CAR-T doses in two patients (UPN1 and UPN3) with chronic lymphocytic leukemia (CLL). Disease burden and CAR-T doses are varied by 0.5-, 1-, 1.5-, and two-fold of their baseline values. The baseline CAR-T doses are 1.1e9 cells (UPN1) and 1.4e7 cells (UPN3), and the baseline disease burdens are 4.14e9 cells (UPN1) and 9.3e8 cells (UPN3). Figure reused from Hardiansyah and Ng with permission

clinical studies can be designed to confirm the correlation between disease burden and CRS and to explore disease burden-based CAR-T dosing to mitigate CRS.

4 Case Study 3: A Modular Whole-body QSP Platform of the Cancer Immunity Cycle to Identify Biomarkers and Optimal Drug Combinations for Cancer Immunotherapy

4.1 Study Objective

Immune checkpoint blockade is another type of immunotherapy. Immune checkpoints are molecules expressed on the surface of immune cells that maintain self-tolerance and modulate immune response through downstream signaling pathways (Pardoll 2012). They are also used by cancer cells as an immune evasion mechanism to create an immuno-suppressive tumor microenvironment and inhibit T cell activation in the lymph nodes. Through the blockade of immune checkpoint molecules, immune checkpoint inhibitors (ICIs) aim to restore the functions of immune cells and facilitate antitumoral response in patients. The common targets of ICIs are cytotoxic T-lymphocyte-associated protein 4 (CTLA-4), programmed cell death protein 1 (PD-1), and its ligand PD-L1. Due to their promising effect observed in clinical trials, ICIs have become the standard-of-care treatment for multiple cancer types, such as melanoma, lung cancer, triple-negative breast cancer, and head and neck carcinoma. Nonetheless, the efficacy of ICIs varies among different cancer types. Although tumor mutational burden and PD-L1 expression are found to be correlated with response to anti-PD-(L)1 therapies in some cancer types (Yarchoan et al. 2017; Davis and Patel 2019), there is a need to identify predictive biomarkers in the context of specific cancer types, especially those with low overall response rate to ICIs, and to improve the efficacy of ICIs via combination therapy. To address this gap, a modular and expandable QSP platform model has been developed as detailed below.

4.2 Model Design

Since immune checkpoints have inhibitory effects on both the tumor microenvironment and lymph nodes, it is important to investigate the cancer-immune cell interactions in both compartments. To this end, Milberg et al. (Wang et al. 2019; Milberg et al. 2019) built a QSP model to describe the cancer-immunity cycle with extensive mechanistic detail, which was later developed into a modular QSP platform, QSP-IO (Jafarnejad et al. 2019; Sové et al. 2020). The latest version of the platform consists of ten modules, each of which describes the dynamics of a major cellular/molecular species in the model, including cancer cells, T cells (Teff, Treg, and helper T cell), antigen-presenting cells, tumor antigens, immune checkpoints, therapeutic agents (e.g., ICIs, T cell engagers), myeloid-derived suppressor cells (MDSCs), and tumor-associated macrophages (TAMs) (Wang et al. 2022, 2023; Anbari et al. 2023; Arulraj et al. 2023; Ippolito et al. 2023). The modular structure of the model has proven its capability to serve different study objectives. Particularly, some modules can be called more than once to account for different clones of cancer cells, T cells, tumor antigens, etc. In addition, new modules developed for drugs of interest, when incorporated into the QSP model, can be used to simulate combination therapies with ICIs (Wang et al. 2020, 2021; Ma et al. 2020a, b) or novel ICIs with more complex mechanisms of activation, such as a masked antibody (Ippolito et al. 2023). When integrated with an agent-based model, the spatial QSP model (spQSP) can also investigate spatial heterogeneities in the tumor microenvironment during treatment (Gong et al. 2021; Zhang et al. 2021; Ruiz-Martinez et al. 2022; Nikfar et al. 2023).

One of the challenges in mechanistic modeling is generating hypotheses for causal mechanisms based on the limited experimental/clinical observations (Baker

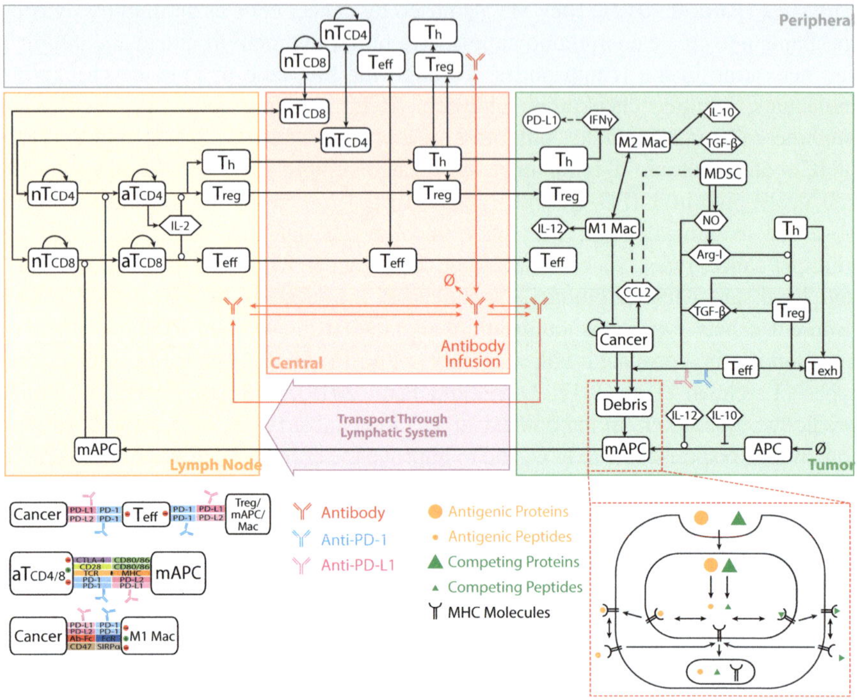

Fig. 5 QSP-IO model diagram. The model contains four main compartments (central, peripheral, tumor, and tumor-draining lymph node) and immune synapse compartments (where immune checkpoint interactions occur), which together describe the cancer-immunity cycle. The effect of cytokines and immune checkpoints are approximated by Hill functions. nT, naïve T cell; aT, activated T cell; NO, nitric oxide; Arg-I, arginase I; Treg, regulatory T cell; Teff, effector T cell; Th, helper T cell; Mac, macrophage; mAPC, mature antigen presenting cell. Cytokine degradation and cellular clearance, while included in the model, are omitted in the figure

et al. 2018). Figure 5 shows all the mechanisms incorporated into the QSP-IO model. Here, we discuss the major mechanisms included in the model. First, the growth of cancer cells can be described by a variety of models, such as exponential, logistic, and Gompertz models. When fitting different models to tumor growth kinetics in 94 animals (including 833 measurements and 3 tumor models), Vaghi et al. concluded that the Gompertz model best predicted the tumor growth and the age of tumors in mice (Vaghi et al. 2020). Notably, the performance of the Gompertz model depended on the initial tumor size, which corresponded to the volume of the injected tumor and was fixed to 1 mm^3. For clinical translation, Vaghi et al. suggested an initial tumor volume of 10^{-6} mm^3 (i.e., a single cancer cell), while Garcia-Cremades et al. set the initial tumor volume to the measured tumor size at diagnosis in a translational model of pancreatic and ovarian tumor growth in human (Garcia-Cremades et al. 2019). Another consideration is the effect of angiogenesis, which can be reflected by a time-dependent tumor-carrying capacity. Hahnfeldt et al. proposed a modified Gompertz model to describe the induction and inhibition of tumor vasculature, which were used in multiple translational models to simulate tumor growth in humans (Hahnfeldt et al. 1999). More recently, Sayama et al. developed an angiogenesis-driven tumor growth model that explicitly incorporated the effect of mitogen-activated protein kinase (MAPK) signaling (Sayama et al. 2021; Zhang et al. 2023). Due to the lack of clinical measurements of uncontrolled tumor growth, the performance of various growth models in predicting tumor kinetics in humans and their therapeutic implications require further investigation (West and Newton 2019; Ghaffari Laleh et al. 2022). In QSP-IO, the cancer module allows switching between the Gompertzian and the logistic tumor growth kinetics to accommodate various study goals and data available (Wang et al. 2023).

Further, it was found that T cell activation occurs in three stages: transient encounters with APCs, stable contacts with APCs and cytokine production, and rapid proliferation (Mempel et al. 2004). The first step is modeled as native T cells are activated at a rate estimated by the ratio of mature APCs and native T cells. Once activated, T cells secrete IL-2 and transition to the proliferation stage when they become functional T cells (e.g., Teff, Treg). The number of T cell divisions during rapid proliferation is found to be the linear sum of the effect of three signals: T cell receptor (TCR) upon recognition of peptide-major histocompatibility complex (MHC), co-stimulation, and inflammatory stimuli. The peptide-MHC (pMHC) complex is expressed on APCs upon antigen uptake and processing. In the antigen module of QSP-IO, this process was adapted from a multiscale model of immunogenicity for therapeutic proteins by substituting the antigenic and competing proteins with tumor neoantigens and tumor-associated self-antigens, respectively (Chen et al. 2014). Assuming that tumor antigens are released from cancer cells at the same rate as the overall cancer cell death, the module describes the internalization, degradation, MHC binding of tumor antigens, and the translocation of the pMHC complex. Then, the total number of TCR bound with pMHC complexes was approximated by the "kinetic proofreading with limited signaling" model, which was found to best describe the experimental data, and used to estimate the number of T cell divisions by TCR via a Michaelis-Menten kinetics (Lever et al. 2014). Next, the interactions among CD80, CD86, CD28, and CTLA-4 were incorporated by the checkpoint module (Jansson et al. 2005), where the number of T cell divisions by co-stimulation is estimated by a Hill function of bound CD28 on the surface of APCs in lymph nodes. Alternatively, the ratio between bound CD28 and bound CTLA-4 in the immune synapse can be incorporated into the Hill function to account for the regulatory effects of CTLA-4 signaling. Lastly, the number of T cell divisions by inflammatory stimuli is estimated by a Hill function of IL-2.

When modeling interactions among receptors, ligands, and therapeutic antibodies, binding affinity, avidity, and cross-arm binding efficiency should be considered. Binding affinities are the strength of interactions between receptors and ligands and are equal to the ratio of dissociation and association rates, which depend on the type of interaction and external physical properties and can be measured by in vitro experiments (Jarmoskaite et al. 2020; Erlendsson and Teilum 2021). Binding avidity, or functional affinity, accounts for cooperativity and multivalency when describing the changes in the rate of association and dissociation between two interacting molecules (Erlendsson and Teilum 2021). For example, CTLA-4 and CD80 often exist as homodimers on the surface of T cells and APCs, respectively. As a result, the rate of association between CTLA-4 and CD80 with a total of 4 free binding sites should be 4 times higher than that between CTLA-4 and CD80 monomers. It is thus important to note how binding affinity is measured in experimental studies by which model parameters are estimated. Moreover, cross-arm binding efficiency is a property of antibodies that describes the strength of cross-arm binding relative to monovalent binding. Harms et al. defined cross-arm binding as the binding between the free arm of a bivalent antibody and its target when the other arm is already bound, and they showed how mechanistic modeling of cross-arm binding can inform therapeutic antibody designs (Harms et al. 2014).

4.3 Model Parameterization and Applications

The QSP-IO platform has 255 parameters with 10 modules, including 1 cancer clone, 1 T cell clone, 1 tumor neoantigen clone, and 1 tumor compartment that represents a primary or a metastatic tumor. Among all the parameters, less than 10% of the parameters are tumor-specific parameters. These are estimated by data on the cancer type of interest. Other parameters are mostly related to the immune system and are estimated by data from experimental studies and healthy individuals. When considering 1 primary and up to 3 metastatic lesions (i.e., 4 tumor compartments), the model was expanded to have 737 parameters with 5 cancer clones, 8 neoantigens, and 8 effector T cell specificities (Arulraj et al. 2023). Notably, cross-reactivity of effector T cells is not accounted for in this model.

The platform has been applied to multiple cancer types that have received limited survival benefits from currently FDA-approved immunotherapies to predict the efficacy of newly discovered drugs and novel drug combinations via in silico virtual clinical trials. In silico virtual clinical trials aim to simulate tumor response to treatments of interest in virtual patients represented by parameter variations. In most cases, tumor-specific parameters were varied around their baseline values within a physiologically reasonable range to account for inter-patient heterogeneity. Recently, as patient-level data became available, additional algorithms were applied to further select virtual patients that better mimic the actual patient cohort (Allen et al. 2016). In addition, by integrating transcriptomic data from metastatic breast tumors into the model, Arulraj et al. predicted the clinical response of metastatic lesions to PD-1 inhibition and proposed the responder inclusion score (RIS) to identify patients with the maximized probability of response (Arulraj et al. 2023, 2024).

Figure 6 shows an example of the QSP-IO model predictions from an in silico clinical trial of a PD-1 inhibitor with 1 primary breast tumor compartment and 3 metastatic tumor compartments (i.e., 2 lung metastases and 1 “other” metastatic tumor). Parameters of the metastatic tumor compartments were estimated so that the relative abundance of immune cells with respect to the primary tumor compartment

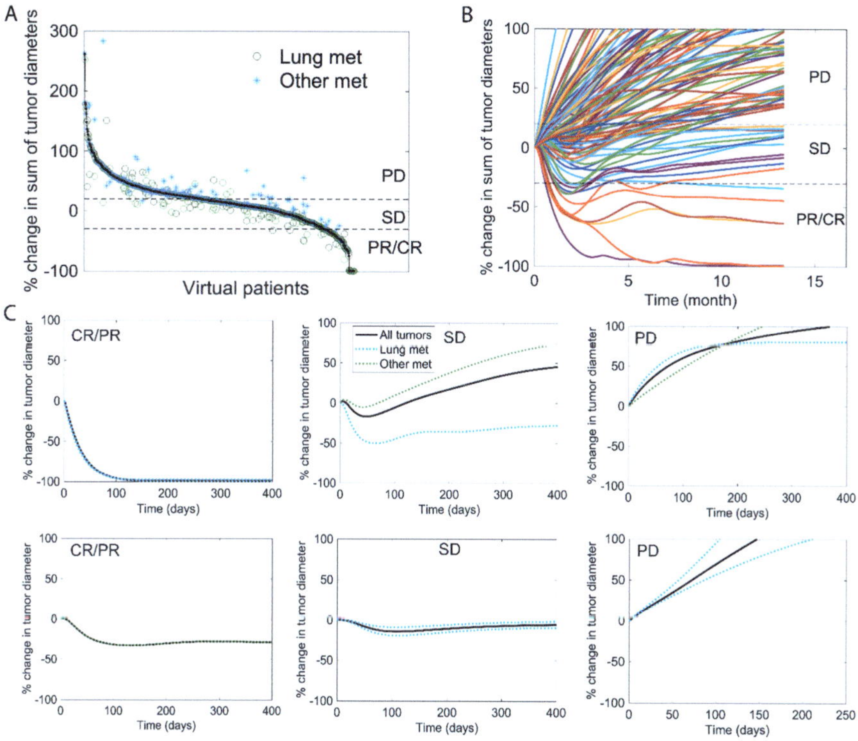

Fig. 6 Virtual clinical trial simulation of anti-PD-1 monotherapy in metastatic triple-negative breast cancer. (**a**) Waterfall plot showing the percentage change in the sum of tumor diameters of all simulated lesions (black dots) in individual virtual patients together with the response of individual tumors (green circles and blue stars for lung and other metastatic tumors, respectively). The horizontal axis represents individual patients. (**b**) Spider plot showing time series of percentage change in the sum of tumor diameters for 100 randomly sampled virtual patients. (**c**) Percentage change in individual metastatic tumor diameters of representative virtual patients with different response status based on RECIST v1.1. CR/PR, complete response/partial response; SD, stable disease; PD, Progressive disease. Figure reused from Arulraj et al. with permission

is consistent with immune cell abundances estimated from RNA-seq data (Siegel et al. 2018) using EPIC (Racle et al. 2017) and quanTIseq (Finotello et al. 2019). Waterfall and spider plots are two common visualizations in clinical analyses showing the best tumor size change from baseline in each patient (Fig. 6a) and the percentage change in tumor size from baseline over time (Fig. 6b). Here, tumor size is the sum of the diameters of each tumor as defined by RECIST v1.1 (Eisenhauer et al. 2009). In addition, Fig. 6c illustrates typical tumor dynamics that fall within each category of clinical response. In this way, the model can make personalized predictions of tumor growth under treatment when parameterized to individual patients' data, laying the groundwork for future precision medicine applications.

5 Discussion

QSP is a highly diverse modeling approach as there is no unified method for model development due to the diversity of available data, model complexity, and study goals (Androulakis 2022). In this chapter, we analyzed 3 cases of QSP model designs in immuno-oncology serving different study objectives and discussed some important considerations when translating experimental findings into model reactions. It is important to note that model predictions with incorporated mechanisms should be consistent with experimental/clinical data, in the learn and confirm cycle, and emerging data may necessitate model modifications reflective of alternative mechanistic hypotheses, as we have seen in the model developed by Ji et al. (2023). In addition, while the mechanistic nature of preclinical QSP models is expected to improve translatability from animal to clinical studies, such applications are limited due to the lack of a comprehensive understanding of the differences between mice and humans. Although allometric scaling was shown to be useful when translating parameter estimates from animals to humans (West et al. 2002; Lindauer et al. 2017; Garcia-Cremades et al. 2019), clinical data are preferred for QSP models of humans, as demonstrated by Hardiansyah and Ng (2019).

Another important consideration is determining model complexity for study goals. The models in the first two case studies are designed with minimized complexity, which suffices to address the objectives of the respective study. For instance, in the model by Ji et al. (2023), IL-10 and IL-2 are assumed to implicitly represent other anti-inflammatory and pro-inflammatory cytokines, respectively. Similarly, Lima et al. proposed multiple candidate models and selected the most simplified model that adequately described the data, therefore lowering the model uncertainty (Lima et al. 2022). On the other hand, the QSP-IO model incorporated dynamics of major immune cell subsets and cytokines in mechanistic detail, which was important to investigate their overall effect on tumor growth during treatment and search for novel predictive biomarkers. For example, the model was extended to a novel immune checkpoint inhibitor that can be conditionally activated by tumor-resident proteases. With the hypothesized mechanisms of action incorporated, the model corroborated the observed patient response rate in a Phase I study of triple-negative breast cancer and suggested predictive biomarkers, which are to be confirmed by upcoming trials (Ippolito et al. 2023). Therefore, the extent of mechanistic details to be included in the model should be determined based on the context and questions to be addressed.

An active field of study in QSP is accounting for inter-patient heterogeneity in model parameterization. To this end, several methods to generate virtual patients and digital twins with different data types have been developed (Niederer et al. 2020; Surendran et al. 2023). While virtual patients can be any model parameterization that falls within the physiologically plausible range and, in some cases, generated to statistically match patient data at the population level, digital twins aim to simulate the behavior of their counterparts in the real world (Chelliah et al. 2021). Undoubtedly, generating a virtual population that resembles the target patient group with specific characteristics (i.e., previous treatment status, cancer type, etc.) is critical to making reliable model predictions, which is especially challenging in oncology research due to 1) limited patient data from clinical trials in public databases and 2) heterogeneity of patient cohorts enrolled by various clinical trials. With the QSP-IO model, the authors have made attempts to generate virtual patient populations that statistically resemble immunogenomic data (Wang et al. 2023; Arulraj et al. 2023; Ippolito et al. 2023). In addition, Wu et al. generated digital twins for patients whose magnetic resonance imaging (MRI) data were collected and processed by an automated pipeline (Wu et al. 2022). More recently, Susilo et al. randomly generated thousands of candidate digital twins for each patient in phase 1 clinical trial and selected 25 digital twins, whose model-predicted tumor size profile best matched the clinical measurement, to investigate dose-response relationships for T cell engagers (Susilo et al. 2023).

In this chapter, we present examples of QSP models designed with a range of mechanistic understanding of biological processes and a small amount of patient data from clinical trials. We believe future QSP impact in immuno-oncology will benefit from clinical/experimental analyses that lead to deepened knowledge of the cancer-immunity cycle (Mellman et al. 2023) and address controversial findings regarding mechanisms of action and resistance mechanisms of immunotherapy drugs, such as immune checkpoint inhibitors (Sharma et al. 2019; Meng et al. 2024). In addition, there is a trend toward integrating multi-omics analyses into QSP model parameterization (Lazarou et al. 2020; Zhang et al. 2021; Wang et al. 2023; Arulraj et al. 2023). This is because bioinformatics analyses (with machine learning-based tools) can provide patient-specific data that are not directly measurable in clinical trials, which leads to lowered uncertainty in model parameterization. With the growing ability to conduct in silico experiments/clinical trials, we envision QSP will gain more presence in model-informed drug development, including dose optimization and correlation/biomarker analyses for patient stratification to determine the optimal treatment for patients.

6 Code Availability

Assumptions, equations, and parameter estimation for models in Case Study 1 and 2 are elaborated by Ji et al. (2023) and Hardiansyah and Ng (2019), respectively. MATLAB code for QSP-IO can be found at https://github.com/popellab/QSPIO-TNBC (Wang et al. 2022) and https://doi.org/10.17632/r46rk4vwdv.1 (Arulraj et al. 2023).

Acknowledgments This study was supported by the National Institutes of Health grant R01CA138264.

References

Allen RJ, Rieger TR, Musante CJ (2016) Efficient generation and selection of virtual populations in quantitative systems pharmacology models. CPT Pharmacometrics Syst Pharmacol 5:140–146. https://doi.org/10.1002/psp4.12063

Anbari S, Wang H, Zhang Y et al (2023) Using quantitative systems pharmacology modeling to optimize combination therapy of anti-PD-L1 checkpoint inhibitor and T cell engager. Front Pharmacol 14:1163432. https://doi.org/10.3389/fphar.2023.1163432

Androulakis IP (2022) Towards a comprehensive assessment of QSP models: what would it take? J Pharmacokinet Pharmacodyn. https://doi.org/10.1007/s10928-022-09820-0

Arulraj T, Wang H, Emens LA et al (2023) A transcriptome-informed QSP model of metastatic triple-negative breast cancer identifies predictive biomarkers for PD-1 inhibition. Sci Adv 9: eadg0289. https://doi.org/10.1126/sciadv.adg0289

Arulraj T, Wang H, Deshpande A et al (2024) Virtual patient analysis identifies strategies to improve the performance of predictive biomarkers for PD-1 blockade. Proc Natl Acad Sci U S A 121(45):e2410911121. https://doi.org/10.1073/pnas.2410911121

Bai JPF, Earp JC, Florian J et al (2021) Quantitative systems pharmacology: landscape analysis of regulatory submissions to the US Food and Drug Administration. CPT Pharmacometrics Syst Pharmacol 10:1479–1484. https://doi.org/10.1002/psp4.12709

Baker RE, Peña J-M, Jayamohan J, Jérusalem A (2018) Mechanistic models versus machine learning, a fight worth fighting for the biological community? Biol Lett 14:20170660. https://doi.org/10.1098/rsbl.2017.0660

Benjamin DJ, Xu A, Lythgoe MP, Prasad V (2022) Cancer drug approvals that displaced existing standard-of-care therapies, 2016-2021. JAMA Netw Open 5:e222265. https://doi.org/10.1001/jamanetworkopen.2022.2265

Braakman S, Pathmanathan P, Moore H (2022) Evaluation framework for systems models. CPT Pharmacometrics Syst Pharmacol 11:264–289. https://doi.org/10.1002/psp4.12755

Bray F, Laversanne M, Sung H et al (2024) Global cancer statistics 2022: GLOBOCAN estimates of incidence and mortality worldwide for 36 cancers in 185 countries. CA A Cancer J Clin 74: 229–263. https://doi.org/10.3322/caac.21834

Butner JD, Dogra P, Chung C et al (2022) Mathematical modeling of cancer immunotherapy for personalized clinical translation. Nat Comput Sci 2:785–796. https://doi.org/10.1038/s43588-022-00377-z

Chelliah V, van der Graaf PH (2022) Model-informed target identification and validation through combining quantitative systems pharmacology with network-based analysis. CPT Pharmacometrics Syst Pharmacol 11:399–402. https://doi.org/10.1002/psp4.12766

Chelliah V, Lazarou G, Bhatnagar S et al (2021) Quantitative systems pharmacology approaches for immuno-oncology: adding virtual patients to the development paradigm. Clin Pharmacol Ther 109:605–618. https://doi.org/10.1002/cpt.1987

Chen DS, Mellman I (2013) Oncology meets immunology: the cancer-immunity cycle. Immunity 39:1–10. https://doi.org/10.1016/j.immuni.2013.07.012

Chen X, Hickling TP, Vicini P (2014) A mechanistic, multiscale mathematical model of immunogenicity for therapeutic proteins: part 1-theoretical model. CPT Pharmacometrics Syst Pharmacol 3:e133. https://doi.org/10.1038/psp.2014.30

Davis AA, Patel VG (2019) The role of PD-L1 expression as a predictive biomarker: an analysis of all US Food and Drug Administration (FDA) approvals of immune checkpoint inhibitors. J Immunother Cancer 7:278. https://doi.org/10.1186/s40425-019-0768-9

De Boer RJ, Perelson AS (2013) Quantifying T lymphocyte turnover. J Theor Biol 327:45–87. https://doi.org/10.1016/j.jtbi.2012.12.025

De Boer RJ, Hogeweg P, Dullens HF et al (1985) Macrophage T lymphocyte interactions in the anti-tumor immune response: a mathematical model. J Immunol 134:2748–2758. https://doi.org/10.4049/jimmunol.134.4.2748

Eisenhauer EA, Therasse P, Bogaerts J et al (2009) New response evaluation criteria in solid tumours: revised RECIST guideline (version 1.1). Eur J Cancer 45:228–247. https://doi.org/10.1016/j.ejca.2008.10.026

Erlendsson S, Teilum K (2021) Binding revisited—avidity in cellular function and signaling. Front Mol Biosci 7:615565. https://doi.org/10.3389/fmolb.2020.615565

Finotello F, Mayer C, Plattner C et al (2019) Molecular and pharmacological modulators of the tumor immune contexture revealed by deconvolution of RNA-seq data. Genome Med 11:34. https://doi.org/10.1186/s13073-019-0638-6

Gadkar K, Kirouac D, Mager D et al (2016) A six-stage workflow for robust application of systems pharmacology. CPT Pharmacometrics Syst Pharmacol 5:235–249. https://doi.org/10.1002/psp4.12071

Garcia-Cremades M, Pitou C, Iversen PW, Troconiz IF (2019) Translational framework predicting tumour response in gemcitabine-treated patients with advanced pancreatic and ovarian cancer from xenograft studies. AAPS J 21:23. https://doi.org/10.1208/s12248-018-0291-9

Ghaffari Laleh N, Loeffler CML, Grajek J et al (2022) Classical mathematical models for prediction of response to chemotherapy and immunotherapy. PLoS Comput Biol 18:e1009822. https://doi.org/10.1371/journal.pcbi.1009822

Gong C, Ruiz-Martinez A, Kimko H, Popel AS (2021) A spatial quantitative systems pharmacology platform spQSP-IO for simulations of tumor-immune interactions and effects of checkpoint inhibitor immunotherapy. Cancers (Basel) 13:3751. https://doi.org/10.3390/cancers13153751

Hahnfeldt P, Panigrahy D, Folkman J, Hlatky L (1999) Tumor development under angiogenic signaling: a dynamical theory of tumor growth, treatment response, and postvascular dormancy. Cancer Res 59:4770–4775

Hardiansyah D, Ng CM (2019) Quantitative systems pharmacology model of chimeric antigen receptor T-cell therapy. Clin Transl Sci 12:343–349. https://doi.org/10.1111/cts.12636

Harms BD, Kearns JD, Iadevaia S, Lugovskoy AA (2014) Understanding the role of cross-arm binding efficiency in the activity of monoclonal and multispecific therapeutic antibodies. Methods 65:95–104. https://doi.org/10.1016/j.ymeth.2013.07.017

Ippolito A, Wang H, Zhang Y et al (2023) Eliciting the antitumor immune response with a conditionally activated PD-L1 targeting antibody analyzed with a quantitative systems pharmacology model. CPT Pharmacometrics Syst Pharmacol 13(1):93–105. https://doi.org/10.1002/psp4.13060

Iyer SS, Cheng G (2012) Role of interleukin 10 transcriptional regulation in inflammation and autoimmune disease. Crit Rev Immunol 32:23–63. https://doi.org/10.1615/CritRevImmunol.v32.i1.30

Jafarnejad M, Gong C, Gabrielson E et al (2019) A computational model of neoadjuvant PD-1 inhibition in non-small cell lung cancer. AAPS J 21:79. https://doi.org/10.1208/s12248-019-0350-x

Jansson A, Barnes E, Klenerman P et al (2005) A theoretical framework for quantitative analysis of the molecular basis of costimulation. J Immunol 175:1575-1585. https://doi.org/10.4049/jimmunol.175.3.1575

Jarmoskaite I, AlSadhan I, Vaidyanathan PP, Herschlag D (2020) How to measure and evaluate binding affinities. eLife 9:e57264. https://doi.org/10.7554/eLife.57264

Ji Y, Madrasi K, Knee DA et al (2023) Quantitative systems pharmacology model of GITR -mediated T cell dynamics in tumor microenvironment. CPT Pharmacometrics Syst Pharmacol 12:413–424. https://doi.org/10.1002/psp4.12925

Kalos M, Levine BL, Porter DL et al (2011) T cells with chimeric antigen receptors have potent antitumor effects and can establish memory in patients with advanced leukemia. Sci Transl Med 3:95ra73. https://doi.org/10.1126/scitranslmed.3002842

Ko K, Yamazaki S, Nakamura K et al (2005) Treatment of advanced tumors with agonistic anti-GITR mAb and its effects on tumor-infiltrating Foxp3+CD25+CD4+ regulatory T cells. J Exp Med 202:885–891. https://doi.org/10.1084/jem.20050940

Kogan Y, Agur Z, Elishmereni M (2013) A mathematical model for the immunotherapeutic control of the Th1/Th2 imbalance in melanoma. Discrete Continuous Dyn Syst Ser B 18:1017–1030. https://doi.org/10.3934/dcdsb.2013.18.1017

Lazarou G, Chelliah V, Small BG et al (2020) Integration of omics data sources to inform mechanistic modeling of immune-oncology therapies: a tutorial for clinical pharmacologists. Clin Pharmacol Ther 107:858–870. https://doi.org/10.1002/cpt.1786

Leonov V, Mogilevskaya E, Gerasimuk E et al (2023) CYTOCON: the manually curated database of human in vivo cell and molecule concentrations. CPT Pharmacometrics Syst Pharmacol 12: 41–49. https://doi.org/10.1002/psp4.12867

Lever M, Maini PK, Van Der Merwe PA, Dushek O (2014) Phenotypic models of T cell activation. Nat Rev Immunol 14:619–629. https://doi.org/10.1038/nri3728

Lima EABF, Wyde RAF, Sorace AG, Yankeelov TE (2022) Optimizing combination therapy in a murine model of HER2+ breast cancer. Comput Methods Appl Mech Eng 402:115484. https://doi.org/10.1016/j.cma.2022.115484

Lindauer A, Valiathan C, Mehta K et al (2017) Translational pharmacokinetic/pharmacodynamic modeling of tumor growth inhibition supports dose-range selection of the anti-PD-1 antibody pembrolizumab: translational pharmacokinetic/pharmacodynamic modeling. CPT Pharmacometrics Syst Pharmacol 6:11–20. https://doi.org/10.1002/psp4.12130

Ma H, Wang H, Sove RJ et al (2020a) A quantitative systems pharmacology model of T cell engager applied to solid tumor. AAPS J 22:85. https://doi.org/10.1208/s12248-020-00450-3

Ma H, Wang H, Sové RJ et al (2020b) Combination therapy with T cell engager and PD-L1 blockade enhances the antitumor potency of T cells as predicted by a QSP model. J Immunother Cancer 8:e001141. https://doi.org/10.1136/jitc-2020-001141

Macallan DC, Asquith B, Irvine AJ et al (2003) Measurement and modeling of human T cell kinetics. Eur J Immunol 33:2316–2326. https://doi.org/10.1002/eji.200323763

Mellman I, Chen DS, Powles T, Turley SJ (2023) The cancer-immunity cycle: indication, genotype, and immunotype. Immunity 56:2188–2205. https://doi.org/10.1016/j.immuni.2023.09.011

Mempel TR, Henrickson SE, Von Andrian UH (2004) T-cell priming by dendritic cells in lymph nodes occurs in three distinct phases. Nature 427:154–159. https://doi.org/10.1038/nature02238

Meng L, Wu H, Wu J et al (2024) Mechanisms of immune checkpoint inhibitors: insights into the regulation of circular RNAS involved in cancer hallmarks. Cell Death Dis 15:3. https://doi.org/10.1038/s41419-023-06389-5

Messmer BT, Messmer D, Allen SL et al (2005) In vivo measurements document the dynamic cellular kinetics of chronic lymphocytic leukemia B cells. J Clin Invest 115:755–764. https://doi.org/10.1172/JCI23409

Milberg O, Gong C, Jafarnejad M et al (2019) A QSP model for predicting clinical responses to monotherapy, combination and sequential therapy following CTLA-4, PD-1, and PD-L1 checkpoint blockade. Sci Rep 9:11286. https://doi.org/10.1038/s41598-019-47802-4

Niederer SA, Aboelkassem Y, Cantwell CD et al (2020) Creation and application of virtual patient cohorts of heart models. Philos Trans A Math Phys Eng Sci 378:20190558. https://doi.org/10.1098/rsta.2019.0558

Nikfar M, Mi H, Gong C et al (2023) Quantifying Intratumoral heterogeneity and immunoarchitecture generated in-silico by a spatial quantitative systems pharmacology model. Cancers 15:2750. https://doi.org/10.3390/cancers15102750

O'Leary MC, Lu X, Huang Y et al (2019) FDA approval summary: Tisagenlecleucel for treatment of patients with relapsed or refractory B-cell precursor acute lymphoblastic leukemia. Clin Cancer Res 25:1142–1146. https://doi.org/10.1158/1078-0432.CCR-18-2035

Pardoll DM (2012) The blockade of immune checkpoints in cancer immunotherapy. Nat Rev Cancer 12:252–264. https://doi.org/10.1038/nrc3239

Racle J, De Jonge K, Baumgaertner P et al (2017) Simultaneous enumeration of cancer and immune cell types from bulk tumor gene expression data. eLife 6:e26476. https://doi.org/10.7554/eLife.26476

Ruiz-Martinez A, Gong C, Wang H et al (2022) Simulations of tumor growth and response to immunotherapy by coupling a spatial agent-based model with a whole-patient quantitative systems pharmacology model. PLoS Comput Biol 18:e1010254. https://doi.org/10.1371/journal.pcbi.1010254

Sadelain M, Brentjens R, Rivière I (2013) The basic principles of chimeric antigen receptor design. Cancer Discov 3:388–398. https://doi.org/10.1158/2159-8290.CD-12-0548

Sayama H, Marcantonio D, Nagashima T et al (2021) Virtual clinical trial simulations for a novel KRASG12C inhibitor (ASP 2453) in non-small cell lung cancer. CPT Pharmacometrics Syst Pharmacol 10:864–877. https://doi.org/10.1002/psp4.12661

Sermer D, Elavalakanar P, Abramson JS et al (2023) Targeting CD19 for diffuse large B cell lymphoma in the era of CARs: other modes of transportation. Blood Rev 57:101002. https://doi.org/10.1016/j.blre.2022.101002

Sharma A, Subudhi SK, Blando J et al (2019) Anti-CTLA-4 immunotherapy does not deplete FOXP3+ regulatory T cells (Tregs) in human cancers. Clin Cancer Res 25:1233–1238. https://doi.org/10.1158/1078-0432.CCR-18-0762

Siegel MB, He X, Hoadley KA et al (2018) Integrated RNA and DNA sequencing reveals early drivers of metastatic breast cancer. J Clin Invest 128:1371–1383. https://doi.org/10.1172/JCI96153

Sové RJ, Jafarnejad M, Zhao C et al (2020) QSP-IO: a quantitative systems pharmacology toolbox for mechanistic multiscale modeling for Immuno-oncology applications. Clin Pharmacol Ther 9:484–497. https://doi.org/10.1002/psp4.12546

Stefanski HE, Eaton A, Baggott C et al (2023) Higher doses of tisagenlecleucel are associated with improved outcomes: a report from the pediatric real-world CAR consortium. Blood Adv 7:541–548. https://doi.org/10.1182/bloodadvances.2022007246

Sterner RC, Sterner RM (2021) CAR-T cell therapy: current limitations and potential strategies. Blood Cancer J 11:69. https://doi.org/10.1038/s41408-021-00459-7

Stewart DJ, Stewart AA, Wheatley-Price P et al (2018) The importance of greater speed in drug development for advanced malignancies. Cancer Med 7:1824–1836. https://doi.org/10.1002/cam4.1454

Storek J, Lalovic BB, Rupert K et al (2002) Kinetics of B, CD4 T, and CD8 T cells infused into humans: estimates of intravascular: extravascular ratios and total body counts. Clin Immunol 102:249–257. https://doi.org/10.1006/clim.2001.5174

Sun D, Gao W, Hu H, Zhou S (2022) Why 90% of clinical drug development fails and how to improve it? Acta Pharm Sin B 12:3049–3062. https://doi.org/10.1016/j.apsb.2022.02.002

Surendran A, Le Sauteur-Robitaille J, Kleimeier D et al (2023) Approaches to generating virtual patient cohorts with applications in oncology. In: Cesario A, D'Oria M, Auffray C, Scambia G (eds) Personalized medicine meets artificial intelligence. Springer, Cham, pp 97–119

Susilo ME, Li C, Gadkar K et al (2023) Systems-based digital twins to help characterize clinical DOSE–RESPONSE and propose predictive biomarkers in a Phase I study of bispecific antibody, mosunetuzumab, in NHL. Clinical Translational Sci 16:1134. https://doi.org/10.1111/cts.13501

Thoma A, Farrokhyar F, McKnight L, Bhandari M (2010) Practical tips for surgical research: how to optimize patient recruitment. Can J Surg 53:205–210

Turk MJ, Guevara-Patiño JA, Rizzuto GA et al (2004) Concomitant tumor immunity to a poorly immunogenic melanoma is prevented by regulatory T cells. J Exp Med 200:771–782. https://doi.org/10.1084/jem.20041130

Vaghi C, Rodallec A, Fanciullino R et al (2020) Population modeling of tumor growth curves and the reduced Gompertz model improve prediction of the age of experimental tumors. PLoS Comput Biol 16:e1007178. https://doi.org/10.1371/journal.pcbi.1007178

Wang H, Milberg O, Bartelink IH et al (2019) *In silico* simulation of a clinical trial with anti-CTLA-4 and anti-PD-L1 immunotherapies in metastatic breast cancer using a systems pharmacology model. R Soc Open Sci 6:190366. https://doi.org/10.1098/rsos.190366

Wang H, Sové RJ, Jafarnejad M et al (2020) Conducting a virtual clinical trial in HER2-negative breast cancer using a quantitative systems pharmacology model with an epigenetic modulator

and immune checkpoint inhibitors. Front Bioeng Biotechnol 8:141. https://doi.org/10.3389/fbioe.2020.00141
Wang H, Ma H, Sové RJ et al (2021) Quantitative systems pharmacology model predictions for efficacy of atezolizumab and nab-paclitaxel in triple-negative breast cancer. J Immunother Cancer 9:e002100. https://doi.org/10.1136/jitc-2020-002100
Wang H, Zhao C, Santa-Maria CA et al (2022) Dynamics of tumor-associated macrophages in a quantitative systems pharmacology model of immunotherapy in triple-negative breast cancer. iScience 25:104702. https://doi.org/10.1016/j.isci.2022.104702
Wang H, Arulraj T, Kimko H, Popel AS (2023) Generating immunogenomic data-guided virtual patients using a QSP model to predict response of advanced NSCLC to PD-L1 inhibition. npj Precis Onc 7:55. https://doi.org/10.1038/s41698-023-00405-9
West J, Newton PK (2019) Cellular interactions constrain tumor growth. Proc Natl Acad Sci USA 116:1918–1923. https://doi.org/10.1073/pnas.1804150116
West GB, Woodruff WH, Brown JH (2002) Allometric scaling of metabolic rate from molecules and mitochondria to cells and mammals. Proc Natl Acad Sci USA 99(Suppl 1):2473–2478. https://doi.org/10.1073/pnas.012579799
Wong CH, Siah KW, Lo AW (2019) Estimation of clinical trial success rates and related parameters. Biostatistics 20:273–286. https://doi.org/10.1093/biostatistics/kxx069
Wu C, Jarrett AM, Zhou Z et al (2022) MRI-based digital models forecast patient-specific treatment responses to neoadjuvant chemotherapy in triple-negative breast cancer. Cancer Res 82:3394–3404. https://doi.org/10.1158/0008-5472.CAN-22-1329
Yarchoan M, Hopkins A, Jaffee EM (2017) Tumor mutational burden and response rate to PD-1 inhibition. N Engl J Med 377:2500–2501. https://doi.org/10.1056/NEJMc1713444
Zhang S, Gong C, Ruiz-Martinez A et al (2021) Integrating single cell sequencing with a spatial quantitative systems pharmacology model spQSP for personalized prediction of triple-negative breast cancer immunotherapy response. ImmunoInformatics 1–2:100002. https://doi.org/10.1016/j.immuno.2021.100002
Zhang Y, Wang H, Oliveira RHM et al (2022) Systems biology of angiogenesis signaling: computational models and omics. WIREs Mech Dis 14:e1550. https://doi.org/10.1002/wsbm.1550
Zhang Y, Popel AS, Bazzazi H (2023) Combining multikinase tyrosine kinase inhibitors targeting the vascular endothelial growth factor and cluster of differentiation 47 signaling pathways is predicted to increase the efficacy of antiangiogenic combination therapies. ACS Pharmacol Transl Sci 6:710–726. https://doi.org/10.1021/acsptsci.3c00008

Application of Mechanistic Mathematical Modeling to Toxicology: Quantitative Systems Toxicology (QST)

Kylie A. Beattie and Anna Sher

Contents

Abstract

Quantitative systems toxicology (QST) is emerging as an independent field of model-informed drug development (MIDD) with a focus on predicting toxicity endpoints. To enable toxicological predictions, QST models require incorporation of mechanistic details specific to safety applications including the ability to accurately model supratherapeutic doses and appropriately represent safety endpoints. Unique to the field of toxicology, mechanistic knowledge is often described through the use of adverse outcome pathways (AOPs), which formally represent existing knowledge about mechanisms of toxicity. The toxicities represented by QST models can arise from exaggerated or adverse pharmacological effects of engaging the drug's intended target (on-target toxicity) or from adverse events due to modulation of additional targets beyond the

K. A. Beattie (✉)
GSK, Stevenage, UK
e-mail: kylie.x.beattie@gsk.com

A. Sher
GSK, Waltham, MA, USA

E. L. Bradshaw, M. E. Spilker (eds.), *Quantitative Systems Pharmacology*,
Handbook of Experimental Pharmacology 289, https://doi.org/10.1007/164_2025_758

primary target (off-target toxicity). In cases of on-target toxicity, QST models can be considered as a type of Quantitative Systems Pharmacology (QSP) model that incorporates safety biomarkers and often includes simulations performed outside the therapeutic dose range to explore potential adverse consequences of exaggerated pharmacology in a pre-clinical or clinical setting. QST models assessing off-target toxicities can be considered distinct from QSP models in that they are typically applicable across molecules of a given modality which can (and often do) have different primary therapeutic targets. Off-target QST models commonly focus on the interrogation of general (e.g. pan-compound) toxicity mechanisms, often within a specific organ system. It can be difficult to categorize a model as purely QSP or QST (given that some models can be considered as both a QSP and a QST model), and therefore, we encourage readers to refer to a model based on its context of use and application. Thus, throughout this chapter, we refer to models as QST models when the context of use is to understand safety-related questions. To illustrate QST modeling approaches, examples of QST model applications for on-target and off-target toxicities at different stages of the drug discovery and development pipeline are presented and discussed. Additionally, contexts of use, triggers, key objectives, and potential impacts of QST models including the types of decisions QST applications can inform across drug discovery and development are reviewed. The chapter concludes with an overview of key challenges and future perspectives in the field of QST.

Keywords

Quantitative systems toxicology · Safety · Toxicity · Toxicodynamics · Toxicokinetics

1 Mechanistic Safety Modeling

Major causes of attrition within the pharmaceutical industry, particularly at phase 2 and phase 3 of drug development, continue to be attributed to lack of drug efficacy and the presence of drug safety concerns (Waring et al. 2015). Model-informed drug development (MIDD) approaches, such as Quantitative Systems Pharmacology (QSP) modeling for quantitative, mechanistic understanding of disease, (patho)-physiological pathways, and pharmacological interactions of drugs with biological systems, have the potential to increase the probability of success (PoS) and/or probability of technical and regulatory success (PTRS) by optimizing target identification and validation, dosing regimen, biomarker and population selection throughout drug discovery and development stages. Such mechanistic models can also be applied to address questions arising post-approval. Toxicological interactions of drugs with biological systems are explored through the application of mechanistic safety modeling in the field of toxicology, known as Quantitative Systems Toxicology (QST), which has the potential to improve PoS and/or PTRS by reducing toxicity-related attrition during drug development.

Quantitative studies of toxicity pathways and mechanisms exist at an intersection of systems biology, toxicology, and chemistry. QST models therefore have applications across a number of different industries including pharmaceutical, petroleum, agrochemical, cosmetics, food and drink, tobacco, as well as in the development of household products. Within the pharmaceutical industry, QST can play a role in the development of different medicinal products and may be referred to by different terminology, for example, as part of model informed vaccine development approaches (Desikan et al. 2024), where modeling of vaccines toxicity is more commonly referred to as 'vaccine reactogenicity' modeling.

QST models are highly relevant in the context of animal testing paradigms and related legislation associated with the refinement, reduction and replacement (3Rs) of animal use in safety testing with additional potential to impact experimental study designs and understanding of translation of toxicity findings or concerns (Harrell et al. 2024). To impact these areas, QST models integrate classical toxicology with mechanistic mathematical and computational approaches and quantitative measurements of molecular, biological, and physiological functions (Sturla et al. 2014; Bloomingdale et al. 2017; Hartung et al. 2017). This chapter will focus on QST approaches that aid in identifying and understanding mechanisms of toxicity and can be used to assess hazard and risk associated with toxic effects and adverse outcomes of an exogenous or endogenous biological or chemical intervention on biological systems across pre-clinical or clinical stages of drug discovery and development.

Over the past decade, QST has become an increasingly recognized independent field with multiple consortia and initiatives focused on advancing the development of QST model applications in the pharmaceutical industry. These include (i) comprehensive in vitro Proarrhythmia Assay (CiPA), a US Food and Drug Administration (FDA)-led initiative for predicting cardiotoxicity (Strauss et al. 2019) which was started in 2013, (ii) Innovative Medicines Initiative 2 (IMI2) TransQST, an academia and industry initiative co-funded by the European Union with a focus on developing QST models in four key organ systems: cardiovascular, liver, kidney and gastrointestinal (Ferreira et al. 2020) started in 2017, (iii) the DILI-sim Initiative, focused on development of a QST model for predicting drug-induced liver injury (Watkins 2022) started in 2011, and (iv) the RENAsym® Consortium, a pre-competitive industry partnership focused on modeling renal injury (web1) launched in 2018. Despite these active efforts to develop new QST modeling capabilities, QST remains a relatively young field. Nonetheless, even as a relatively new discipline, QST is becoming a recognized approach that is part of MIDD (Marshall et al. 2023). Overall, current state-of-the-art QST modeling spans different therapeutic areas with differing maturity levels. For instance, cardiac electrophysiology and DILI modeling are relatively mature with examples of QST modeling being used routinely as part of safety assessment strategies. On the other hand, mathematical modeling of gastrointestinal toxicity is less mature with fewer current applications (Beattie et al. 2024).

From a mathematical modeling perspective, QST models, similar to QSP models, vary in size from less than ten to a few thousand variables and can be represented as ordinary differential equations (ODEs), partial differential equations (PDEs), agent-based modeling or other. Similar to QSP, a QST model has species-specific parameterizations and represents key physiological processes of interest with relevance to the toxicity of concern typically in a given organ system. Assumptions related to drug or target modulation effect on the biological system and parameterization of the drug effect often include a means to translate the drug concentrations and effects across species, especially when used in a predictive capacity in a pre-clinical context of use. Calibration, validation, and verification frameworks are as essential to QST model development as they are for QSP model development (as outlined in 'A Framework for Quantitative Systems Pharmacology Model Execution' (Sokolov et al. 2025)). General MIDD guidance documents incorporating QSP and QST models are currently being developed as part of International Council for Harmonisation (ICH) M15 (Marshall et al. 2023; web2), and by the European Medicines Agency (EMA) (web3).

Categorizing a model as a QSP or QST model can be difficult; therefore, we encourage readers to refer to a model based on its context of use and application. Furthermore, while QSP models typically focus on representing pharmacology related to the primary target of the molecule being modeled, QST models can be developed for both on-target and off-target toxicities. On-target toxicity refers to exaggerated or adverse pharmacological effects resulting from drug interaction with the drug's intended target. Off-target toxicities refer to adverse effects caused by modulation of targets other than the intended drug target. QST models and applications for understanding both types of toxicities are elaborated on further in this chapter.

Adverse outcome pathways (AOPs) and AOP networks can be considered as the blueprints for some on-target or off-target QST models. AOPs allow a toxicologist to map out, at different biological scales of organization, the mechanistic details of a particular toxicity outcome, starting with a molecular initiating event, leading to any number of known intermediatory key events, which ultimately lead to an adverse outcome at the whole organism scale (Ankley and Edwards 2018; web4; web5). An AOP can form the foundation for a QST model, and similarly, a QST model may be able to help explore potential hypotheses and further refine putative AOPs. Characterizing the key events in an AOP often helps identify a series of biomarkers most informative for that key event and, therefore, facilitates identifying pathway components and predicted endpoints to include in the QST model. Moreover, review of the proposed AOP or AOP network can provide additional justification to measure these select biomarkers experimentally, thus generating valuable data for QST model calibration and/or validation. Within the field of QST, the AOP is equivalent to the mapping out of biological knowledge to create a conceptual diagram of a QSP model, as described in the atopic dermatitis example in 'What Is QSP and Why Does It Exist?: A Brief History' (Fig. 1 in Friedrich (2024)). However, in contrast to AOPs in risk assessment and collating toxicology knowledge, the QSP model mapping exercise performed for pharmacological efficacy assessment often takes place informally, with no current standardized methodology or universally recognized terminology.

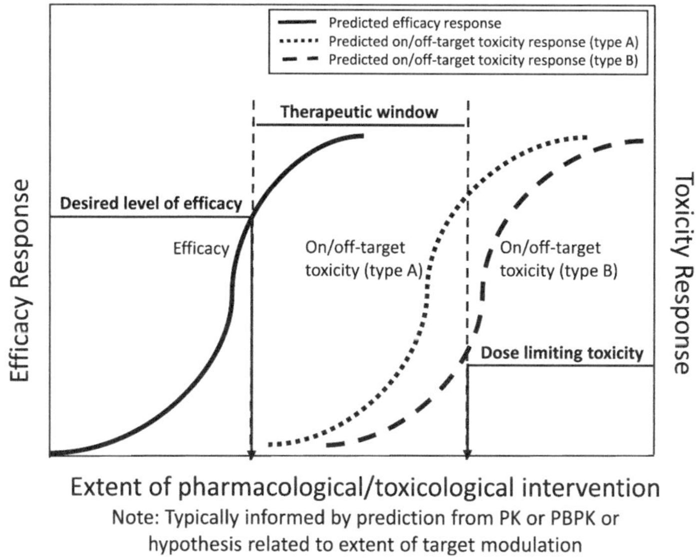

Fig. 1 Illustration of a therapeutic window. The y-axes denote the pharmacological (efficacy) or toxicological (toxicity) response. The x-axis represents the extent of on-target or off-target intervention either as the dose administered, drug concentration or degree of target modulation. Black solid, square-dotted, and dashed curves exemplify predictions of efficacy, and two distinct on-and/or off-target toxicities, respectively. Note there may be numerous toxicity curves for both on-and off-target toxicities, and these may be in different locations and/or order to the examples shown here. The weight of different curves to the overall safety margin may also vary depending on the severity and monitorability of the particular toxicity, with these factors influencing what is considered to be the dose limiting toxicity. The therapeutic window is shown as the range between the concentrations of minimum desired efficacy and of predicted dose limiting toxicity. PK pharmacokinetic, PBPK physiologically based pharmacokinetic

QST models can be utilized to inform the therapeutic window of a compound, providing information about the safety margins for various toxicities of concern. Figure 1 shows an adapted version of a classical illustration of a therapeutic window, where the y-axes show the pharmacological or toxicological responses and the x-axis represents the extent of the pharmacological or toxicological intervention, where the latter could be in the form of dose administered, plasma exposure, tissue exposure, or degree of target modulation. As shown in Fig. 1, the therapeutic window (defined as the window between where the minimal desired level of efficacy is achieved and a level of unacceptable toxicity is anticipated to occur) informs on the difference between efficacious and toxic drug concentrations or target modulation which quantitatively facilitates internal and/or regulatory decision making related to the progression of a target or an asset. Overall QST modeling aims to provide additional quantitative evidence to support decision making alongside the traditionally used metrics of toxicological assessment such as the no observed adverse effect level (NOAEL) and lowest observed adverse effect level (LOAEL).

QST models can quantitatively inform the toxicity component of the therapeutic window, where toxicity profiles for both on- and off-target toxicities can impact the shape and size of the therapeutic window. It is important to note that not all toxicities are equal and some may have more weight than others when evaluating a safety margin. To define the full model-based predicted therapeutic window, an accompanying model prediction of an efficacy biomarker or endpoint is also required, which highlights the potential for a close interplay between, and/or combined use of, QST and QSP models.

A key difference between pharmacological/efficacy and toxicological/safety modeling is the type of data and knowledge that model components are focused on and derived from, including toxicity-specific data and knowledge, in safety modeling. Throughout this chapter, we use the terms of pharmacology and efficacy as well as of toxicology and safety interchangeably; however, readers should acknowledge multiple potential contexts of use of these terms. For example, endpoint predictions related to efficacy or safety may not be the only outputs of QSP and QST models, as these can also include mechanistic or biomarker information. Figure 2 presents key components of pharmacological and toxicological quantitative modeling, highlighting commonalities and differences, as well as the overlap in some contexts of use. Toxicokinetic (TK) and physiologically based toxicokinetic

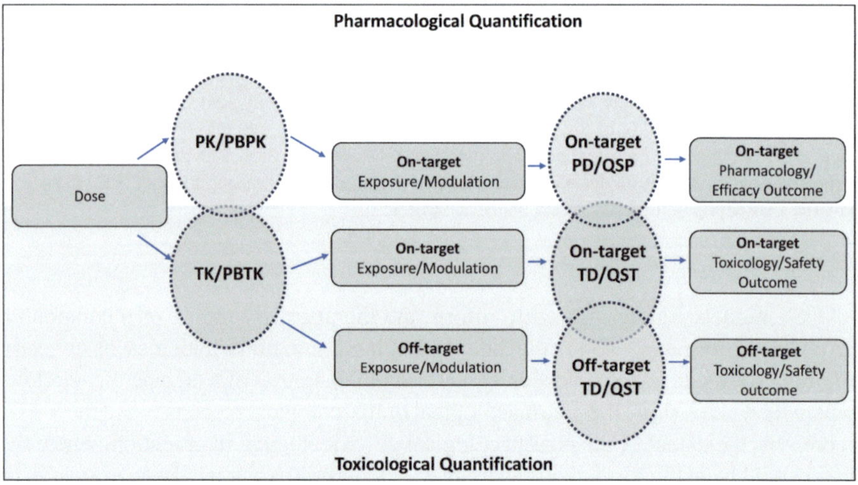

Fig. 2 Comparison of key components of mechanistic modeling of pharmacological and toxicological effect. The figure highlights the connection of the analogous components as well as the overlap of mechanistic pharmacological and toxicological modeling when predicting on-target safety outcomes. PK pharmacokinetic, TK toxicokinetic, PBPK physiologically based pharmacokinetic, PBTK physiologically based toxicokinetic, PD pharmacodynamic, TD toxicodynamic, QSP quantitative systems pharmacology, QST quantitative systems toxicology. The arrows included in the diagram are illustrative, and as described in the text, there may be more interconnections between the various model components in different contexts of use. Note that in this figure, we refer to the outputs of the QSP and QST models as pharmacology/efficacy and toxicology/safety outcomes, respectively, as described in the main text

(PBTK) models quantitatively describe the relationship between dose and drug exposure in plasma or at the site of action. In the absence of using a TK or PBTK model, some contexts of use employ a drug concentration range of interest or, alternatively, assumptions related to target modulation may also be appropriate. The latter are particularly relevant for contexts of use involving preliminary exploration with QST models. Toxicodynamic (TD) or mechanistic systems toxicology models (QST) predict toxicity outcomes of interest such as biomarkers and/or clinical scores. On-target TD/QST models are likely to share similarities with PD/QSP models given the common primary mechanism of action of the therapeutic; however, the two models may be focused on different cell systems and manifestations of pathway modulation for toxicology and pharmacology application. Beyond common pharmacokinetic components, off-target TD/QST models may differ significantly from the on-target mechanistic models given their focus on alternative mechanisms from the therapeutic's primary target. Additionally, while Fig. 2 highlights the usefulness of both TD and QST models, the mechanistic nature of QST models more readily permits extrapolation beyond acquired datasets to impact questions that may not be suited for TK/TD models.

QST applications are growing across drug discovery and development with clear trends of increased use in both pre-clinical and clinical stages of drug development to complement and advance traditional and novel experimental approaches to understanding safety (Beattie et al. 2024). In this regard, QST plays a key role in integrating experimental and computational approaches with mechanistic knowledge to gain new insights into toxicity. The next two sections focus on QST examples of on-target and off-target mechanistic safety modeling, respectively.

2 On-Target QST Modeling

QST models used to explore on-target toxicities are often QSP models with extensions or modifications to allow for simulations of safety endpoints either under normal or in exaggerated pharmacology scenarios. Multiple examples of QST models that are QSP or semi-mechanistic PK/PD models with a focus on safety outcomes can be found in oncological applications (e.g. Friberg et al. 2002; Craig 2017; Collins et al. 2017; Chen et al. 2020a, b; Mackey et al. 2021) where mechanistic models are employed to inform the on-target side effects of chemotherapy such as prediction of the probability of, and the severity of, neutropenia or of increased risk of bleeding due to thrombocytopenia. Additional examples exist in other therapeutic areas including hemostatic and immune system disorders where prothrombin deficiency increases the risk of bleeding or hypogammaglobulinemia is associated with an increased risk of infections (e.g. Nayak et al. 2015).

Below we discuss a mechanistic model developed by Mackey et al. (2021) (Fig. 3) that suggests an optimal frequency of drug administration to reduce the toxic side effects of chemotherapy in childhood acute lymphoblastic leukaemia (ALL) patients. This case study is an example of a QST model application to

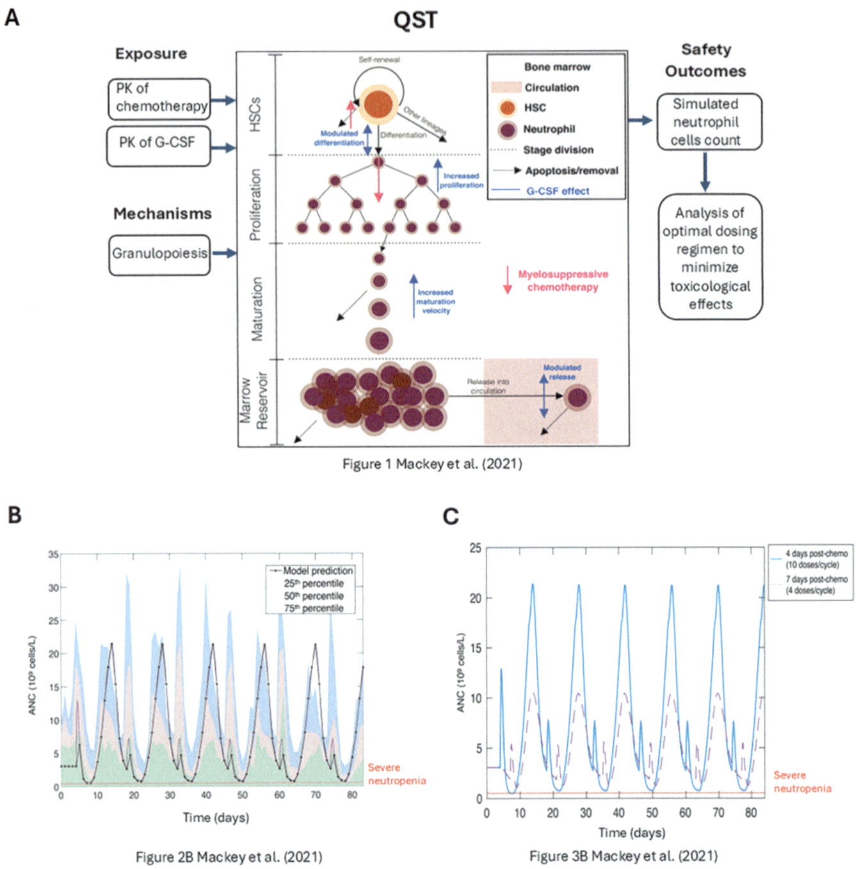

Fig. 3 An on-target QST application example: summary of Mackey et al. (2021) model. Model inputs include clinical PK/PD data and a mechanistic description of granulopoiesis. The model evaluates the risk of on-target toxic side effects based on the timing of drug-induced cytotoxic chemotherapy. The model aims to rationalize and personalize the clinical therapeutic strategy based on predicted safety biomarkers of neutropenia and neutrophilia. Panel A (Figure 1 from Mackey et al. (2021)) shows a QST model schematic. Panel B (Figure 2B from Mackey et al. (2021)) shows model predictions at clinical sampling points (black) compared to quartile results (blue: 75% quartile, pink: median, green: 25% quartile) from the CHOP14 trial of the German high-grade non-Hodgkin's lymphoma study group. Panel C (Figure 3B from Mackey et al. (2021)) shows model predictions of neutrophil oscillations when granulocyte-colony stimulating factor (G-CSF) is given 4 days (blue) and 7 days (purple) post-chemotherapy with severe neutropenia level shown as red solid line. As shown, the model predicts that timing administration of G-CSF 6-7 days post-chemotherapy (instead of 4 days) better controls the neutrophilic nadir. ANCs absolute neutrophil counts, HSCs hematopoietic stem cells, PK pharmacokinetics. Reproduced from Mackey et al. (2021)

mitigate myelosuppressive effects of cytotoxic chemotherapy in childhood ALL patients.

The authors analysed neutrophil counts from the childhood ALL cohort of 286 Caucasian children diagnosed with ALL at the Sainte-Justine University Health Center (Montreal, Canada) and identified resonance in neutrophil count occurring in lymphoma patients undergoing cyclic chemotherapy. The authors then leveraged a QST model to predict a new post-chemotherapy scheduling regimen of prophylactic treatment with granulocyte-colony stimulating factor (G-CSF) in ALL pediatric patients to reduce the risk of hematological toxic side effects such as chemotherapy-induced neutropenia given the regularity of neutropenia patients experienced. In brief, the Mackey et al. (2021) model (Fig. 3a) combines their previously developed QSP model of granulopoiesis with PK/PD models of chemotherapy and of G-CSF (Craig 2017; Craig et al. 2015, 2016) to provide mechanistic insight and suggest a clinically actionable schedule to reduce neutropenia and neutrophilia effects. Relevant biology includes mechanisms of hematopoiesis, neutrophilia and neutropenia, as well as the pathophysiology associated with ALL and myelosuppression. The model leverages data of neutrophil counts following cytotoxic chemotherapy with G-CSF prophylactically administered to mitigate the risk of neutropenia. The PK of the chemotherapy was defined by a 4-compartment model. The cytotoxic drug effect of the chemotherapy (i.e. the PD of chemotherapy) was assumed to affect the hematopoietic stem cells (HSC) death rate and the expansion of neutrophil progenitors. The PK of G-CSF was modeled by two ODEs to describe binding and unbinding kinetics of G-CSF to maturing bone marrow neutrophils, mature reservoir-based neutrophils and circulating neutrophils. The mechanism of action of G-CSF (i.e. the PD of G CSF) mechanistically linked G-CSF with HSC and neutrophil-lineage cells. Calibration and validation of the model included the use of childhood ALL cohort data (Fig. 3b), in vitro studies, and literature data. The incorporation of a mechanistic granulopoiesis model into the PK/PD models predicted that delaying supportive G-CSF therapy by 6-7 days after chemotherapy would reduce the risk of myelosuppressive effects by controlling both the neutropenic nadir and the neutrophilic rebound. QST modeling outputs provided on-target toxicity quantification by defining the onset of chemotherapy-induced toxicity and the authors proposed a clinically actionable schedule for pediatric patients where each patient may receive the most efficacious and least toxic effects of chemotherapy. Model impact has not yet been realized in clinical practice, but according to the authors of the paper, clinical work is ongoing.

This modeling work demonstrated that the length of chemotherapy cycles (with and without G-CSF) has an important influence on the severity of the neutrophil nadir. While limitations of this work are that clinical confirmation of neutrophil resonance has yet to be reported and that the neutrophil resonance pattern was identified only in 26% of patients in the ALL cohort studied, this QST example is critical in highlighting the need for continued assessment of chemotherapy schedules using quantitative approaches to minimize toxic side effects.

This model exemplifies the interchangeability of QSP and QST terminology, where Mackey et al. (2021) refer to their model as a QSP model (and not a QST model) even though in this application the outcomes of interest are on-target toxicity (in this case represented as neutropenia and neutrophilia). Thus, given its context-of-use and its focus on safety endpoints, it is recommended to categorize this model as a QST model.

This case study demonstrates how QST models can be applied to minimize on-target toxicological liabilities and inform optimal dosing regimens. This could have significant implications for optimal trial design and/or optimal drug regimens post-approval at the individual level. Additionally, the above case study illustrates how QST modeling can provide unique contributions to the weight of evidence for decision making in pharmaceutical and clinical practice through an understanding of exaggerated pharmacology and mitigation of toxicological effects.

3 Off-Target QST Modeling

The previous section illustrates the application of QST modeling to address an on-target safety question. Additionally, QST models can be used to investigate off-target toxicity concerns. In this section, we summarize a QST modeling example applied to understand liabilities associated with drug-induced liver injury (DILI).

DILIsym® is a QST model which can be used to predict changes in liver biomarkers to inform on the potential for DILI (Watkins 2019). DILIsym® includes three key components: (i) a mathematical model of liver biochemistry, (ii) a PBPK/TK model used to represent the liver concentration of the compound which drives liver toxicity in the model, and (iii) a series of model assumptions relating to the mechanisms of toxicity for each compound. The model of liver biochemistry is parameterized for the species being modeled (including, for example, representation of species differences in normal liver physiology) since DILIsym® has the ability to model both pre-clinical species (rat, mouse and dog) and human (Watkins 2020). Species-specific PBPK models for each compound are used to inform the predicted liver exposure for each compound under different dosing regimens. Species-specific response to the toxicity mechanisms represented in DILIsym® are also modeled. The mechanisms of toxicity include inhibition of bile acid transport, generation of oxidative stress and mitochondrial dysfunction. In vitro data produced for a compound to be modeled is used to identify the appropriate mechanisms of toxicity to be represented in the DILIsym® model and then to quantitatively represent the exposure-response relationship for each mechanism included in the model. The DILIsym® model, integrating together the predicted liver concentrations and the in vitro data informing on mechanisms of toxicity within the mathematical model of liver biochemistry, is then used to simulate different dosing scenarios to explore changes in predicted liver biomarkers in different regimens. Figure 4 summarizes the key components of the DILIsym® model including the steps required to build a model (panel A) and example results of fitting to in vitro data to inform toxicity mechanisms in the model (panel B).

Woodhead et al. (2022) applied the DILIsym® model to explore the DILI liability for a series of compounds with the same target, the calcitonin gene-related peptide

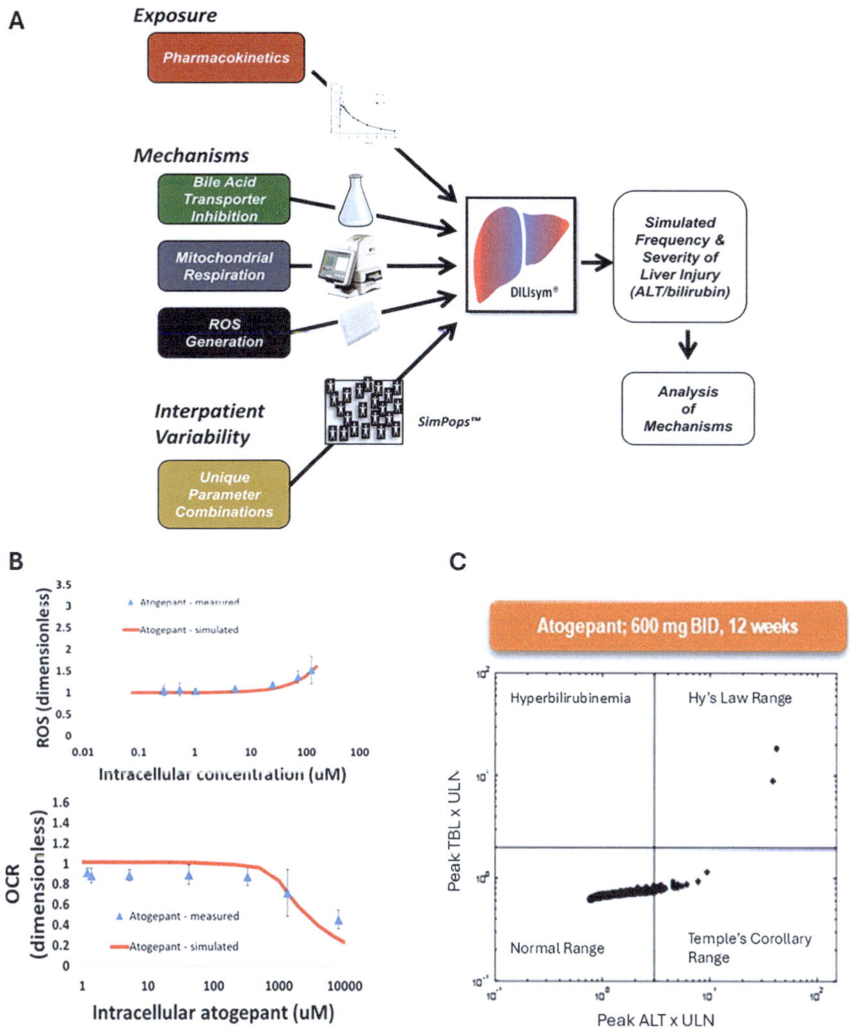

Fig. 4 An off-target QST application example: summary of Woodhead et al. (2022) DILIsym modeling. Panel A shows key model inputs and outputs with examples of mechanisms of toxicity included in the DILIsym® model. The three components described in the text are shown in Panel A, with the compound PBPK and in vitro data information on mechanisms of toxicity being integrated with the mathematical model of liver biochemistry to perform DILIsym model simulations. The case study described in this chapter only includes application of the model to human species. Model inputs include concentration-time profiles from the PBPK model and in vitro data informing on mechanisms of toxicity which are used to derive model parameters. Panel B shows example calibration results of fitting to in vitro data to inform toxicity mechanisms in the model. Panel C shows an example of model outputs with model simulations in the form of an evaluation of drug-induced serious hepatotoxicity (eDISH) plot which informed the exploration of DILI liability for the CGPR inhibitor atogepant. Reproduced from Watkins (2019); Woodhead et al. (2022). ROS reactive oxygen species, OCR oxygen consumption rate, ALT alanine aminotransferase

(CGRP) receptor. An assessment of DILI liability was prompted following the observation that a compound with the same target (telcagepant) had clinical hepatoxicity findings. The DILI observation in the clinic was not thought to be an on-target effect of CGRP receptor inhibition as (i) CGRP knockout mice did not develop adverse outcomes in the liver and (ii) other CGRP inhibitors such as erenumab were not associated with liver safety concerns (Smith et al. 2020). In this example, DILIsym® was used to explore the liability for off-target toxicities in the liver via a series of mechanisms represented in the QST model. The aim of the study was to compare the hepatoxicity profiles of the four CGPR receptor antagonists (rimegepant, zavegepant, atogepant, and ubrogepant). The availability of clinical data for the CGPR inhibitor talcagepant provided validation data for evaluating the human DILI prediction for that compound and consequently a toxicity profile against which the other CGPR inhibitors could be compared.

In vitro study results for each compound were used to parameterize the mechanisms of toxicity represented in the model. Clinically measured liver enzyme data for telcagepant were used to validate DILIsym® model predictions before simulating human liver biomarkers for the other four CGPR receptor antagonists. For telcagepant, simulation results suggested that hepatoxicity was driven by the combined mechanisms of electron transport chain (ETC) inhibition and bile acid transport inhibition, while for the other compounds, where changes in clinical liver biomarkers were predicted, only a single mechanism was identified as the main contributor to the predicted hepatotoxicity profile (specifically either reactive oxygen species (ROS) generation or ETC inhibition).

Figure 4 (panel C) shows example modeling results (in the form of an evaluation of drug-induced serious hepatotoxicity (eDISH) plot) which informed the assessment of DILI liability for the CGPR inhibitor atogepant. The example shown is for dosing 600 mg BID for 12 weeks, and the eDISH plot summarizes predicted changes in alanine aminotransferase (ALT) and bilirubin across a simulated population.

The percentage of the simulated population with ALT changes greater than 3x the upper limit of normal are also calculated. Predictions for the four next-generation CGPR receptor antagonists were then compared with those for telcagepant and were predicted to have significantly less DILI liability than telcagepant, providing evidence to support their progression to later stage clinical trials. Modeling led to the hypothesis that either an adaptive response to ROS, which is not included in DILIsym®, or mitochondrial biogenesis could prevent serious injury from occurring from the next-generation CGRP receptor antagonists—providing further mechanistic insight. The outcome of this QST model application was to inform the clinical progression decision for the series of related compounds (Woodhead et al. 2022; web6). Insights from this exercise could also inform further mechanistic follow-up studies (e.g. to explore ROS and mitochondrial biogenesis hypotheses) or suggest modification to standard clinical protocol designs to further explore DILI liability. The key added value of the QST model application to explore off-target toxicities in this case study is the ability to use clinical information already available to understand potential for DILI for compounds still in development. Having the ability to gain these insights at this earlier stage of drug development and having modeling evidence, which can add to the weight of evidence for decision making in pharmaceutical development, can be invaluable in selecting the compounds with the highest probability of being safe and efficacious in the clinic. The ability to explore toxicity responses under different dosing scenarios, as in this example, allows for the selection of the regimen which achieves efficacy while minimizing the associated predicted toxicity. This case study is also an example of where QST and QSP models could be used in conjunction as there would be the necessity to confirm that the compound is predicted to be efficacious with a proposed dosing regimen where the safety liability has been explored.

4 QST Applications and Impacts in Toxicology

The specific questions addressed by QST, the trigger for these questions, as well as potential QST model impact(s) will vary depending on the stage of drug discovery and development at which a QST model is developed and applied as well as its context of use. Table 1 highlights some common QST model contexts of use, model triggers, key modeling objectives, and model impacts. It also provides an overview of the types of decisions QST applications can inform across drug discovery and development.

During the earliest stages of drug discovery, the application of a QST model may be triggered by a particular flag from review of the target biology. For example, there may be known safety concerns associated with the target's mechanism of action or a precedence with competitor molecules with the same target. At the stage of target identification, there may be limited information or knowledge about the intended modality and limited information on drug concentrations in plasma or relevant tissues. QST modeling can be used to directly explore potential consequences of target modulation on downstream events or to investigate hypothetical PK profiles. These early applied QST models may have high uncertainty in model predictions; however, complementary modeling techniques such as performing uncertainty quantification and sensitivity analyses for the key model outputs mean that these QST approaches can generate qualitative or initial quantitative insights to support progression decisions related to target selection and validation.

In lead discovery and optimization stages, additional compound-specific information typically becomes available which can be used to inform QST modeling. Such information may arise from in vitro experiments, including high-throughput in vitro experimental screens that generate evidence to support lead compound advancement. For toxicity mechanisms that are well understood, such as cardiac ion channel inhibition leading to disruption of cardiac electrophysiology, high-throughput screening data informing on specific key events can be integrated in QST models to give an early indicator of potential toxicity liability. In such circumstances, the QST model can be incorporated as a routine screen, applied to all relevant compounds where the toxicity is of concern as soon as model input data becomes available (e.g. Mohr et al. 2022). This may then allow for modeling applications in a medium to high-throughput manner, with the potential for an

Table 1 Illustrative examples of QST model contexts of use, triggers, objectives, and potential impacts of model outcomes across the different stages of drug discovery and development

	Stages of drug discovery and development							
	Target validation	Lead discovery and optimization		Pre-clinical testing		First time in human		Further clinical development
Context of use	Gain confidence in rationale of a target	Inform modality selection	Identify mechanism(s) of toxicity	Inform first time in human studies including human dose prediction		Inform biomarker additions to clinical trial		Inform patient population selection
Model trigger	Target flag	Routine screen	Phys chem property	In vivo toxicity finding		Clinical toxicity finding		
Model scope/ objective	Understand target safety liability	Predict toxicity liability	Understand pre-clinical toxicity finding	Translate target safety liability	Inform clinical trial design	Support compound progression	Inform clinical trial design	Interpret clinical toxicity finding
Model impact	Inform target progression	Inform termination decision	Inform prioritization decision	Inform internal resourcing decision	Inform clinical trial design	Support regulatory submission		Impact label or support in-licensing

The listed illustrative examples of model triggers, context of use, objectives and impacts are not limited to the stage these are listed under in the table as these can occur at various stages of drug discovery and development, spanning from the earliest stages of drug discovery, when a potential target is being identified and validated, through pre-clinical and clinical development of an asset, to post-approval drug administration and ultimately to post-marketing drug surveillance. (This table is an extended version of Table 1 in Beattie et al. (2024))

automated workflow where model simulations are generated as soon as ion channel data is generated. Model outputs can be also used in decision making alongside high-throughput in vitro data, for example, providing evidence that integrates the effects of multiple ion channel blocks on cardiac electrophysiology as opposed to relying on interpretation of results from blockage of each individual ion channel, where it may be challenging to foresee the combined effect from the data alone. For other QST modeling applications performed at this stage, which may require a more bespoke modeling approach or more involved pre-processing or fitting of data, the level of throughput described may not be feasible due to resource constraints or lack of necessary mechanistic knowledge.

As compounds progress through pre-clinical development, QST models can play an integral part of safety assessment strategies and facilitate quantitative decision making to support compound progression, prioritization, and/or termination decisions. This type of QST modeling approach is most often used for screening off-target toxicities, where a specific toxicity mechanism represented in the model may be applicable to multiple targets, thus being applicable across portfolio projects, typically for assets of the same modality types. Examples of screening type models include platform models such as the DILIsym® model (discussed in detail in the previous section) as well as mathematical models of cardiac electrophysiology which integrate ion channel screening data described in the paragraph above (e.g. Mirams et al. 2011; Beattie et al. 2013).[1] When in vivo toxicity data is generated, this can be used alongside in vitro data to further refine and validate QST model predictions. There may also be opportunities during pre-clinical development to generate bespoke data to support QST model building or validation by adding biomarker endpoints to a toxicity study protocol or by designing specific in vitro experiments to explore a hypothesized toxicity mechanism and provide quantitative data to model that mechanism. In pre-clinical development, QST models are particularly used to: (1) understand mechanisms of toxicity when there is a concern (either related to the target or from an experimental finding) and (2) understand the translation of the toxicity from in vitro and/or in vivo studies to humans. Similar to QSP applications, QST applications are used to enable risk assessment to support progression of a compound to clinical development and to address specific questions in different stages of early drug discovery and development, which may not be practical or feasible with in vitro or in vivo experiments alone. QST during pre-clinical development stages can contribute to defining and optimizing planning of clinical trial design, establishing a safe clinical dose and regimen, and identifying specific populations, which may be more susceptible to a particular toxicity.

In clinical development and as clinical trials progress, emerging clinical data can be used to further refine and calibrate QST models. Clinical trial observations can be compared to prospective model predictions made during pre-clinical development to

[1] Note that the highlighted above publications pre-date the terminology of 'QST', so while the models are referred to as 'mathematical models' or 'biophysical models' within the publications, nowadays these models would be categorized as 'QST' models, adopting more recent terminology.

evaluate model performance and support model refinement to facilitate future QST translational activities as well as QST efforts at clinical stages and beyond. QST can be leveraged to support understanding of safety associated with combination dosing regimens. QST efforts at clinical stages of development can also focus on ensuring that the compound will be safe in the wider post market population and that the label accurately reflects the risks associated with the compound. Similar to QSP, QST can facilitate business development decisions on in-licensing a drug by providing additional quantitative toxicological confidence in mechanistic rationale. QST model applications can be triggered for the first time during clinical development, for example, following adverse findings during clinical trials. Moreover, if there is a precedent for toxicity events from other compounds with the same target, it is important to explore this potential liability and to the fullest extent possible develop an understanding of the toxicity mechanism(s) to de-risk the toxicity. With the execution of an effective end-to-end QST strategy, this latter activity would take place earlier in drug development with the aim to avoid clinical trial toxicity findings. A QST model developed during clinical stages can both facilitate the progression of (or sometimes contribute to termination of) a given compound through clinical development and provide insights for developing other compounds targeting the same mechanism.

The life cycle of a compound continues beyond clinical development stages and approval by a regulatory body. For example, the marketed drug could become part of a combination strategy with a novel compound, requiring safety assessment to evaluate the risk of adverse events. Another example relates to a marketed drug being used in the clinic with a need to identify ways to minimize the drug's toxic side effects. Such situations are common in indications such as oncology or diseases where the need for any cure or slowing of disease progression is of higher precedence than the presence of non-life threatening toxic side effects. Notably, QST applications throughout the different stages of drug discovery and development necessitate an appropriate level of model qualification or validation that is commensurate with the significance of the decision they are intending to impact. QST-driven decisions require an evaluation of the confidence in the model predictions given the data and knowledge informing it. It is also important to carefully consider the specific weight of evidence that the model prediction has on the decision relative to other data and considerations. Hence, similar to QSP modeling, inclusion of uncertainty quantification (see 'A Framework for Quantitative Systems Pharmacology Model Execution' by Sokolov et al. (2025)) is an important aspect of QST modeling.

To summarize, predictions from QST models can contribute to the evidence in quantitative hazard and risk assessments related to adverse outcomes across the discovery, pre-clinical and clinical stages of drug development, where safety-related questions QST models can be used to inform include the following: (i) identifying and understanding the mechanisms of toxicity; (ii) exploring time and dose-dependence of toxicities; (iii) exploring the potential for reversibility of toxicity; (iv) understanding how in vitro or in vivo toxicity measures translate to clinical scenarios; (v) exploring toxicity liabilities under different dosing regimens to inform on design of clinical dosing protocols; (vi) exploring characteristics of individuals most susceptible to toxicity, and many others.

5 QST Considerations, Challenges and Future Perspectives

5.1 Considerations for Modeling Safety

Pharmacological and toxicological modeling exhibits both similarities and some unique distinctions (Fig. 2). In select scenarios, a distinction may not be necessary. For example, a PBPK model developed to describe the biodistribution of a compound in the rat may be able to describe the TK measured in a rat toxicity study when supratherapeutic doses are administered. In other scenarios, this may not be the case, and modifications to models used to predict therapeutic doses may be required to adequately describe behaviour observed at higher doses. As a result, the modeler needs to carefully consider the relevance of the following: i) tailoring the PK/PBPK model to the safety context, such as the incorporation of potential saturable mechanisms at supratherapeutic doses, ii) accounting for deviations from normal biological function that may arise when exaggerated pharmacology occurs, and iii) capturing the potential interplay between these two factors (PK and biological functions). These aspects are supported by the distinction made between terms such as PK and TK, PBPK and PBTK, PD and TD, QSP and QST. The section below outlines some key considerations relevant to safety modeling.

Until recently QST model development has mainly focused on modeling individual organs with specific mechanisms of toxicity. A fundamental limitation of this approach is that it does not capture any interactions between organs. Not accounting for this cross-talk, which may play a key role in the development of an adverse outcome, could result in inaccurate model predictions by single organ QST models. These limitations are acknowledged by QST modelers, and further research in this area is expected.

Feedback mechanisms can have important effects on both drug concentrations and the physiological system at high drug concentrations. For instance, if a drug causes damage to an organ tissue, the tissue may activate feedback mechanisms to repair or limit further damage. These feedback mechanisms, which may only be relevant when toxicity begins to manifest, can influence both the TK and the TD. Feedback is not limited to individual organ tissues, and toxicity in one organ system may have consequences on the function of connected organ systems.

QST models are often parameterized to represent a healthy individual. This may be an appropriate assumption for replicating dosing in early-stage clinical trials such as phase 1 in indications where initial dosing is in healthy volunteers. However, in later clinical trials or indications where early clinical trials dose directly into diseased patients (e.g. oncology), QST models that do not represent the patient population diseased state may be less informative. This situation may arise from QST models developed to support screening capability, particularly for off-target toxicities. This contrasts to QSP models which are typically built specifically for the disease population of interest. In the future, it is expected that more QST models will represent both healthy and patient disease states.

Importantly, not all toxicities are equal. Numerous factors such as whether toxicity is considered adverse, monitorable, or reversible are considered when defining the therapeutic window of a new drug alongside the acceptable risk-benefit profile for the indication in the specific patient population. The default assumption made when applying any QST model is to consider the worst-case scenario. Given predictions ultimately concern patient safety, a conservative approach is usually taken.

For the considerations highlighted here, it is conceivable that these differences in assumptions in therapeutic and toxic scenarios could be built into a single model with specific effects triggered when the concentrations are within a certain range or surpass a certain level. However, since models are typically constructed for a specific context of use, this is not often done in practice, unless a particular context of use includes both efficacy and safety components.

5.2 Key QST Challenges

Many challenges in QST modeling mirror those faced by QSP modeling. These include limited pre-clinical and clinical data for model calibration and validation (e.g. often insufficient longitudinal data), high variability of experimental data, limited knowledge of the disease and drug related pathways, and imperfect translation of models across experimental systems (e.g. especially, from in vitro to in vivo) and across species (e.g. critically, pre-clinical to clinical translation).

Not as relevant to QSP, a challenge in QST modeling is that the occurrence of toxic effects can be rare, resulting in limited clinical data for model credibility assessment. Therefore, QST model validation often involves comparison to in vivo pre-clinical data, the translation of which to clinical outcomes is imperfect on its own. The scarcity of such data further emphasizes the importance of simulations used to explore diverse populations. Incorporating population diversity, whenever possible, should be considered a part of best practices in QST model development and validation.

Similarly, a notable challenge in QST modeling is that experimental readouts for toxicity, such as histopathology endpoints, are frequently recorded only at terminal endpoints. This limitation necessitates changes in study design to provide richer data without compromising statistical power. One solution is staggering the termination of animals in studies to capture multiple time points, which not only benefits modeling but may also offer additional insights into toxicological data. Such an approach again highlights the need of close integration and interpretation of data across in vitro, in vivo and in silico studies to increase the PoS and ultimately PTRS of a particular target and/or compound. This can only be achieved through close collaboration and communication between modelers, biologists, clinicians, chemists, and other drug discovery or development project team members.

Another critical challenge in QST modeling is ensuring that model validation is appropriately aligned with the significance of the decision it is intended to inform. As discussed in Sect. 4, different types of models used at different stages in the drug development process will require different validation strategies. For this approach to succeed, there needs to be suitable data sets with appropriate biomarkers that are indicators of a particular toxicity type or informative of the target mechanism included in the model. However, in practice, interpreting biomarkers alone is often confounded by various factors, such as disease state, individual activities, and external influences. These factors affect measurements, making it challenging to use these data to validate a computational model. Moreover, toxicities may occur infrequently, idiosyncratically, or after a very long dosing period, all of which present additional challenges for QST model validation.

5.3 QST Future Perspectives

As with QSP, defining the context of use, aligning modeling efforts with key questions from biologists and clinicians, and clearly communicating modeling findings along with model assumptions and limitations will ensure QST models stay relevant and maximally useful and help facilitate multidisciplinary collaboration between modelers and non-modelers. The latter are key stakeholders in advocating for QST modeling to impact internal decision making within pharmaceutical industries. Despite QST modeling limitations and challenges, there are many opportunities described in this chapter, both technical and strategic for the QST field and its modelers to explore.

Selected examples of immediate opportunities for QST to impact drug discovery and development include: (i) drug discovery applications of QST to identify compounds with the highest PTRS, (ii) continued leveraging of existing QST models to bridge into the clinical development space including regulatory submissions (e.g. DILIsym® QST modeling off-target toxicity case study), (iii) increased use of QST during clinical development to inform dosing regimens (dose, frequency, sequence) in monotherapies and/or combinations, (iv) use of QST model results to influence drug labelling, (v) pro-active integration of QST approaches within pharmaceutical drug discovery and development pipelines to revisit safety assessment of previously rejected compounds (e.g. application of the CiPA model for further investigating hERG blocking compounds) since some compounds may actually be safe and viable and thus could proceed with further drug development, (vi) incorporation of feedback loops across different organ systems (e.g. heart and brain, liver, and kidney) both representing the normal physiological feedback and the impact of toxicity effects in one organ system on a connected organ. Inclusion of this feedback within a model will more accurately reflect the reality of a toxicological response and lead to more accurate QST model predictions.

QST, along with QSP, is part of the wider mathematical mechanistic modeling discipline of MIDD used to support pharmaceutical discovery and development, all of which fields can thrive by being closely allied. Toxicology focused questions can provide interesting and challenging mathematical problems that both utilize and drive advances in numerical analysis, artificial intelligence and machine learning (AI/ML), computational algorithms, and other related fields. Increasingly, integration of omics data within QSP and QST models is also being explored. Questions related to model identifiability, parameter estimation, global sensitivity analysis, virtual patient generation, uncertainty quantification, numerical methods and techniques, data storage, model curation, storage and reproducibility, programming languages, and others will always form an important foundation for the QST field as they do within other mechanistic modeling fields. Alignment, standardization, and solutions to some of these challenges shared across the mechanistic modeling fields will be highly beneficial for the advancement of mechanistic systems modeling approaches including QST.

Within the pharmaceutical industry, frameworks for mechanistic mathematical modeling and tools such as PBPK, PBTK, QSP, QST will inevitably share common methodologies and similar approaches as outlined in the ICH M15 draft guidance (Marshall et al. 2023; web2). As a result, these frameworks would benefit from learnings across these related disciplines and, where appropriate, reaching consensus amongst different modeling communities as part of the MIDD roadmap. Despite commonalities, QST is emerging as an independent field separate from QSP (Bloomingdale et al. 2017; Beattie et al. 2024). The focus of the model application and its context of use are the primary drivers that differentiate QSP and QST leading to QST's emergence as a separate field. The remit of QST will be to quantitatively study mechanisms by which either biological or chemical entities may produce unwanted adverse effects including on-target toxicity (where there are overlaps with QSP as outlined in Fig. 2) as well as in assessing off-target adverse events. Both QSP and QST are expected to become an integrative part of digital twin, precision medicine, virtual clinical trials, healthcare decision making (e.g. during pandemics), and other innovative emerging applications of MIDD in drug development and medicine. To support QST's impact in the pharmaceutical industry, a key objective of QST is to gain increasing acceptance by regulatory bodies within the next few decades. Ideally, QST approaches would become an integral part of regulatory body submissions for new drugs, similar to the current acceptance of PBPK/PD and PK/PD models.

Key takeaways of this chapter include: (1) QST is an end-to-end MIDD tool both within the pharmaceutical industry and other healthcare applications supporting decisions in early target evaluation, target and lead prioritization and optimization, biomarker selection, clinical trial design, interpretation of clinical trial results, and in combination therapy predictions; (2) QST model feasibility assessment is the first step of QST model development and includes assessment of the question of interest, data availability, toxicological, biological and clinical knowledge (often including outlining an adverse outcome pathway), resource availability, anticipated outcomes/deliverables, and timelines; (3) the strategy for QST model validation should be appropriately aligned with the significance of the decision it is intended to inform and QST-specific challenges in this regard may be considered as general MIDD guidelines are formulated; and (4) QSP and QST share many commonalities in terms of methodologies and approaches. Key distinctions of QST from QSP can be found in the application focus and the contexts of use.

QST as an independent field is in its infancy; and while there are many opportunities, significant challenges remain. Various QST modeling applications are at different levels of maturity, which is reflected in the context of use of these models. While this chapter focused on applications of QST to drug discovery and development questions, similar QST approaches and challenges are relevant to understand toxicity liability in other industries including petroleum, agrochemicals, cosmetics, food and drink, tobacco, as well as in the development of household products. Future QST success relies on modelers working in multidisciplinary environments alongside biologists, toxicologists, and clinicians, where modelers' focus on accelerating the development of relevant and useful fit-for-purpose QST models to increase QST model use and impact on decision making.

Acknowledgements Special thanks to Morgan Craig (University of Montreal, Canada) for review of the on-target QST case study, Brett Howell (GSK, USA) for review of the off-target QST case study and Anna Kirpichnikova (University of Stirling, UK) for reviewing and providing feedback on the draft chapter. Additionally, we thank the anonymous reviewers for their comments as the chapter was being drafted. We are also deeply grateful to the editors Mary Spilker (Pfizer Inc., USA) and Erica Bradshaw (Genesis Therapeutics, USA) for their expertise, insightful contributions and support, and being instrumental in enhancing the quality of this chapter.

References

Ankley GT, Edwards SW (2018) The adverse outcome pathway: a multifaceted framework supporting 21st century toxicology. Curr Opin Toxicol 9:1–7

Beattie KA et al (2013) Evaluation of an in silico cardiac safety assay: using ion channel screening data to predict QT interval changes in the rabbit ventricular wedge. J Pharmacol Toxicol Methods 68(1):88–96

Beattie KA et al (2024) Quantitative systems toxicology modeling in pharmaceutical research and development: an industry-wide survey and selected case study examples. CPT Pharmacometrics Syst Pharmacol 13(12):2036–2051

Bloomingdale P et al (2017) Quantitative systems toxicology. Curr Opin Toxicol 4:79–87

Chen W et al (2020a) A physiological model of granulopoiesis to predict clinical drug induced neutropenia from in vitro bone marrow studies: with application to a cell cycle inhibitor. J Pharmacokinet Pharmacodyn 47:163–182

Chen W et al. (2020b) Predicting chemotherapy-induced neutropenia and granulocyte colony–stimulating factor response using model-based in vitro to clinical translation AAPS J 22:1–14

Collins TA et al (2017) Translational modeling of drug-induced myelosuppression and effect of pretreatment myelosuppression for AZD5153, a selective BRD4 inhibitor. CPT Pharmacometrics Syst Pharmacol 6(6):357–364

Craig M (2017) Towards quantitative systems pharmacology models of chemotherapy-induced neutropenia. CPT Pharmacometrics Syst Pharmacol 6(5):293–304

Craig M et al (2015) Neutrophil dynamics during concurrent chemotherapy and G-CSF administration: mathematical modelling guides dose optimisation to minimise neutropenia. J Theor Biol 385:77–89

Craig M, Humphries AR, Mackey MC (2016) A mathematical model of granulopoiesis incorporating the negative feedback dynamics and kinetics of G-CSF/neutrophil binding and internalization. Bull Math Biol 78:2304–2357

Desikan R et al (2024) A quantitative clinical pharmacology-based framework for model-informed vaccine development. J Pharm Sci 113(1):22–32
Ferreira S et al (2020) Quantitative systems toxicology modeling to address key safety questions in drug development: a focus of the TransQST consortium. Chem Res Toxicol 33(1):7–9
Friberg LE et al (2002) Model of chemotherapy-induced myelosuppression with parameter consistency across drugs. J Clin Oncol 20(24):4713–4721
Friedrich C (2024) What is QSP and why does it exist?: a brief history. Handb Exp Pharmacol. https://doi.org/10.1007/164_2024_733
Harrell AW et al (2024) Endeavours made by trade associations, pharmaceutical companies and regulators in the replacement, reduction and refinement of animal experimentation in safety testing of pharmaceuticals. Regul Toxicol Pharmacol 152:105683
Hartung T et al (2017) Systems toxicology: real world applications and opportunities. Chem Res Toxicol 30(4):870–882
Mackey MC et al (2021) The timing of cyclic cytotoxic chemotherapy can worsen neutropenia and neutrophilia. Br J Clin Pharmacol 87(2):687–693
Marshall S et al (2023) Model-informed drug development: steps toward harmonized guidance. Clin Pharmacol Ther 114(5):954–959
Mirams GR et al (2011) Simulation of multiple ion channel block provides improved early prediction of compounds' clinical torsadogenic risk. Cardiovasc Res 91(1):53–61
Mohr M et al (2022) Accurate in silico simulation of the rabbit Purkinje fiber electrophysiological assay to facilitate early pharmaceutical cardiosafety assessment: dream or reality? J Pharmacol Toxicol Methods 115:107172
Nayak S et al (2015) Using a systems pharmacology model of the blood coagulation network to predict the effects of various therapies on biomarkers. CPT Pharmacometrics Syst Pharmacol 4(7):396–405
Smith B et al (2020) Mechanistic investigations support liver safety of ubrogepant. Toxicol Sci 177(1):84–93
Sokolov V et al (2025) A framework for quantitative systems pharmacology model execution. Handbook of experimental pharmacology. https://doi.org/10.1007/164_2024_738
Strauss DG et al (2019) Comprehensive in vitro proarrhythmia assay (CiPA) update from a cardiac safety research consortium/health and environmental sciences institute/FDA meeting. Ther Innov Regul Sci 53(4):519–525
Sturla SJ et al (2014) Systems toxicology: from basic research to risk assessment. Chem Res Toxicol 27(3):314–329
Waring MJ et al (2015) An analysis of the attrition of drug candidates from four major pharmaceutical companies. Nat Rev Drug Discov 14(7):475–486
Watkins PB (2019) The DILI-Sim initiative: insights into hepatotoxicity mechanisms and biomarker interpretation. Clin Transl Sci 12(2):122–129
Watkins PB (2020) DILIsym: quantitative systems toxicology impacting drug development. Curr Opin Toxicol 23:67–73
Watkins PB (2022) Quantitative systems toxicology and drug development: the DILIsym experience. Systems medicine. Springer US, New York, pp 181–196
web1 https://www.simulations-plus.com/resource/renasym-a-qst-model-to-predict-drug-induced-kidney-injury/
web2 https://database.ich.org/sites/default/files/ICH_M15_EWG_Step2_DraftGuideline_2024_1031.pdf
web3 https://www.ema.europa.eu/en/documents/scientific-guideline/concept-paper-development-guideline-assessment-reporting-mechanistic-models-used-context-model-informed-drug-development_en.pdf
web4 https://ntp.niehs.nih.gov/whatwestudy/niceatm/comptox/ct-aop/aop
web5 https://www.epa.gov/chemical-research/adverse-outcome-pathways
web6 https://www.nature.com/articles/d41573-022-00093-3
Woodhead JL et al (2022) Comparing the liver safety profiles of 4 next-generation CGRP receptor antagonists to the hepatotoxic CGRP inhibitor telcagepant using quantitative systems toxicology modeling. Toxicol Sci 188(1):108–116

The Role of Cross-Institutional and Interdisciplinary Collaboration in Defining and Executing a Quantitative Systems Pharmacology Strategy

Paolo Vicini and Piet H. van der Graaf

Contents

Abstract

The application of quantitative systems pharmacology (QSP) has enabled substantial progress and impact in many areas of therapeutic discovery and development. This new technology is increasingly accepted by industry, academia, and solution providers, and is enjoying greater interest from regulators. In this chapter, we summarize key aspects regarding how effective collaboration among institutions and disciplines can support the growth of QSP and expand its application domain. We exemplify these considerations through a selection of successful cross-institutional or cross-functional collaborations, which resulted in reuse, repurposing, or extension of QSP modeling results or infrastructure, with important and novel results.

Keywords

Collaboration · Data sharing · Modeling and simulation · Pharmacodynamics · Pharmacokinetics · Quantitative systems pharmacology · Software

P. Vicini
Confo Therapeutics, Ghent, Belgium

P. H. van der Graaf (✉)
Certara, Canterbury, UK

University of Leiden, Leiden, The Netherlands
e-mail: p.vandergraaf@lacdr.leidenuniv.nl

E. L. Bradshaw, M. E. Spilker (eds.), *Quantitative Systems Pharmacology*,
Handbook of Experimental Pharmacology 289, https://doi.org/10.1007/164_2024_736

Modern biomedical science has improved the quality of life worldwide. However, several diseases remain a challenge, both due to their intrinsic complexity and the increased demand placed on emerging technology to accelerate and improve the probability of success in drug development. Continued progress now requires new ways of cross-institutional and interdisciplinary collaboration to meet patients' needs. A complete and rigorous understanding of biological mechanisms and pathways, coupled with sophisticated modeling and predictive tools, has become crucial to the discovery and development of novel therapeutics, to maximize their impact on human health. As a predominantly quantitative science, quantitative systems pharmacology (QSP) (Sorger et al. 2011) relies on effective collaboration among multiple, diverse disciplines to gather, analyze, display, and use data to develop actionable computational models of biology and pharmacology. We have seen in recent years a significant rise in such collaborations, which have demonstrably increased the effectiveness of QSP in multiple applications and therapeutic areas.

Several pharmaceutical companies with integrated research and development (R&D) infrastructures have recently developed rigorous quantitative frameworks based on pharmacological and pharmacokinetic-pharmacodynamic (PK-PD) principles to assess the viability of their internal programs. These include the three pillars of survival at Pfizer (Morgan et al. 2012) and the 5R (the right target, patient, tissue, safety, and commercial potential) framework at AstraZeneca (Cook et al. 2014). The overall purpose was to improve what was perceived as a major issue in the development of new therapeutics, namely, the success rate of phase 2 programs, which historically has been low due to the lack of efficacy of new molecules at doses that show acceptable tolerability in the clinic. Remarkably, these quantitative frameworks have led to significant improvement in the successful clinical development of new entities. For example, Pfizer's end-to-end success rate (from first-in-human to regulatory approval) increased from 2% in 2010 to 21% in 2020 (Fernando et al. 2022), by using a reformulated Signs of Clinical Activity (SOCA) paradigm inspired by the three pillars analysis, while AstraZeneca's success rates from candidate drug nomination to phase 3 completion improved from 4% in 2005–2010 to 19% in 2012–2016 (Morgan et al. 2018). Speed to decision-making is a common trait of these and other innovation programs aimed at improving the flow of new medicines and decreasing the financial and time burden of clinical failure, see e.g., the Chorus model at Eli Lilly (Owens et al. 2015). This is not only a large pharma concern: arguably, small- to medium-size biotech companies have an even more urgent need for efficient decision-making and optimal use of venture or public funds in the pursuit of their programs. As the skills and knowledge base required to implement these approaches do not reside in a single discipline, close-knit collaboration among units in a drug R&D program bringing to bear a multidisciplinary viewpoint has become paramount.

Model-Informed Drug Development (MIDD) (EFPIA MID3 Workgroup et al. 2016) has proven to be a powerful tool in obtaining deeper biological understanding and improved communication (at all levels) of commercial priorities. It is a cornerstone of quantitative modeling approaches and underpins many of the examples in the primary references cited above. Regulatory acceptance of quantitative methods has contributed to accelerating their adoption in the business, as shown for MIDD by its mention in the Prescription Drug User Fee Act (PDUFA) VI (Wang et al. 2019) and, specifically for QSP, via an increase in its use in regulatory submissions to the FDA over time [see Zineh (2019) for a review of the 2013–2018 period and Bai et al. (2021) for a review including the 2013–2020 period, recently extended to 2023 in Bai et al. (2024)]. While MIDD approaches are rooted in the established science of PK-PD and its contribution to drug development, they also employ novel approaches such as mechanistic modeling, real-world evidence, and of course systems pharmacology. Physiologically-based pharmacokinetics (PBPK) is probably the best example of mechanistic models that made the transition “from academic curiosity to regulatory acceptance” (Jamei 2016), thus paving the way for an entirely new class of approaches. Broadly accepted regulatory sciences now include precision medicine/pharmacogenomics, exposure/response modeling, and simulation, in addition to physiologically-based pharmacokinetic modeling (Zineh 2019). In addition to MIDD’s inroads in regulatory sciences, as a highly technical discipline (especially true for QSP) embedded in a business environment, MIDD practitioners must successfully interface with the other components of the drug discovery and development enterprise to realize their potential impact on business deliverables.

What these new methods have in common is their reliance on an increasing number of data sources, as also discussed in detail in Chapter “Role of Data in Development and Application of Quantitative Systems Pharmacology Models”. Traditional pharmacokinetic, pharmacodynamic, biomarker, and immunogenicity datasets are often not sufficient to satisfy the needs of these mechanistic quantitative methods, and often measurements at a higher resolution and/or more mechanistic detail are required to inform these approaches. Naturally, the availability of high-resolution or custom data sets requires increased coordination among data users and data providers. In the case of integrated big pharma companies this would be between R&D business units and support lines [“a collaborative network structure and empowered matrix leadership,” as remarked in Visser et al. (2014)], and, in the case of smaller biotech firms or academic researchers, ensuring data at the appropriate resolution are gathered would entail close collaboration with either contract research organizations (CROs) or multiple academic centers. Increased coordination is of course not an end in itself, and at their best, these collaborations include harmonized experimental design and data collection, as well as coordinated interpretation of data once gathered. Against this background, a quantitative systems pharmacology model is an ideal framework for data gathering, compilation, and interpretation, acting as a repository of institutional knowledge. The importance of communication and integration of functional areas has also been mentioned by others (Bradshaw et al. 2019).

With all that being said, the usage of QSP approaches could increase further from its current state. One way to approach this is through increased efficiency. Strategies and recommendations have been formulated on best practices to maximize the use and reuse of QSP models, so that time is spent tailoring a model to new and broader applications, rather than validating and reconstructing previous results (Cucurull-Sanchez et al. 2019). Additionally, we believe that to be successful and to maximize the impact of QSP models, collaborations aimed at increasing quantitative understanding of drug development programs must include clinical trial data. The addition of these data, combined with the integration of a priori knowledge of biology along with pharmacological perturbations of the intact system, can improve our understanding of the system and the model used to represent it. Such a collaboration integrating publicly available "omics" human data and clinical data from trials is mentioned in Fernando et al. (2022), but there are of course many others. Despite the potential challenges involving intellectual property, especially at the early stages of development, quantitative, translational data in the preclinical space can also be very informative for QSP model development. Effective sharing of these data can be accomplished by their inclusion in the published record, when feasible, as it is done with other data types.

While QSP models leverage diverse datasets to achieve the necessary mechanistic representation of physiology and pharmacology, sometimes perceived data gaps in QSP model development can be solved by adjusting the level of detail of the model or by changing their intended scope. It is not always required to design specific experiments for quantifying model parameters, however crucial to the simulation they may initially appear. It is better sometimes to employ techniques such as sensitivity analysis or more comprehensive simulations to measure the impact of certain parametric or structural assumptions on the system behavior. This is illustrated by Franssen et al. (2023) where the impact of the model parameter representing the initial number of antigen-specific naïve T cells on a predictive model of therapeutic protein immunogenicity was assessed. Sometimes what appears to be a bioanalytical limitation is in fact a modeling limitation: in the same article, it is put forward that the lack of availability of absolute anti-drug antibody (ADA) concentrations can be overcome by developing computational models of bioanalytical ADA assays, so that the model prediction can then be directly compared with the assay output. Regardless, close collaboration between modeling scientists and bioanalytical experts can be very valuable, as it sometimes generates important insights into the behavior of the system and motivates detailed experimental assessment of key system properties that can have a major impact on the understanding of system dynamics.

Execution of a QSP strategy is an important area where increased synergies are still possible. There is considerable heterogeneity in how academia, industry, and CROs tackle the realization of a QSP model. Academic efforts are of high quality (the modern instance of PBPK, for example, started as an academic research project), but sometimes suffer from a lack of model documentation and ways to improve reproducibility. This can be linked to the academic model development environment, which traditionally has relied on a transient population of graduate students and postdoctoral fellows, a situation which has much improved due to an emphasis by peer review requiring the inclusion of reproducible model code whenever possible. In addition to the above, models developed in academia tend to focus on systems biology, as opposed to systems pharmacology, and thus their effective deployment in drug R&D, where the emphasis is on understanding pharmacological modes of action, may not be straightforward. In industry, the strategy is often formulated internally with QSP experts and ideally in a cross-functional R&D team involvement. Execution can then take place internally and/or externally with support from academia and/or CROs. CROs interface with their customers to address their stated strategy, and thus normally do not retain the intellectual property or the broad usage rights that their clients have. However, they can sometimes facilitate or even lead cross-industry/cross-institutional consortia, when precompetitive collaborations are feasible. QSP model development efforts can transcend independent institutional work and become effective cross-institutional collaborations. While such examples are increasing in number, there remains a certain aspect of serendipity about it. More consortia directed to a specific, hitherto intractable problem, or an increase in precompetitive collaborations resulting in cross-institutional publications would be beneficial to the discipline.

To increase QSP's impact, we posit that QSP is a powerful technology that has not yet found its "killer application," or the area of discovery and development where it is poised to make its highest impact; not simply as a one-off but influencing and changing the field consequently. QSP models have been applied to many steps in the drug discovery and development cascade: from the assessment of novel targets to clinical study design and analysis, supporting biomarker development, assessment of new modalities, setting potency targets, etc. No technology, however, can credibly solve issues from early discovery to late-stage development and everything in between. When QSP was initially being recognized as a discipline, it was thought that its major area of application would be translational research, or the interface between discovery and development. That remains a very natural application for a mechanistic model incorporating both biology and pharmaceutical intervention. For example, human dose projection and the efficient design of dose escalation schemes, with particular attention to starting doses, is something where QSP can have, and is already having, a measurable impact. A challenge for the field will be to identify areas of focus where impact is maximum, or where QSP proves indeed indispensable. Note that this is not just a matter of resourcing, but also external visibility. Thinking about PBPK and its impact on regulatory practice, this is highest in the investigation of drug-drug interactions (DDI). This does not mean that there are no other applications of PBPK, just that DDI is its "killer application," from which the technology can branch out in other areas of R&D. There are of course several factors underpinning the success of PBPK in the regulatory space: some of these can be generalized to QSP and for others it is not so straightforward. While the narrower focus on DDI and special populations was, in our opinion, a critical factor, PBPK also benefited from a better scientific and mechanistic understanding of in vitro-in vivo correlations in its field of application. Moreover, PBPK modeling has a broader potential application domain: the primary enzymes involved in the metabolism of small molecule drugs are common between different new molecular entities. This maximizes the potential for model reuse in a way that may not straightforwardly apply to the heterogeneous biological pathways and disease states that QSP models target. This of course negatively impacts the investment required for reuse, which may be one of the reasons why QSP has so far not identified its main area of emphasis. We argue that a clear, focused, and unambiguous statement of where impact can be directed would support the future of the discipline and its broad uptake.

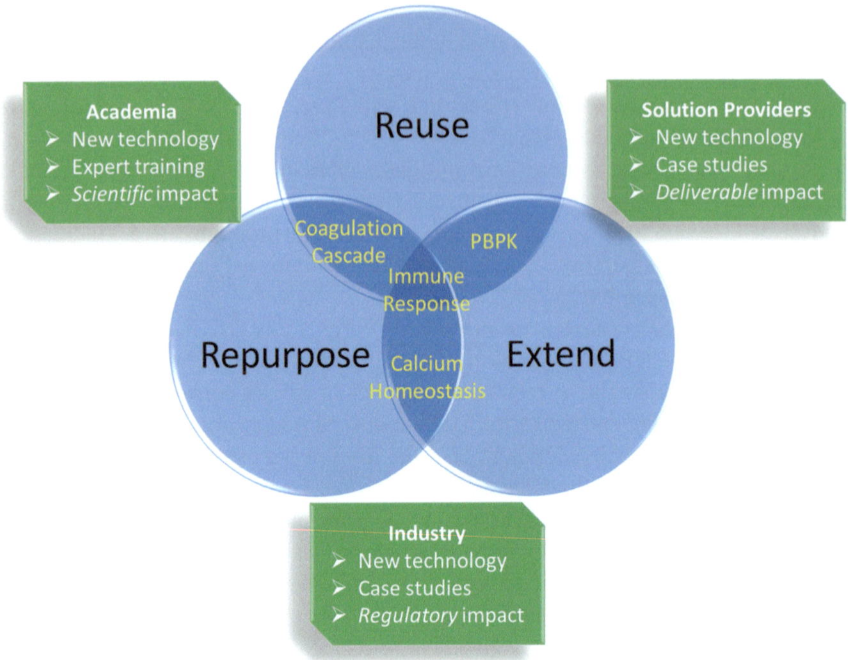

Fig. 1 The figure shows selected QSP models and the features they represent (among Reuse, Repurpose, and Extend) based on their impact since publication. The role of academia, solution providers (e.g., contract research organizations) and industry is also highlighted

What follows is a selection of successful cross-institutional or cross-functional collaborations in QSP. These examples, graphically summarized in Fig. 1, resulted in reuse, repurposing, or extension of QSP modeling results or infrastructure, with important and novel outcomes. The list is of course not exhaustive, but we chose these examples to highlight best practices as well as potential pitfalls and factors to watch for. PBPK approaches would, of course, belong to this list, but they have been extensively reviewed elsewhere (Jamei 2016). Also, PBPK models deal with a very specific problem (the absorption, distribution, metabolism, and excretion of xenobiotics in the intact organism) which has been studied for a long time (Teorell 1937). Since the development of PBPK models has been so unique, its generalization to modern QSP approaches is difficult. While it has been argued that they squarely belong in QSP (Jamei 2020), this is still a subject of debate in the QSP field.

It is important to note that we mean "collaboration" in an extended sense. In fact, some of these collaborative efforts were asynchronous, or occurred independently of the originators of the QSP modeling effort. Moreover, when looked at after some time has elapsed since their publication or first disclosure, the serendipitous nature of these collaborations is clearer: rather than emerging from a well-thought and planned strategy, they often resulted from the reuse and extension of existing QSP modeling frameworks, as dictated by specific programs and needs. These examples provide important lessons on how such investigations can be conducted more deliberately.

Example 1 (Cross-Institutional Collaboration): Immune Response Modeling (Reuse, Extension)
In 2014, a two-part report was published outlining the theoretical development and example applications of an integrated model of the immune response meant to aid in the prediction of immunogenicity of therapeutic proteins (Chen et al. 2014a, b; parts 1 and 2). The model arose from a cross-functional research project supported by the Pfizer Worldwide Research and Development postdoctoral training program. Anti-drug antibodies (ADA) are an unwanted endogenous immune response to therapeutic proteins, and more in general to biologics. Due to the difficulty in predicting the extent of ADA response in vivo, they can be a risk factor for new biological entities. The model was developed to ultimately integrate available information and enable forward prediction of immunogenicity for individual programs. Notably, in addition to model diagrams, tabulated parameter (reference) values, lists of differential equations, and use cases, which are relatively standard in scientific modeling articles, the actual model code was published as well. The model code being released with the publication resulted in increased efficiency for reuse and meant that prospective model users would not have to invest time and resources in reconstructing the model calculations based on tabulated values and the differential equation diagrams in their own computing environments. Others have discussed how the availability of model code is an essential component of reproducibility (Kirouac et al. 2019). However, the availability of code per se is not a guarantee of reproducibility or ultimate uptake by the community. Most open-source projects (outside of QSP) have been successful because of strong community advocacy and collaborative ownership by a relatively small group of committed individuals. Such a proactive community has not fully coalesced around QSP yet, and building on published models remains difficult, in part because of an overall lack of curation and quality control (these topics are discussed more in detail in Chapter “A Framework for Quantitative Systems Pharmacology Model Execution”). There are, however, promising signs, and curated model code repositories such as BioModels (Malik-Sheriff et al. 2020) are of course a first step in the right direction.

In Kierzek et al. (2019), a consortium approach is described that takes the originally published Chen model (Chen et al. 2014a, b; parts 1 and 2) as the basis for the development of an integrated IG (immunogenicity) Simulator, which extends the original approach by embedding the immune response description in a PBPK context. The IG Consortium was led by Certara and started in 2017, with members including several leading pharmaceutical companies. The original intent was to provide research and development practitioners of new therapeutic proteins with a tool to better understand and mitigate the immunogenicity risk of their molecules. Following the coronavirus disease 2019 (COVID-19) pandemic, the model was repurposed for vaccination dosing strategies (Giorgi et al. 2021). This was possible as the underlying functional components of the immune response model are the same whether the intervention is a therapeutic protein immunogen or a vaccine. The model was evaluated (Franssen et al. 2023) and calibrated on several examples and datasets

(32 clinical datasets, 3 vaccines, 8 dosing regimens, and 10 biomarkers) (van der Graaf 2022) and has shown the importance of dose and dosing frequency to optimize the antibody response following vaccination. While this had always been an empirical science, based on "trial and error," the modeling approach has demonstrated that more rigorous dose optimization in vaccination is possible. Of note, complementary modeling efforts have been deployed in the development of antivirals (Dai et al. 2021; Rao et al. 2023), contributing to trial design and dose selection (Singh et al. 2022).

If we posit that this body of work constitutes "successful" QSP model development and prospective utilization, with significant impact, we can list a few key components of that success:

- Original model code availability was a contributing factor to the model being repurposed. The availability of the original code means that the reproducibility of the model should not be an issue, and it facilitates the repurposing of the modeling framework beyond its initial domain of application. As previously mentioned, the BioModels database has emerged as an authoritative resource for QSP model code (Malik-Sheriff et al. 2020).
- Availability of multi-level biomedical data and sharing of these data also facilitates the repurposing of the model. Successful prospective application (over time) of the same quantitative modeling framework to multiple distinct data sets, once they become available, increases confidence in the model structure and resulting predictions. This can be considered an important variable influencing usage patterns (Braakman et al. 2022). As the model application grows in scope, the link between QSP models and clinical endpoints can also be established or improved, which arguably is crucial to continued success.
- Crisp, clearly stated questions to be addressed, such as optimal dose and dosing regimen, help focus development efforts and improve the final product. These questions can shift over time, with the model being continuously adapted to answer a growing set of relevant problems in its application area. This immune response modeling case study is a good example of model development and extension carried out in multiple environments, with the original model being developed in industry by Pfizer, where the original question was immunogenicity of biotherapeutics, and the redeployment in a different context by a CRO, Certara, where the question became the study of vaccine immune responses.
- Organizational constructs can be a significant barrier. Paradoxically, model repurposing is sometimes more straightforward outside of the institution where it was originally developed, as the original model developers may have left the company or research laboratory, or (even more likely) business priorities may have shifted.
- The opportunities for repurposing sometimes become clear only after a certain time has elapsed from model publication or availability, either because of new business priorities or the emergence of new, unanticipated questions to be addressed. Often, this is just due to the typical pace of progress of scientific investigation, and not to specific institutional obstacles.

Example 2 (Cross-Discipline Collaboration): Coagulation Cascade Modeling (Reuse, Repurpose)
The coagulation cascade is a good example of the prospective utility of QSP approaches. It is a very intricate network of interconnected parts, and its complexity makes it difficult to predict the outcome of any therapeutic interventions without a QSP-based approach. Accordingly, the coagulation cascade has been a popular objective of modeling efforts, both in the systems biology as well as the quantitative pharmacology community.

A recent systematic review (Chung et al. 2023) has identified some issues in the field of coagulation. Chiefly, irreproducible code, which is a frequent and serious impediment to the progression of the discipline, as discussed among other obstacles in Kirouac et al. (2019), and lack of coordination and linkage between models that do not show a clear intent to build on previous findings. This is unfortunate in general, because a major advantage of QSP is that models can be extended and/or reused, with every contribution building on the previous one in a measurable progression of knowledge. Key challenges that have hindered progress include a lack of validation for some of the representations of the pathways and the inability to build on previous modeling efforts, either because the code is not available or because reproducibility cannot be assured. Despite that, QSP models of coagulation have proven successful in mechanistic investigation, specifically to increase understanding of in vivo mechanisms and the mechanism of action of treatments.

If we thought of the field of coagulation cascade modeling as having some opportunities for improvement, these would be some of the lessons learned:

- Most of the published QSP models for coagulation have been developed independently and in relative isolation. Incentives or requirements to share the model computer code or pseudocode, as several academic journals have already introduced, would continue to improve coordination, and facilitate building novel modeling findings upon previous efforts. However, obstacles in this regard may include the need to maintain a degree of control on present and future QSP research, e.g., for prospective publication, or protecting intellectual property, and sharing the original code could conceivably compromise these activities to some extent. Most of the time a balance between these (sometimes) contrasting needs must be found.
- It can be speculated that the complexity of the coagulation cascade and the high number of components, and their interconnections, may have hindered model development and/or use by QSP groups operating in the private sector. Given the challenges in reusing and repurposing existing models, the role of a QSP model to investigate new therapies affecting the coagulation cascade may have been diminished due to, e.g., the requirements of tight timelines and uncertainty in the specific mechanism of action of novel interventions. Moreover, while many of these models reproduce in vitro assays well, their impact and accuracy in vivo are still somewhat limited. Increased availability of clinical data may contribute to addressing this limitation.

- When comparing systems biology and systems pharmacology models, there is clearly an unmet need for models that capture the signals arising from pharmacological intervention. This highlights a philosophical difference in how these models are assembled: systems biology models attempt to recapitulate the basic biology, while systems pharmacology models take advantage of pharmacological perturbations to explore a range of behaviors of the intact system. A more systematic approach to understanding the pharmacology of the coagulation cascade would be beneficial to increase the applicability of QSP to this important problem.

Example 3 (Asynchronous Collaboration): Calcium Homeostasis Modeling (Repurpose, Extension)
One of the most often cited examples of QSP modeling is the calcium homeostasis and bone remodeling construct by Peterson and Riggs (2010). The original model development began in 2005, and by 2012 it had been extended from calcium and bone homeostasis to changes in bone mineral density (Peterson and Riggs 2012). Since then, the model has been further extended and applied to a variety of situations and experimental designs [see e.g., Stodtmann et al. (2021)], including a "watershed" moment (Peterson and Riggs 2015) when in 2014 it was included in a United States FDA advisory meeting for review of a Biological License Application, as described in Khurana et al. (2019) and commented in Riggs (2019).

For the QSP calcium homeostasis model, it is instructive to focus for our purposes on how its usage changed over time, as opposed to the scientific accomplishments it facilitated, which have been covered well elsewhere Riggs (2019). While initially motivated by a single drug development program (denosumab), it was extended to other programs by several investigators, beyond the original group of innovators. Moreover, it became the first QSP model to inform a regulatory recommendation. Regulatory impact is probably the best possible and the most visible outcome for a QSP model. Similarly, influencing and determining regulatory decisions was an important contributing factor to success in the early days of pharmacometrics, before population pharmacokinetic analyses became expected elements of new drug applications.

Because this project has been ongoing for longer than a typical QSP modeling project, and in fact it was one of the earliest contributions in this area, there are several lessons to be gained:

- It is possible for a single QSP model to impact several areas of discovery and development: choice of therapeutic targets, selection of patient subpopulations and potential indications, dose optimization, and trial design. Beyond the specifics of the modeling exercise, this example shows the discipline's potential breadth of influence on many aspects of modern drug development. This breadth of applications is similar to what was discussed earlier regarding Example 1 (immune system modeling).
- Many QSP modeling efforts have focused on the role of biomarkers, in both the preclinical and clinical settings. The calcium homeostasis model shows the

importance of going beyond biomarkers (like C-terminal telopeptide, CTX), towards measurable clinical endpoints (like bone mineral density and fracture risk) that are broadly accepted and may be used for regulatory purposes.
- Many drug development programs can make use of a predictive approach to bone mineral density that can be extended to incorporate the effects of treatment. This meant that the model was applied to a relatively large number of programs and has been subject to several external validation efforts. Incidentally, we see this kind of "collaborative validation," exemplified by the uses of this model over the years since it was initially proposed, as an expedient avenue to building consensus from a multidisciplinary and cross-organizational group of experts, the QSP modelers and their collaborators, about the biology of the system being studied and its pharmacology. This kind of "peer validation" can give rise to a consensus model that captures the current state of the art in the understanding of the underlying biology and physiology. While this may be proven wrong at any subsequent point in time, it provides at least an actionable consensus built on expert opinion and a solid way forward. Availability of the model code was, once again, an important factor in broad acceptance.

A 2016 review (Musante et al. 2016) identified three critically important themes for the success of QSP as a discipline: (1) building effective teams, (2) building mutually beneficial cross-sector collaborations, and (3) building a community. While it is sometimes unclear what "progress" would entail in each one of these, we can make some conclusions as to whether the general approach to QSP has changed since its introduction. We will discuss now the status of cross-institutional collaboration in keeping with these three critical themes.

Building effective teams is meant in a general sense, including collaborations across academic institutions, solution providers, the drug R&D industry, and regulatory agencies. QSP integration within industry project teams, e.g., the multidisciplinary teams that work on specific drug R&D projects, is paramount. Many of the disciplines on these teams, e.g., biology, chemistry, drug safety, drug metabolism and pharmacokinetics (DMPK), and clinical pharmacology, are required partners for optimal QSP model development and execution. A critical task for teams is to evaluate appropriate and consequential questions that can be addressed with QSP models. As far as fostering the right kind of talent that can propel QSP forward, two key aspects that were identified included training programs as well as a renewed and increased emphasis on teamwork. While there are a few very successful academic labs that train scientists to develop QSP models, the majority of working QSP scientists are still drawn from complementary disciplines, e.g., engineering, (applied) mathematics, and (quantitative / systems) biology and physiology. Since the QSP "toolbox" is so heterogeneous, it is difficult to formulate training modules that would have universal applicability. Another factor that makes this complex is that the available technology is changing rapidly, see for instance the emergence of machine learning and artificial intelligence, which now is impacting pharmacometrics and likely soon will be integrated with QSP (Janssen et al. 2022) (see Chapters "Integrating QSP and ML to Facilitate Drug Development and

Personalized Medicine" and "Future Directions for Quantitative Systems Pharmacology" for additional insights). A virtuous circle between academia and industry has been proposed to overcome some of these difficulties (Zhang et al. 2022), likely leaving to industry a deep understanding of the needs and requirements of the drug R&D business cycle. Solutions providers (CROs) also have an important role to play in this, as they will have to continue to push the boundaries of available technology, at the same time ensuring its direct applicability to the R&D business model, to stay relevant.

Building mutually beneficial cross-sector collaborations is another very important theme, one where the disconnect between the academic and private sector priorities starts to become acute. The main issue with measuring "success" in this theme is that the value systems are disconnected. The priority for the private sector (both industrial R&D and solution providers) is to contribute to success, i.e., progress or termination, which are equally valid measures of success, in research projects leading to new therapeutics. Accordingly, the measure of success for QSP applications is clear and unambiguous. For any consultant, for example, there is a clear link between an R&D project and the scientific contribution the consultant is asked to provide. This is not necessarily true in academia, where the measures of success, or rather the criteria on which a research program, and researcher, are evaluated are often measured on publication, training, and fundraising metrics. Because publication of the results of an academia-industry collaboration can be subject to business review, and the training component can be elusive and not explicitly part of a collaboration, fundraising can assume an outsized role from the point of view of academic institutions. Moreover, a collaboration for a drug R&D program can be successful for the industrial partner if it contributes to decision-making, but perhaps not for an academic partner if it does not result in a publication, for example. In other words, the criteria for productive collaborations may be misaligned between the entities involved in industrial-academic partnerships.

The last theme involves *building a community*, and arguably this is where most progress has been made. A number of professional societies have now established vibrant and active QSP-oriented focus groups, and these play a crucial role in contributing to scientific conference programming, organizing workshops and symposia whose proceedings can then be published, increasing the visibility of the contributors and their accomplishments. These cross-institutional groups are also making inroads to involve academics more deeply in drug R&D pursuits, but a productive dialog is still needed. It is encouraging though to see more examples of successful cross-institutional partnerships [some important examples are listed in Bradshaw et al. (2019)], spanning the repurposing of published academic models by industry in their own programs, explicit partnerships between academia and industry, and pre-competitive consortia which also involve solution providers. These efforts often include data sharing, which we have already highlighted as a crucial step in the validation, and thus the adoption of QSP models. Of course, such collaboration becomes more feasible if a QSP model is focused on general mechanisms of efficacy and toxicity and not on asset-specific mechanisms of action. However, despite the formidable obstacles, largely due to subject protection and intellectual property in sharing patient-level biomedical data for example, this is an area where progress can be made with careful planning.

In conclusion, the future of QSP is likely bright, as the problem it tackles cannot be solved with other means, and objective obstacles to its adoption, such as computing power, are becoming less and less of a hindrance. A significant increase in multi-institutional collaborations has effectively shown that QSP can be deployed in multiple applications and therapeutic areas. We feel that significant progress has been made in selected precompetitive collaborations and the development of effective data and code-sharing efforts, and those must continue to ensure long-term success. The discipline is learning by doing, and synergies between academia, CROs, and industry should continue. These cross-institutional collaborations are a testimony to the value that QSP models can bring to drug discovery and development and have delivered much success. Continuing obstacles to effective cross-institutional cooperation however remain, including the need for appropriately trained practitioners, misalignment of goals and value systems, and shared measures of success.

Acknowledgments The comments of the anonymous referees and of the Editors on a previous version of this chapter are gratefully acknowledged.

References

Bai JPF, Earp JC, Florian J, Madabushi R, Strauss DG, Wang Y, Zhu H (2021) Quantitative systems pharmacology: landscape analysis of regulatory submissions to the US Food and Drug Administration. CPT Pharmacometrics Syst Pharmacol 10(12):1479–1484. https://doi.org/10.1002/psp4.12709. Epub 2021 Nov 3

Bai JPF, Liu G, Zhao M, Wang J, Xiong Y, Truong T, Earp JC, Yang Y, Liu J, Zhu H, Burckart GJ (2023) Landscape of regulatory quantitative systems pharmacology submissions to the U.S. food and drug administration: an update report. CPT Pharmacometrics Syst Pharmacol 13(12):2102–2110. https://doi.org/10.1002/psp4.13208. Epub 2024 Oct 18

Braakman S, Pathmanathan P, Moore H (2022) Evaluation framework for systems models. CPT Pharmacometrics Syst Pharmacol 11(3):264–289. https://doi.org/10.1002/psp4.12755. Epub 2022 Jan 10

Bradshaw EL, Spilker ME, Zang R et al (2019) Applications of quantitative systems pharmacology in model-informed drug discovery: perspective on impact and opportunities. CPT Pharmacometrics Syst Pharmacol 8(11):777–791. https://doi.org/10.1002/psp4.12463

Chen X, Hickling TP, Vicini P (2014a) A mechanistic, multiscale mathematical model of immunogenicity for therapeutic proteins: part 1-theoretical model. CPT Pharmacometrics Syst Pharmacol 3(9):e133. https://doi.org/10.1038/psp.2014.30

Chen X, Hickling TP, Vicini P (2014b) A mechanistic, multiscale mathematical model of immunogenicity for therapeutic proteins: part 2-model applications. CPT Pharmacometrics Syst Pharmacol 3(9):e134. https://doi.org/10.1038/psp.2014.31

Chung D, Bakshi S, van der Graaf PH (2023) A review of quantitative systems pharmacology models of the coagulation cascade: opportunities for improved usability. Pharmaceutics 15(3): 918. https://doi.org/10.3390/pharmaceutics15030918

Cook D, Brown D, Alexander R et al (2014) Lessons learned from the fate of AstraZeneca's drug pipeline: a five-dimensional framework. Nat Rev Drug Discov 13(6):419–431. https://doi.org/10.1038/nrd4309

Cucurull-Sanchez L, Chappell MJ, Chelliah V, Amy Cheung SY, Derks G, Penney M, Phipps A, Malik-Sheriff RS, Timmis J, Tindall MJ, van der Graaf PH, Vicini P, Yates JWT (2019) Best practices to maximize the use and reuse of quantitative and systems pharmacology models: recommendations from the United Kingdom quantitative and systems pharmacology network. CPT Pharmacometrics Syst Pharmacol 8(5):259–272. https://doi.org/10.1002/psp4.12381. Epub 2019 Mar 22

Dai W, Rao R, Sher A, Tania N, Musante CJ, Allen R (2021) A prototype QSP model of the immune response to SARS-CoV-2 for community development. CPT Pharmacometrics Syst Pharmacol 10(1):18–29. https://doi.org/10.1002/psp4.12574. Epub 2020 Dec 21

Van der Graaf. Rapid repurposing of a QSP platform to guide COVID-19 vaccine development. AAPS PharmSCI360, October 16–19, 2022. Available at: https://www.certara.com/app/uploads/2022/10/Piet-van-der-Graaf_Rapid-Repurposing-of-a-QSP-Platform-to-Guide-COVID-19-Vaccine-Development.pdf. Accessed 25 May 2023

EFPIA MID3 Workgroup, Marshall SF, Burghaus R et al (2016) Good practices in model-informed drug discovery and development: practice, application, and documentation. CPT Pharmacometrics Syst Pharmacol 5(3):93–122. https://doi.org/10.1002/psp4.12049

Fernando K, Menon S, Jansen K et al (2022) Achieving end-to-end success in the clinic: Pfizer's learnings on R&D productivity. Drug Discov Today 27(3):697–704. https://doi.org/10.1016/j.drudis.2021.12.010

Franssen LC, Swat MJ, Kierzek AM, Rose RH, van der Graaf PH, Grimm HP (2023) Learn-confirm in model-informed drug development: assessing an immunogenicity quantitative systems pharmacology platform. CPT Pharmacometrics Syst Pharmacol 12(2):139–143. https://doi.org/10.1002/psp4.12887

Giorgi M, Desikan R, van der Graaf PH, Kierzek AM (2021) Application of quantitative systems pharmacology to guide the optimal dosing of COVID-19 vaccines. CPT Pharmacometrics Syst Pharmacol 10(10):1130–1133. https://doi.org/10.1002/psp4.12700. Epub 2021

Jamei M (2016) Recent advances in development and application of physiologically-based pharmacokinetic (PBPK) models: a transition from academic curiosity to regulatory acceptance. Curr Pharmacol Rep 2(3):161–169. https://doi.org/10.1007/s40495-016-0059-9. Epub 2016 Apr 14

Jamei M (2020) Where do PBPK models stand in pharmacometrics and systems pharmacology? CPT Pharmacometrics Syst Pharmacol 9(2):75–76. https://doi.org/10.1002/psp4.12493. Epub 2020 Jan 26

Janssen A, Bennis FC, Mathôt RAA (2022) Adoption of machine learning in pharmacometrics: an overview of recent implementations and their considerations. Pharmaceutics 14(9):1814. https://doi.org/10.3390/pharmaceutics14091814

Khurana M, Zadezensky I, Lowy N, Roman D, Guettier JM, Li L, Florian J, Sahajwalla CG, Sinha V, Mehrotra N (2019) Use of a systems pharmacology model based approach toward dose optimization of parathyroid hormone therapy in hypoparathyroidism. Clin Pharmacol Ther 105(3):710–718. https://doi.org/10.1002/cpt.1200. Epub 2018 Oct 22

Kierzek AM, Hickling TP, Figueroa I et al (2019) A quantitative systems pharmacology consortium approach to managing immunogenicity of therapeutic proteins. CPT Pharmacometrics Syst Pharmacol 8:773–776

Kirouac DC, Cicali B, Schmidt S (2019) Reproducibility of quantitative systems pharmacology models: current challenges and future opportunities. CPT Pharmacometrics Syst Pharmacol 8(4):205–210. https://doi.org/10.1002/psp4.12390. Epub 2019 Mar 3

Malik-Sheriff RS, Glont M, Nguyen TVN, Tiwari K, Roberts MG, Xavier A, Vu MT, Men J, Maire M, Kananathan S, Fairbanks EL, Meyer JP, Arankalle C, Varusai TM, Knight-Schrijver-V, Li L, Dueñas-Roca C, Dass G, Keating SM, Park YM, Buso N, Rodriguez N, Hucka M, Hermjakob H (2020) BioModels-15 years of sharing computational models in life science. Nucleic Acids Res 48(D1):D407–D415. https://doi.org/10.1093/nar/gkz1055

Morgan P, Van Der Graaf PH, Arrowsmith J et al (2012) Can the flow of medicines be improved? Fundamental pharmacokinetic and pharmacological principles toward improving Phase II survival. Drug Discov Today 17(9–10):419–424. https://doi.org/10.1016/j.drudis.2011.12.020

Morgan P, Brown DG, Lennard S et al (2018) Impact of a five-dimensional framework on R&D productivity at AstraZeneca. Nat Rev Drug Discov 17(3):167–181. https://doi.org/10.1038/nrd.2017.244

Musante CJ, Abernethy DR, Allerheiligen SR, Lauffenburger DA, Zager MG (2016) GPS for QSP: a summary of the ACoP6 Symposium on quantitative systems pharmacology and a stage for near-term efforts in the field. CPT Pharmacometrics Syst Pharmacol 5(9):449–451. https://doi.org/10.1002/psp4.12109. Epub 2016 Sep 17

Owens PK, Raddad E, Miller JW et al (2015) A decade of innovation in pharmaceutical R&D: the Chorus model. Nat Rev Drug Discov 14(1):17–28. https://doi.org/10.1038/nrd4497

Peterson MC, Riggs MM (2010) A physiologically based mathematical model of integrated calcium homeostasis and bone remodeling. Bone 46(1):49–63. https://doi.org/10.1016/j.bone.2009.08.053. Epub 2009 Sep 2

Peterson MC, Riggs MM (2012) Predicting nonlinear changes in bone mineral density over time using a multiscale systems pharmacology model. CPT Pharmacometrics Syst Pharmacol 1(11):e14. https://doi.org/10.1038/psp.2012.15

Peterson MC, Riggs MM (2015) FDA advisory meeting clinical pharmacology review utilizes a quantitative systems pharmacology (QSP) model: a watershed moment? CPT Pharmacometrics Syst Pharmacol 4(3):e00020. https://doi.org/10.1002/psp4.20. Epub 2015 Mar 12

Rao R, Musante CJ, Allen R (2023) A quantitative systems pharmacology model of the pathophysiology and treatment of COVID-19 predicts optimal timing of pharmacological interventions. NPJ Syst Biol Appl 9(1):13. https://doi.org/10.1038/s41540-023-00269-6

Riggs MM (2019) When learn/confirm leads to expand/understand: the expanding role of quantitative systems pharmacology in the betterment of therapeutics development. Clin Pharmacol Ther 105(3):558–560. https://doi.org/10.1002/cpt.1287. Epub 2018 Dec 18

Singh RSP, Toussi SS, Hackman F, Chan PL, Rao R, Allen R, Van Eyck L, Pawlak S, Kadar EP, Clark F, Shi H, Anderson AS, Binks M, Menon S, Nucci G, Bergman A (2022) Innovative randomized Phase I study and dosing regimen selection to accelerate and inform pivotal COVID-19 trial of Nirmatrelvir. Clin Pharmacol Ther 112(1):101–111. https://doi.org/10.1002/cpt.2603. Epub 2022 May 4

Sorger PK, Allerheiligen SRB, Abernethy DR, Altman RB, Brouwer KLR, Califano A et al (2011) Quantitative and systems pharmacology in the post-genomic era: new approaches to discovering drugs and understanding therapeutic mechanisms. An NIH White Paper by the QSP Workshop Group; October 2011

Stodtmann S, Nader A, Polepally AR, Suleiman AA, Winzenborg I, Noertersheuser P, Ng J, Mostafa NM, Shebley M (2021) Validation of a quantitative systems pharmacology model of calcium homeostasis using elagolix Phase 3 clinical trial data in women with endometriosis. Clin Transl Sci 14(4):1611–1619. https://doi.org/10.1111/cts.13040. Epub 2021 May 7

Teorell T (1937) Studies on the diffusion effect upon ionic distribution: II. Experiments on ionic accumulation. J Gen Physiol 21(1):107–122. https://doi.org/10.1085/jgp.21.1.107

Visser SA, de Alwis DP, Kerbusch T, Stone JA, Allerheiligen SR (2014) Implementation of quantitative and systems pharmacology in large pharma. CPT Pharmacometrics Syst Pharmacol 3(10):e142. https://doi.org/10.1038/psp.2014.40

Wang Y, Zhu H, Madabushi R, Liu Q, Huang SM, Zineh I (2019) Model-informed drug development: current US regulatory practice and future considerations. Clin Pharmacol Ther 105(4):899–911. https://doi.org/10.1002/cpt.1363

Zhang T, Cho CR, Bonate PL (2022) Perspectives on training quantitative systems pharmacologists. CPT Pharmacometrics Syst Pharmacol 11(6):669–672. https://doi.org/10.1002/psp4.12783. Epub 2022 Apr 7

Zineh I (2019) Quantitative systems pharmacology: a regulatory perspective on translation. CPT Pharmacometrics Syst Pharmacol 8(6):336–339. https://doi.org/10.1002/psp4.12403

Part III

Current Challenges and Future Directions

Future Directions for Quantitative Systems Pharmacology

Birgit Schoeberl, Cynthia J. Musante, and Saroja Ramanujan

Contents

B. Schoeberl
Biomedical Research, Novartis, Cambridge, MA, USA

C. J. Musante
Pfizer Inc., Cambridge, MA, USA
e-mail: cynthia.j.musante@pfizer.com

S. Ramanujan (✉)
Genentech Inc, South San Francisco, CA, USA
e-mail: ramanujan.saroja@gene.com

E. L. Bradshaw, M. E. Spilker (eds.), *Quantitative Systems Pharmacology*,
Handbook of Experimental Pharmacology 289, https://doi.org/10.1007/164_2024_737

Abstract

In this chapter, we envision the future of Quantitative Systems Pharmacology (QSP) which integrates closely with emerging data and technologies including advanced analytics, novel experimental technologies, and diverse and larger datasets. Machine learning (ML) and Artificial Intelligence (AI) will increasingly help QSP modelers to find, prepare, integrate, and exploit larger and diverse datasets, as well as build, parameterize, and simulate models. We picture QSP models being applied during all stages of drug discovery and development: During the discovery stages, QSP models predict the early human experience of in silico compounds created by generative AI. In preclinical development, QSP will integrate with non-animal "new approach methodologies" and reverse-translated datasets to improve understanding of and translation to the human patient. During clinical development, integration with complementary modeling approaches and multimodal patient data will create multidimensional digital twins and virtual populations for clinical trial simulations that guide clinical development and point to opportunities for precision medicine. QSP can evolve into this future by (1) pursuing high-impact applications enabled by novel experimental and quantitative technologies and data types; (2) integrating closely with analytical and computational advancements; and (3) increasing efficiencies through automation, standardization, and model reuse. In this vision, the QSP expert will play a critical role in designing strategies, evaluating data, staging and executing analyses, verifying, interpreting, and communicating findings, and ensuring the ethical, safe, and rational application of novel data types, technologies, and advanced analytics including AI/ML.

Keywords

Clinical trial simulation · Digital twins · Machine learning and artificial intelligence · Microphysiological systems · Modeling and simulation · Quantitative systems pharmacology · Virtual patients

1 Introduction

Since the 2011 publication of the landmark white paper by a National Institutes of Health working group on Quantitative and Systems Pharmacology (QSP) (Sorger et al. 2011), QSP has matured as a discipline and become widely accepted in drug discovery and development. The 2011 description of QSP as: "an approach to translational medicine that combines computational and experimental methods to elucidate, validate and apply new pharmacological concepts to the development and use of small molecule and biologic drugs," for understanding "in a precise, predictive manner, how drugs modulate cellular networks in space and time and how they impact human pathophysiology" still applies. More than a decade later, academia has increased focus on research, application, and training in QSP, most major pharmaceutical companies and even small biotech companies frequently employ QSP in drug design and clinical development, and health authorities have been increasingly reviewing and guiding the use of QSP. Although the scope of approaches and applications included in QSP is continuously evolving, the originally identified need to integrate knowledge from different disciplines remains. At the time the white paper called out the integration of pharmacology, systems biology, mathematics, and biochemistry. Today, we see the need to integrate new data types as well as the disciplines of data science and machine learning and artificial intelligence (ML, AI) (Aghamiri et al. 2022; Ribba 2023; Terranova et al. 2023).

Indeed, the present moment finds us in a time of seismic changes that broadly impact medicine and healthcare and will require us to bring all such approaches to bear. Rapid advancements in technology and computational power are drastically increasing the amount and speed of data generated, which in turn drives the need for novel analytics methods including but not limited to AI. Novel biological technologies, such as single-cell RNAseq, spatial transcriptomics, or CRISPR/Cas9 gene editing have spurred the design and subsequent success of new drug modalities such as cell and gene therapy and RNA-based therapies and vaccines (Porto et al. 2020; Wang et al. 2023). Development of complex in vitro systems such as organs-on-a-chip are promising to improve bench-to-bedside translation and reduce animal use (Ingber 2022). Experimental approaches such as spatial, single-cell, and high-content imaging techniques provide extraordinarily rich information on the underlying biology. Collection and organization of larger-scale clinical data, real-world evidence, and digital measurements from wearables and other decentralized technologies offer unprecedented information on patient biology, disease, and behavior. Notably, ML and AI approaches that infer and learn from large, multi-modal datasets to make predictions, drive decisions, and more recently create new data (e.g., images, text, molecular formulas, protein sequences) are upending fields and industries across the board (Hasselgren & Oprea 2023). These exciting developments lead to new financial, ethical, and regulatory considerations, which the biopharmaceutical industry together with the health authorities need to navigate and reconcile.

At the same time, the biopharmaceutical industry is grappling with multiple pressures one the drug development process. Long development cycles and high failure rates continue to pose a significant challenge (Harrison 2016). The increasing pressure on drug pricing and the growing size and bargaining power of payors in the United States further push the pharmaceutical industry to develop transformative, novel therapies in a shorter period and with a greater probability of success. Regulatory agencies meanwhile are revisiting topics including dose justification and accelerated approval with an eye towards greater stringency. They are actively encouraging model-informed drug development (MIDD) as an approach to support clinical plans (Bradshaw et al. 2019; J. Zhao et al. 2022). This combination of scientific and policy pressures is driving disruptive change in the field of biomedicine.

In this environment, regulators, researchers, and drug developers are increasingly embracing quantitative approaches that harness the abundance of knowledge and

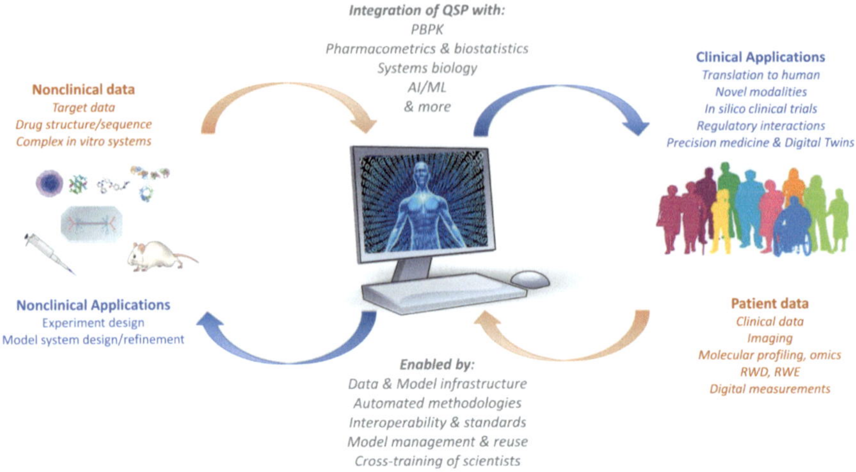

Fig. 1 Future Vision of QSP: Preclinical and clinical data from diverse sources are integrated into hybrid QSP platforms that integrate multiple modeling methodologies to drive high-impact applications in biomedical research and development. © *Saroja Ramanujan (reproduced with permission)*

data to *efficiently and robustly* advance novel medicines and clinical strategies. Quantitative scientific disciplines, including QSP, are already evolving in response to this need, incorporating methodologies such as machine learning to meet the challenge. In this evolving landscape, the field of QSP has a unique role. By providing an interpreatable and biologically meaningful quantitative framework that can integrate diverse and disparate data and knowledge, QSP enables us to propose and test mechanistically-based hypotheses and to contextualize experimental and clinical results to drive the development of novel medicines. In particular, we see QSP as an essential quantitative approach with the means to bridge different biological scales: (i) from the level of single cells, to a single organ or full body, (ii) the time scales of minutes and hours to weeks or months, (iii) across species, from mouse to man.

Figure 1 illustrates a vision of the future where QSP capitalizes on novel technologies and data sources to drive progress in key areas, including: (1) reducing the reliance on animal models; (2) optimizing the design of novel therapeutics in the human disease context including for new modalities; (3) increasing efficiency and probability of technical success in various aspects of translational and clinical development; and (4) driving precision medicine. These applications will necessitate foundational enabling efforts: methods to integrate and utilize the various rich data sources mentioned while keeping track of data lineage; advanced modeling infrastructure, standardization, and automation to support efficiency, reproducibility, data integrity, and credibility; and perhaps most importantly the holistic adoption of different modeling techniques including AI/ML. At present, QSP often involves manual collection and integration of disparate small to medium-size datasets from a multitude of preclinical and clinical studies into an expert-designed model. This "artisanal" process is not only highly subjective but also expensive and time-consuming. Approaches and technological platforms that easily integrate different datasets and model frameworks and modalities similar to Novartis's Data42 platform are needed to fully leverage non-clinical and clinical data (Eichler et al. 2022). This will require the development of approaches to data quality and reliability assessment. Although AI/ML approaches are not yet capable of generating biologically meaningful de novo QSP models that are predictive of multi-phenotypic mechanistic outcomes, we can envision a state where QSP leverages more algorithm-driven approaches to integrate multiscale, multisource (molecular, clinical, digital) datasets into hybrid data driven and mechanistically informative quantitative models. In this scenario, QSP practitioners could leverage standardized and automated frameworks for model generation, simulation, and analysis, and combine it with expert curation and verification to help drive preclinical science, clinical development, and precision medicine.

Without a doubt, AI will increasingly help QSP modelers find, prepare, and merge disparate data, as well as to build and parameterize models. More intriguingly, the rapid growth in generative Artificial Intelligence that goes beyond data-driven prediction to the creation of novel content offers new opportunities for the field of QSP. For example, perhaps generative AI engines trained on a wealth of prior models could be used to "generate" novel models that seamlessly integrate different modalities to address novel data and scenarios. We picture applications of such a QSP platform during all stages of drug discovery and development. During drug discovery, novel therapeutic candidates are generated by AI and will be translated to the human/clinical context by QSP models that predict clinical endpoints long before entering actual human studies. If this vision becomes reality, the "design, make, test, and analyze" (DMTA) cycles will be significantly reduced, and fewer candidates will be tested in animal experiments. During clinical development, different modeling approaches and multimodal patient data will be leveraged together to create equally multidimensional digital twins and populations for clinical trial simulation and to guide precision medicine, including individualized treatment decisions in a mechanistically *explainable* manner. We envision that in the future, QSP will capitalize on the advances in technology and analytics to radically increase the efficiency and success rate of creating efficacious and safe treatments for the patients who need them. To do this successfully, a close partnership with other quantitative disciplines will be essential, echoing the original vision set forth in 2011 by Sorger *et al.* Furthermore, given the implications to human health and safety, especially when replacing animal testing or defining clinical plans, engineering, computational, and QSP experts will play a critical role in developing and vetting the modeling approaches, ensuring the quality and relevance of inputs and data, interrogating and assessing results for plausibility and veracity, and ultimately determining safe and meaningful ways to leverage the learnings. In this chapter, we describe how QSP can evolve into this future by (1) pursuing high-impact applications enabled by novel lab technologies and data types; (2) integrating closely with analytical and computational advancements; and (3) increasing efficiencies through automation, standardization, and model reuse. While this list is by no means exhaustive, it highlights how immediate and mid-term efforts can pave the way towards this long-term vision.

2 Driving High Impact: Near and Long-Term Opportunities for QSP in Drug Development

In the following sections, we will discuss five areas that will lead to the expansion of QSP and the associated key enablers (Fig. 2). With the FDA Modernization Act 2.0, which became law in December 2022, the United States Food and Drug Administration (FDA) stated that new approach methodologies (NAMs) may be used to reduce animal use. This presents an important immediate role and opportunity for QSP to help translate results from experimental NAMs, such as organs-on-a-chip, into the human context via integrated model predictions. A second area, where we see an opportunity is to further expand discovery applications in early feasibility assessment and guidance of optimal drug design especially when there is a disconnect between pharmacokinetics and pharmacodynamics. During the clinical development process, QSP models will grow in importance to inform and justify dose selection given the mechanistic understanding of the drug and disease incorporated into the models as an orthogonal but complementary approach to more traditional exposure-response analyses. As a third opportunity for the expansion of QSP, we will discuss in silico clinical trial simulations to accelerate clinical development by reducing the clinical trial burden. A fourth critical area for QSP is in the support of regulatory submissions and the use of model simulations to supplement clinical data. However, use in the regulatory context requires verification and qualification of a QSP model for its intended use, which will require additional effort by model developers as well as regulatory agencies. Precision Medicine and the selection of predictive biomarkers is the fifth area where we see an increasing application of QSP models. In this context, we will discuss the concept of digital twins.

For the field of QSP to reach its maximal impact in the five areas outlined above, we argue that the advancement of hybrid modeling techniques as well as the ability of the QSP practitioners to build models more efficiently, will be key to increase adoption (Fig. 2).

2.1 Novel Assessment Methods and Reduced Reliance on Animal Models

The frequent failure of animal models to predict therapeutic responses in humans is one contributor to the high failure rate in drug development (Bart van der Worp et al. 2010). QSP is well suited to interpret and leverage animal data in the context of human disease understanding to improve translation, as was done for example to successfully predict the clinical response rate to a novel ERK inhibitor in BRAF mutant colorectal cancer (Kirouac et al. 2017). Meanwhile, the well-recognized

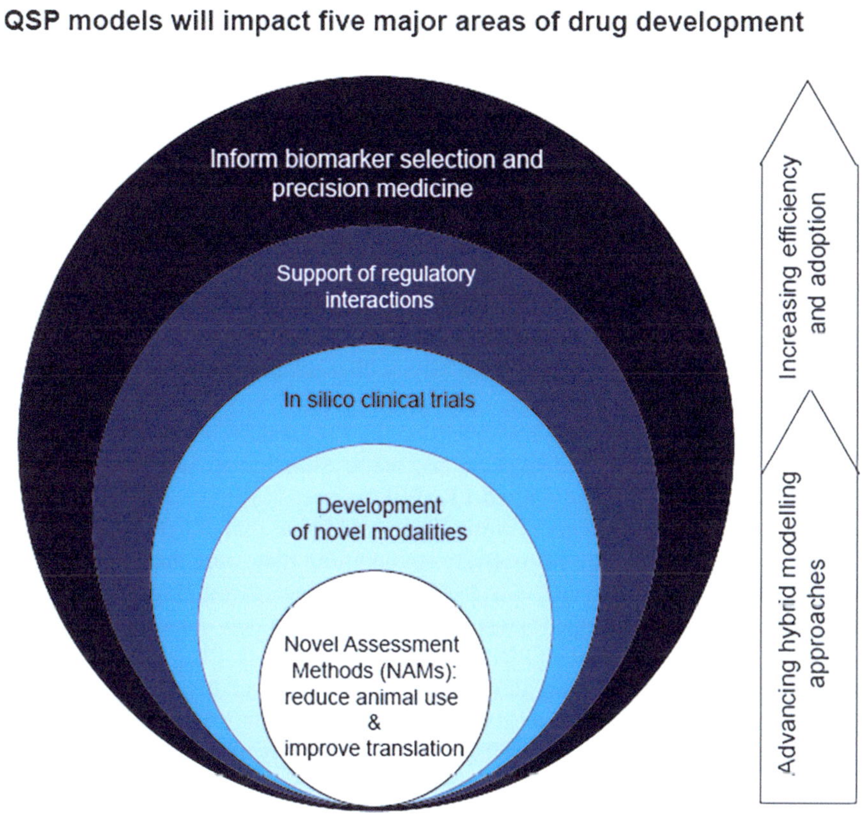

Fig. 2 Five areas that will lead to the expansion of the field of QSP and key enablers

disconnect between animal and human biology (Junhee Seok et al. 2013), combined with the need and responsibility to reduce animal use in biological and pharmaceutical research, have converged to an emphasis on reducing reliance on animal models. This has fueled a focus on in vitro, ex-vivo (non-animal) new approach methodologies (NAMs). With the FDA Modernization Act, the agency has been encouraging "the thoughtful and directed development of NAMs that will contribute to safety and efficacy assessments of pharmaceuticals" (Avila et al. 2023). Although frequently used to refer to in vitro laboratory systems, "NAMs" also include non-animal in silico approaches to predicting human safety and efficacy (e.g., Ram et al. 2022).

Experimental NAMs range from complex culture systems that recapitulate different aspects of in vivo cellular and tissue architecture and behavior, to organ-on-a-chip models, and even connected multi-organ systems that reproduce physiology and pathophysiology with greater fidelity (Herland et al. 2020; Ingber 2022). The use of human-derived cell lines, especially primary material in these microphysiological systems (MPS) holds potential for improving translation

(Benam et al. 2022). In silico, and especially QSP model NAMs can also focus and reduce the extent of animal experimentation. However, additional synergies could be achieved by coupling mechanism-based modeling with MPS to inform biomedical research (Cirit & Stokes 2018; Maass et al. 2019; Taylor et al. 2019). With their ability to measure various biological responses under controlled conditions and perturbations, MPS can probe multi-component interactions and mechanisms, thus synergizing naturally with QSP modeling; QSP models can in turn be used to capture, interpret, translate, and integrate data on biological functions, molecular and cellular distribution, and biomarker responses from the experimental systems into whole organism representations and predictions of human physiology and pathophysiology.

We see great potential in combining MPS with mechanistic modeling to improve human translation and especially inform on predictive safety. Herland et al. demonstrated that they were able to obtain similar PK parameters to what is observed clinically for IV administered cisplatin and orally administered nicotine in humans by combining physiologically based PK (PBPK) models with data generated with either a fluidically linked bone marrow, liver, and kidney chip or a fluidically linked gut, liver, and kidney chip, respectively. In addition, they were able to model the bone marrow toxicity observed with cisplatin in the bone marrow chip (Herland et al. 2020). More recently, Keuper-Navis et al. demonstrated how multi-organ on chip models combined with in silico PBPK modeling, can provide PK parameters during the preclinical phase of drug development that will lead to improved clinical translation (Keuper-Navis et al. 2023). Since efficacy and toxicity are driven by drug exposure at the target, we see a growing opportunity to extend efforts such as these for QSP models of predictive safety (e.g., Quantitative Systems Toxicology (QST); see Chapter "Toxicology" for more details). QSP models combined with PBPK models that focus on intercompartmental transport and biodistribution may prove especially useful as we translate from MPS systems into the clinic (Bloomingdale et al. 2017; Watkins 2020). As a natural extension, we can picture how the increased use of liver-on-chip models can be used to further advance our ability to parameterize mechanistic QSP models and to predict liver toxicity (Deng et al. 2019; Mcaleer et al. 2019).

QSP and MPS could also synergize in the study of human disease and translation of novel drugs. For example, intestinal organ-on-a-chip models (Guo et al. 2023) could be used to measure the impact of perturbations on cytokine, cellular, and functional responses to advance or validate existing QSP models of inflammatory bowel disease (IBD) (Balbas-Martinez et al. 2020; Rogers et al. 2021). Similarly, while translational QSP models of novel T cell engagers or T cell therapies might currently rely on preclinical animal observations (Betts et al. 2019; Betts & van der Graaf 2020; Flowers et al. 2023), in the future, they could leverage and integrate T cell infiltration, drug distribution, and cytotoxicity measured in immunocompetent human tumor MPS. The use of patient-derived material in fact opens the door to individualized treatment. An example where QSP signaling models were calibrated for individual patients based on a microfluidic high throughput biopsy drug screen is described below in the context of Digital Twins.

Significant progress has been made in advancing MPS. However, most organ chips are low throughput, but have high content and need the expertise to obtain high-quality data. Even though use of MPS is still far from common practice, we see them and QSP models as key approaches to improve clinical translation and reduce animal use. For MPS systems to reach their full potential, developers must demonstrate and validate the predictive performance of MPS and resolve issues related to availability, reproducibility, and standardization. Workshops were recently held by the FDA and Innovation and Quality Microphysiological Systems (IQ MPS) to address these issues and should involve the QSP modeling community in the future (Baran 2022). While MPS models continue being developed and validated to gain wider use, QSP is already actively used to translate data obtained from existing in vitro systems such as cocultures, tumor spheroids, and ex vivo assays, laying the groundwork for greater integration to fully leverage and translate data from MPS moving forward.

2.2 Development of Novel Modality Therapeutics

A major advancement in the field of medicine in the past decade has been the development and success of novel treatment modalities. These include natural extensions of prior formats, as with multi-specific antibodies and antibody drug conjugates (ADCs), but also newer approaches such as protein degraders (proteolysis targeting chimeras, PROTACs, and molecular glues), radioligands, cell therapy, RNA-based medicines, and gene therapy. The ability of QSP to leverage preclinical data and mechanistic knowledge for clinical prediction before the accumulation of significant clinical experience is especially suited to inform the design and development of such novel therapeutics.

Mechanistic modeling approaches have been adopted for such molecules due to the complexity of PK-PD-efficacy relationships. Drugs may demonstrate disconnects between exposure and response due to mechanism of action (PROTACs and covalent inhibitors), tissue-targeting (radioligands and ADCs), induction or deletion/silencing of protein production (mRNA vaccines, gene therapies), long-lasting effects (siRNA therapies administered every 6 months), or expansion of living drugs in the case of cell therapies. We believe for these new therapies QSP models that explicitly address the mechanisms underlying complex behaviors will be critical to inform the full drug discovery and development process.

Many mechanistic models have been published in recent years, focusing on informing discovery and optimization of PROTACs (Haid & Reichel 2023; Stein et al. 2023; H. Zhao & Narjes 2023). With the increased number of PROTACS in clinical development (Békés et al. 2022), the field is ripe for the use of QSP models and their application to inform clinical development in the near future. Recently, Claas et al. showed that a reduced model based on mechanistic principles leads to different predictions of the extent of protein degradation in patients compared to an empirical indirect response model, suggesting a potential advantage of QSP models over traditional models in such contexts. In addition, the authors showed that the reduced model with interpretable parameters allowed the prediction of in vivo degradation from in vitro degradation kinetics, once translational parameters were learned from tool compound data (Claas et al. 2023).

Seamless integration of PBPK and QSP will be critical in the development of novel treatment modalities as evident from recent examples where PBPK models developed for ADCs delivered to mice were able to inform on tissue targeting in humans (Chang et al. 2022; Kletting et al. 2019). For radioligand therapy (RLT), a recently developed PBPK model for ^{177}LuPSMA was extended to account for tumor growth inhibition by solely using the PET/CT and modeling to predict the PSMA-positive tumor volume after RLT (Kletting et al. 2019). We can envision a further mechanistic extension of this work in the future as combination therapies are considered.

Nucleic acid medicine is a burgeoning therapeutic area that can benefit from the use of QSP. Antisense oligonucleotides (ASOs) are promising therapeutic agents for a variety of neurodegenerative and neuromuscular disorders, and Monine et al. (2021) have developed a PBPK model for intrathecal administration of ASOs in non-human primates. As this novel anti-Tau ASOs proceed through clinical development, we see a great opportunity to combine the scaled PBPK model with a mechanistic QSP model to inform dose selection. The approval of four small interfering RNA (siRNA) products in the past few years has demonstrated unequivocally the therapeutic potential of this novel modality. The application of PBPK/QSP models for RNAi therapeutics is in its infancy. A minimal PBPK model for fitusiran, a GalNAc-conjugated siRNA targeting liver antithrombin, was published by Ayyar et al. (2021). The model bridged multiple PK-PD data sets in preclinical species (mice, rats, monkeys) and successfully captured reported plasma pharmacokinetics and antithrombin knockdown in a phase I ascending-dose study. As siRNA therapeutics are considered for targets outside the liver, we believe PBPK/QSP models will be essential. In addition, when targets are pursued with ASOs and siRNA like lipoprotein(a) (Lp(a)), we believe QSP models are essential to understand potential differences between the various modalities, especially when pharmacological and biological mechanisms lead to qualitative differences in PK-PD-efficacy relationships (Thau et al. 2024). mRNA vaccines and gene therapy have also seen recent clinical successes across indications from infectious disease and now cancer therapy for mRNA to genetic and rare diseases for gene therapy. QSP has been applied in these cases to explore the effect of therapeutic properties and treatment parameters on efficacy (Chen et al. 2014; Selvaggio et al. 2021; Zheng et al. 2021). As was highlighted briefly in Chapter "QSP Model Execution and Partnership in Academia vs Industry vs CROs, Touching on How to Partner/Communicate, Repurposing of Models and Academic/Industry Alliances", a mathematical model for an mRNA vaccine was created by adapting and extending a multiscale model of immunogenicity, opening the door for greater use of quantitative approaches in vaccine discovery and development.

The need for QSP to address therapeutics with complex mechanisms centered on cellular dynamics will only become more important moving forward. We have already seen the impact in the area of immune-cell stimulating agents for molecular design, dose/regimen optimization, and biomarker prediction (Betts & van der Graaf 2020; Hosseini et al. 2020; Ma et al. 2020; Qi et al. 2023; Susilo et al. 2023). More recently, cell therapy has been an emerging area of application. The kinetics, persistence, and efficacy of chimeric antigen receptor T-cell (CART) therapies are difficult to anticipate, especially since preclinical experimental models are often not relevant and clinical behavior depends heavily on patient variables such as antigen levels, immune fitness, and cellular-infusion-material characteristics. As such, QSP efforts have begun to focus on CART therapy, especially in hematological malignancies (Hardiansyah & Ng 2019; Kirouac et al. 2023, 2024), as discussed in the second case study in Chapter "Applications of Quantitative Systems Pharmacology in Cancer Drug Development: Target Identification, Dose Optimization, and Efficacy Prediction". Kirouac et al. used QSP to simulate cell dynamics and tumor response, and in parallel used machine learning to demonstrate the importance of cell product characteristics, specifically transcriptional patterns, for the prediction of response. T cell therapy efforts have now expanded to other approaches that exploit tumor-infiltrating-lymphocytes and TCR-engineered T cells and have moved into solid tumor indications. In the latter context, Joslyn et al. (2024) investigated the cellular phenotypes and properties governing the persistence of TCR-engineered T-cell treatment of solid tumors by creating and analyzing patient-matched QSP digital twins. Going forward, we see increasing use of QSP models to quantitatively understand multi-specific immune cell engagers, proliferative cytokines, and cellular therapies, especially as these modalities are considered for solid tumors where there can be greater barriers to tumor access and assessment, and as these modalities are applied to indications beyond oncology (Mougiakakos et al. 2021). Furthermore, integrating these QSP models with MPS models as discussed earlier will be extremely valuable, as animal models are often not as relevant for the study of novel "TCR-like" T cell engagers and cell therapies.

2.3 In Silico Clinical Trials

Over the past decade, the FDA has established modeling and simulation (M&S) as a regulatory science priority, coinciding with the explosive growth in data generation, data science, machine learning, and model-based technologies. The FDA has highlighted QSP and mechanism-based models as a component of the MIDD Pilot Program (Friedrich 2016; J. Zhao et al. 2022) and has reported the increased use of QSP in regulatory submissions (Bai et al. 2021; Zineh 2019). Additionally, the FDA's Project Optimus has emphasized the need for more rigorous dose justification and exploration in oncology clinical development, before approval (Rodney & Banerji 2024). In this context, the ability to leverage in silico methods as part of a "totality of evidence" approach can help address regulatory scrutiny without prohibitively exhaustive exploration of dose/regimen in randomized clinical trials (Venkatakrishnan & van der Graaf 2021). QSP presents a valuable approach here, and can bridge preclinical knowledge/data, clinical biomarkers, and mechanistic understanding with dose/regimen exploration to focus clinical trials and justify clinical strategy and dose/regimen. As an example, Hosseini et al. used a translational QSP model of an anti-CD3/CD20 T cell engaging bispecific for in silico evaluation and optimization of a step-up dose regimen to mitigate cytokine release syndrome adverse events observed early in a Phase I study, supporting a protocol amendment (Hosseini et al. 2020). The same model was employed as part of a broader analytics strategy to assess the pivotal dose expansion of mosunetuzumab and, with continued development, to explore recommended Phase 2 doses for other T cells engaging bispecifics.

In silico clinical trial simulation can streamline or reduce the clinical trial burden of a program. For example, disease progression models can be used to simulate "virtual control arms" to inform efficient trial design or to compare with clinical observations from treatment arms in studies without control arms. While disease progression is traditionally modeled using statistical approaches, other modeling approaches are also used (Barrett et al. 2022), and QSP is particularly useful in connecting biological mechanisms with disease progression. Karelina et al. (2017) presented a mechanistic model linking the dynamics of soluble and aggregated amyloid with their impact on the progression of amyloid plaques and subsequent connection to neurotoxicity. The integrated model could simulate responses to amyloid-targeted agents and was used to explore the impact of preventative vs. therapeutic treatment and assess the trial duration required to observe clinical benefit.

In a more acute disease context, mechanistic viral dynamics and QSP models of SARS-CoV-2 have played a key role in advancing our understanding of SARS-CoV-2 to accelerate the development of treatments from the very beginning of the pandemic (Dai et al. 2021; Desikan et al. 2022; Kim et al. 2021; Padmanabhan et al. 2022; Rao et al. 2023). A QSP model of the time course of COVID infection and impact of intervention supported the initial emergency use authorization and subsequent full approval of PAXLOVID™ (nirmatrelvir tablets; ritonavir tablets) for the treatment of mild-to-moderate COVID in adults who are at high risk for progression to severe COVID, including hospitalization or death. The model considered not only the evolution of viral load, but also hospitalization and death, matching observed placebo and treatment responses of viral load and severity. QSP simulations predicting similar efficacy of 5-day and 10-day courses of treatment were used to support the 5-day regimen (Rao et al. 2023), and this application of QSP in their review process was highlighted by the FDA in their annual OCP report (Office of Clinical Pharmacology Report 2023).

A growing and valuable source of data for disease modeling approaches is "real world data" (RWD), defined by the FDA as data "relating to patient health status and/or the delivery of health care routinely collected from a variety of sources" (FDA 2023). These data traditionally contain demographic and clinical data from sources such as electronic health records/systems or payer databases, but also can include molecular, imaging, and digital data. The use of RWD is not new to QSP with the Archimedes model, a statistical network systems model, being a notable early example (Eddy & Schlessinger 2003). With the increased abundance and diversity of such data, it is important that the field actively evaluates and incorporates it into future QSP models, especially as clinical trials are tightly managed and often not representative of behavior and outcomes in the real world.

2.4 Regulatory Interaction Support

The ability to impact dose justification and clinical study design more broadly raises the issue of regulatory interactions. Health authorities are increasingly considering QSP modeling results in regulatory interactions, as in the example discussed above for nirmatrelvir/ritonavir. Regulatory use requires qualification and clear communication of the appropriateness of a QSP model for its intended use. For PBPK models the intended use is framed by a key question, the context of use, and ultimately, the associated risk-based assessment (Cuquerella-Gilabert et al. 2024; Kuemmel et al. 2020; Viceconti et al. 2021). Thus, the risk to patients associated with the proposed use of QSP model-based predictions can be broadly categorized as high, medium, or low risk, depending on the application and potential impact on patients, similar to PBPK models (Ramanujan et al. 2019; Shebley et al. 2018). Since QSP models often involve significant biological uncertainty, it is more challenging to establish rigorous guidelines for data relevance and quality, model building, calibration, validation, and qualification (J. R. Chan et al. 2022). Adoption and consistent use of a common assessment framework will enable easier, objective evaluation during regulatory interactions (Bai et al. 2019; Cucurull-Sanchez et al. 2019; Ramanujan et al. 2019). Model and methodology standardization, publication, and reuse will greatly facilitate this.

2.5 Biomarkers and Precision Medicine

Biomarkers, translational medicine, and precision medicine are rapidly evolving areas of drug development due in part to new clinical biomarker technologies and data types. With its emphasis on integrating quantitative biological measurement data in the context of mechanistic understanding of systems, QSP is highly suited to advancing this area.

2.5.1 High-Content Molecular and Imaging Data

The explosion in data types like -omic, single-cell transcriptomics, multiplexed pathology and high-content imaging, and digital biomarkers opens new opportunities in the space of biomarkers and precision medicine. QSP presents a unique approach to link these data across different scales and connect them with functional activity and clinical measures in a mechanistically relevant framework. This however is likely to require integration of different modeling formalisms to address the different scales and types of data.

Spatial data and dynamics can be addressed by exploiting modeling methodologies such as agent-based models (ABM), which can also capture stochastic events. For example, Ruiz-Martinez et al. presented an immune-oncology QSP model for triple-negative breast cancer coupled with a spatial ABM (Ruiz-Martinez et al. 2022). To enable the integration of high-throughput data, the model blends multiple mathematical modeling frameworks. An ordinary differential equation module simulates tumor progression at the organ and whole-patient levels with parameters that are specifically calibrated for triple-negative breast cancer. The ABM represents either part of a tumor or a coarse-grained approximation of the entire tumor and allows the exploration of spatial tumor evolution in the presence or absence of therapeutic intervention. This and a related paper by Zhang et al. (2021) are examples of how spatial QSP modeling and multiplex digital pathology can be combined for model parameterization. and validation, as well as biomarker discovery.

Systems biology and bioinformatics approaches are well suited to address molecular data such as multi-omics and single-cell profiling. Pathway enrichment analysis, single cell-driven tissue characterization, bulk RNA deconvolution, and network analysis can be incorporated into or used to inform model development and calibration (Brubaker et al. 2019; Meimetis et al. 2024; Paczkowska et al. 2020; Schäfer et al. 2024). Logic-based approaches can be used to adapt systems biology networks into quantitative simulation models that can be integrated into larger QSP models (Balbas-Martinez et al. 2020). Integration of QSP with AI and other analytics approaches is particularly valuable in these high-content data settings, as highlighted by Terranova et al. in their review of future directions in translational medicine (Terranova et al. 2021).

2.5.2 Precision Medicine

The wealth of biomarker data also opens the door to more precision medicine and personalized approaches. In this space, the development and use of "digital twins" is a compelling approach. The concept of the digital twin has been used in many contexts in medicine where an in silico model reproduces behaviors of a specific complex system of interest, be it a device, a process, or a biological entity – be it a cell, tissue, animal, or patient. Chapter "Rare Disease (Niemann-Pick Disease); Highlight Extrapolation from Adults to Pediatrics & FDA Interactions" provides an example of the use of digital twins with a QSP model to assess disease and drug similarity in adult and pediatric patients to support the clinical development of olipudase alfa for the treatment of Acid Sphingomyelinase Deficiency (ASMD).

QSP models currently rely heavily on virtual patient (VP) and population (Vpop) approaches, that are typically used to capture and simulate population or subpopulation behaviors at aggregate levels. However, recent work has begun to focus on capturing the measurements for each individual patient separately, rather than at the population level. These individualized virtual patients reflect baseline biological measurements/properties, treatment regimens, PK and biomarker measurements, and clinical outcomes of a single individual. Individualized VPs thus represent "digital twins" to the clinical patients. Statistical and machine learning digital twins have been used to predict a patient's disease progression from biomarker and clinical measurements (Allen et al. 2021; Bordukova et al. 2024; Katsoulakis et al. 2024). Individual patient parameterizations of population pharmacometric PKPD models can also be considered digital twins with respect to PKPD.

The QSP digital twin is a unique parameterization of a mechanistic QSP model that matches a specific patient's temporal data for relevant measures (PK, biomarkers, clinical endpoints). The QSP digital twin thus uniquely enables one to interrogate biology (markers or mechanisms) which were not explicitly measured but inferred from matching a mechanistic model to available data. Susilo et al. used a QSP model to develop digital twins for individual patients from a Phase I trial of the T-cell engaging anti-CD3/CD20 bispecific antibody, mosunetuzumab (Susilo et al. 2023). These digital twins were then used to simulate and determine population-level behaviors (exposure-response) but also to infer baseline biological characteristics of individual patients (tumor immune infiltration, tumor growth rates) and relate these to post-treatment tumor growth to propose predictive biomarkers of response. Notably, multiple digital twins were used for each individual patient to address the challenge of parametric non-identifiability and biological uncertainty typical in mechanistic QSP models. Venkatapurapu et al. (2022) applied a digital twin strategy for Crohn's disease, generating a library of QSP model-based virtual patients from which a digital twin (or twins) most similar to a clinical patient of interest can be selected. Integrating experimental and model-based approaches, Eduati et al. (2020) generated patient-specific data by perturbing tumor biopsy material obtained from the corresponding patient with various drugs using a high throughput microfluidics approach. The data was used to generate individualized calibrations of a logic-based signaling model, which was then used to predict efficacious combination treatments for the corresponding patient. Interestingly, this example combines advanced in vitro systems with QSP modeling to form a *multi-modal* approach to generating digital twins. While the findings were tested in vitro, others are pursuing clinical validation of QSP-model-based digital twin predictions. Howard et al. (2022) have developed a biophysically-based mechanistic model that uses the structure and vascularity of a patient's tumor (as determined by CT scan) to predict breast tumor progression and shrinkage in response to standard-of-care treatment options as an exploratory tool for surgeons. These examples illustrate how in the future, with appropriate validation of performance and clinical utility, predictive simulation using QSP models and digital twins could help inform individual patient treatment options.

3 Embracing Hybrid Modeling Approaches

Biomedical research and development will increasingly require a diversity of quantitative approaches to address various applications and needs. While QSP by definition will continue to focus on systems-based quantitative approaches, the ever-evolving data and analytics landscape requires that QSP incorporate or integrate with a wide range of modeling methodologies. QSP should be recognized as a discipline that encompasses all methodologies that support the integration of data in a quantitative, biologically interpretable modeling framework, regardless of mathematical formalism or model construction approach ("top-down" vs "bottoms-up"). With time, we predict a high degree of hybridization of the numerous quantitative methodologies will be used across the board.

3.1 PBPK

Well underway is the natural extension of QSP to incorporate PBPK models for small molecules and especially for novel complex modalities. Since 2016, PBPK modeling and simulation has become integral to the development of small molecules (X. Zhang et al. 2020), and PBPK models are routinely included in IND and NDA submissions. PBPK models for small molecules are commonly used to predict drug PK as a function of patient-intrinsic factors like age, organ impairment, pregnancy or patient-extrinsic factors like formulation, food, or drug-drug interactions. PBPK software like GastroPlus®, Simcyp™, or PKsim® in combination with regular workshops between scientists from health authorities, industry, and academia have resulted in regulatory guidelines for PBPK models. These guidelines led to a subsequent increase in the use of PBPK models and the acceptance of in silico predictions in lieu of clinical pharmacology studies by regulators to inform drug labels in the last decade (Shebley et al. 2018; Wu et al. 2021; X. Zhang et al. 2020).

It has become increasingly common to couple PBPK and other mechanistic PK models with mechanistic QSP models of PD. This has been evident in efforts to support the development of complex modalities, especially for tissue-targeted agents. In a recent example, Miyazawa et al. (2024), developed a PBPK-QSP model for a mRNA-loaded lipid nanoparticle (LNP) enzyme replacement therapy for Crigler-Najjar (CN) syndrome. CN syndrome is characterized by the absence of the protein uridine glucuronotransferase (UGT). The minimal PBPK-QSP model investigated LNP-mRNA delivery and biodistribution, the generation of UGT protein, and its catalytic effect on bilirubin to inform dosing regimens. Such integration of PBPK into QSP modeling will increasingly become standard, not only for physiological and pharmacological considerations of PK and biodistribution, but also for addressing safety and efficacy in different organs grounded in a mechanistic understanding of the disease. Alsmadi et al. investigated the impact of irritable bowel syndrome (IBS) and inflammatory bowel disease (IBD) diseases on pravastatin oral bioavailability and pharmacokinetics using a bottom-up PBPK model (Alsmadi et al. 2024). The increased pravastatin oral bioavailability can be explained by a combination of increased pravastatin intestinal permeability and decreased pravastatin gastric degradation resulting in higher exposure. Thiel et al. (2018) developed a PBPK/QSP model to predict the pharmacokinetics (PK) and pharmacodynamics (PD) of COX-2 and 5-LOX inhibitors, which are used to treat inflammatory conditions. Drug-specific PBPK models and existing cellular models of arachidonic acid metabolism and rifampicin-induced CYP activation were coupled. The impact of pre-treatment with rifampicin on the drug clearances of COX-2 and 5-LOX inhibitors and, consequently, on their pharmacological action, i.e., prostaglandins and leukotrienes were evaluated. These examples of combined PBPK/QSP models can be used to predict clinical outcomes and optimize dosing strategies based on mechanistic understanding of drug distribution, drug-target interactions, and disease processes.

3.2 Pharmacometrics and Statistical Outcomes Models

QSP models to date have focused on simulating biological measurements (e.g., tumor growth inhibition and tumor-based response rates) rather than clinical endpoints (e.g., survival), because of their grounding in the mechanism. The use of QSP in the clinical simulation space naturally raises the opportunity and need to go further and to integrate with pharmacometric and statistical modeling approaches. Joshi et al. (2023) describe the three most common ways in which QSP and pharmacometrics models are currently used together: parallel synchronization, cross-informative use, and sequential integration. However, only the last of these actually integrates QSP and pharmacometric models. For example, connecting QSP-model simulated tumor growth inhibition (TGI) with overall survival (OS) using a TGI-OS statistical framework (P. Chan et al. 2021; Claret et al. 2015) would enable simulation from a basic mechanism through to a complex endpoint. However, even this represents low-hanging fruit, and Joshi et al. note that "achieving a true convergence of QSP and pharmacometrics will require new tools and methods that enable greater technical integration." This could enable a future where QSP and pharmacometrics experts work closely together, rather than in parallel.

3.3 Machine Learning

A critical, if not the most pressing, current need in QSP is the efficient incorporation of machine learning approaches to automate and accelerate workflows as well as to effectively leverage and integrate biological knowledge with the rich datasets now available. Machine learning is already being utilized in QSP. In 2021, Aghamire et al. performed a review and analysis of efforts to integrate QSP and ML across indications (Aghamiri et al. 2022). Since then, such examples have only grown. A recent review (T. Zhang et al. 2022) found that the most common current applications of ML in QSP are for parameter estimation/extraction, inference or learning of model structure, dimension reduction (of inputs and of model structure), working with virtual populations, and addressing stochasticity (Myers et al. 2023). Many of these efforts focus on ML-based analysis of biological or QSP-model-derived synthetic data for model development or analysis. Another popular approach reviewed has been the development of ML surrogate models, whereby ML models trained to reproduce QSP-model results are used for more efficient large-scale simulation in computationally expensive efforts such as parameter optimization or virtual population generation and simulation.

Moingeon (2023) considered how AI can enhance QSP by enabling easier integration of data (imaging, medical records, QSAR) to inform and optimize model construction. Historically, model structure determination has often been more "art," based on expert review and integration of disparate literature, knowledge, and data. However, inference of network and model structure through causal ML approaches involving larger data sets (e.g., scRNA or high content imaging) or scientific ML approaches such as physics-informed neural nets and universal ODEs advance this by letting hybrid mechanism-informed methods help identify or enhance QSP model structure. While mechanism-informed ML approaches have occasionally been used with large-scale models, they have primarily been applied to relatively small models. Expanding their use for more complex (e.g., whole disease) models, could significantly reduce the resource demand and scientific complexity of developing, updating, and applying such QSP models, enabling faster progress and broader use. Furthermore, in the clinical setting, thousands to millions of data points are currently collected for each patient. Incorporation of such data can enable for example a QSP model that captures baseline patient biomarker data, PK, dynamic changes in biomarkers, and clinical parameters. ML can help identify previously unappreciated connections/patterns in large datasets that can be investigated mechanistically in QSP models. Furthermore, ML approaches can be used to analyze QSP model-generated synthetic data to explore multivariate parameters/biomarkers driving clinical response. ML analysis of QSP-derived synthetic data has in fact been used previously to propose potential diagnostic biomarkers of treatment response in rheumatoid arthritis, cardiovascular disease, and atopic dermatitis (Go et al. 2024; Ramanujan et al. 2016). Mechanistic underpinnings of these markers can then be explored in the context of the QSP model.

Certainly, a burgeoning area of work in the machine learning field pertains to natural language processing and generative AI large language models (LLMs). Use of such approaches to interrogate wide bodies of knowledge and data and infer mechanism and causality from literature and data are already in progress and can be used to inform model design as described above (Feder et al. 2022). Additionally, efforts are already underway to create biological LLM inquiry tools (e.g., Lee et al. 2020), especially by focused training of open source LLMs. These can extend beyond the ingestion of learnings and knowledge from scientific sources to the ingestion of information from published datasets and databases and can be enhanced to enable standard analyses. Thus, making large datasets more accessible and integrating them with scientific knowledge can revolutionize QSP by enabling efficient integration of large amounts of information and reducing the manual overhead in determining mechanisms and constructing QSP models. Ultimately, LLMs and other AI capabilities can be used to support QSP modelers in framing model topology, parameterizing, and interacting with models. Such efforts are already underway in various modeling disciplines including pharmacometrics.

Figure 3 maps the various machine learning applications in QSP onto a generalized workflow for QSP model development and application, illustrating the potential broad utility of diverse current ML approaches in QSP efforts. Machine learning may be used in any or all these QSP areas, integrating with other methodologies, but requires expert supervision and review, especially in areas such as data curation, model reduction, and prediction assessment to ensure the reliability of the work.

While the examples discussed are obvious areas of promise, marrying evolving machine learning advances with mechanistic QSP will likely enable many more exciting approaches such as generative AI-driven model design or other applications

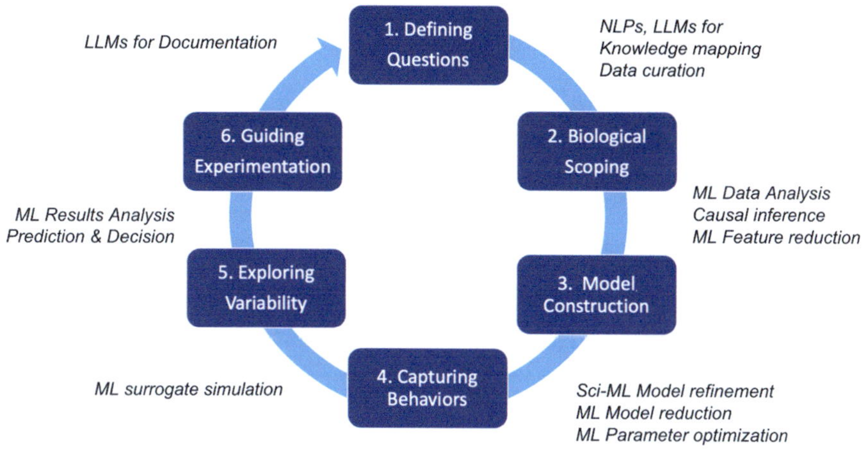

Fig. 3 Examples of how existing machine learning approaches can enable the QSP workflow

yet to be anticipated. We again emphasize however the important role of QSP experts even in the use of evolving automated and AI technologies. For example, knowledge-mining results, inferred networks, reduced models, and simulation predictions must be reviewed for biased, artifactual, dubious, or even nonsensical results and appropriate modifications made. While still predicated on the current systems pharmacology skillset of the experts, this will require greater familiarity with data science and ML approaches along with their strengths and liabilities.

4 Increasing Efficiency and Adoption

4.1 Continued Adoption of QSP

In Fig. 4, we show how four fundamental analytics capabilities that are currently being used in the pharmaceutical industry differ in their: 1) adoption, from early adoption in the case of AI/ML to routine application in the case of exposure-response models, 2) amount of data needed and 3) current impact on drug development. AI/ML models need large amounts of data that contain features that have predictive value and are collected reliably using a similar method or assay at scale (thousands to millions). On the other end of the spectrum, exposure-response models and biostatistical models used in preclinical and clinical development to support translation, dose selection, trial designs, or subgroup analyses typically need quality preclinical and clinical data from controlled trials with usually tens to thousands of patients. PBPK models typically utilize pharmacokinetics data collected for a given compound and incorporate in vitro and in vivo ADME data for the model parametrization and verification. The application of PBPK models in small molecule drug

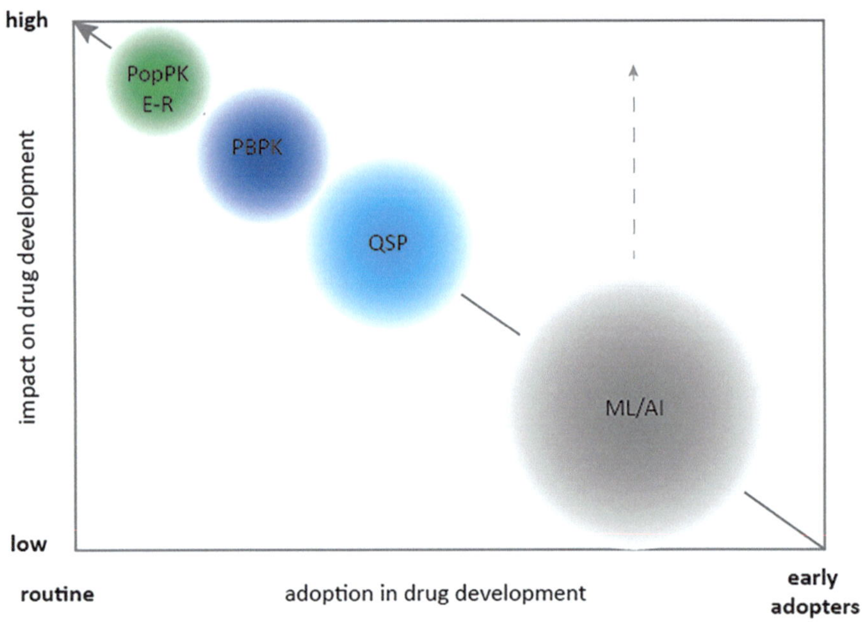

Fig. 4 The x-axis reflects the technology's adoption in drug development, from routine to early adopters, while the y-axis reflects each technology's impact on drug development currently. The size of the circle is indicative of the amount of data required (not to scale). The dashed line indicates the anticipated growing impact of AI/ML on the drug development

development has become standard in recent years. Currently, we see QSP models between PBPK models and AI/ML in terms of adoption, data use and impact on drug development. QSP models are able to integrate mechanistic kinetic data from multiple disparate data-sources across biological scales, i.e., in vitro, in vivo, pre-clinical and clinical. In addition, in-house data most often will be combined with data retrieved from publications.

Looking ahead, the discipline of QSP will further gain in importance. We believe that QSP will significantly benefit from and complement ML/AI models. Early considerations of a sound target product profile require a mechanistic drug and disease understanding. This understanding is essential to benchmark novel therapies to existing standard-of-care or competitor therapies in development and to guide clinical development strategy as well as investment decisions. Here we see a great opportunity to benchmark molecules designed by generative AI with QSP models in silico. With the rising importance of precision medicine, QSP models could become a key enabler of personalized medicine, providing an integrated framework to connect pharmacokinetics and predictive biomarkers with disease biology and clinical endpoints.

4.2 Key Enabling Infrastructure

QSP model development and use require significant time and effort, involving expert review of literature and data and often manual and subjective design of model topology. Computational expense further increases time and costs. This has limited the wider and more "on-demand" use of QSP, to those where advanced planning is feasible. And even as the use of QSP models in the regulatory arena continues to grow, the absence of uniformly adopted model communication and assessment strategies, requires a significant effort by the regulatory reviewers during the review process.

For the field to become as widely accepted as pharmacometrics, bioinformatics, or even AI, two aspects in particular of QSP modeling need to be addressed: (1) model building, training, and validation need to become faster, and (2) model communication, assessment, and reuse need to be enabled by more standardized approaches, to build credibility with collaborators, regulators, and the wider scientific community. If both are fulfilled, we will see an increase in speed and impact. Here, we argue, lie the fundamental driving forces that will define the future of QSP as an essential discipline alongside the well-established disciplines of pharmacometrics and biostatistics and the rapidly growing field of AI.

As key enabling infrastructure, we see a need to implement data and meta-data traceability combined with the associated model version control and documentation. The traceability of internal and external data requires increased data stewardship and more liberal, or "FAIR" (findable, accessible, interoperable, reusable) data access standards and policies across many organizations.

However, infrastructural needs also exist in the model development and management space to allow continuous improvements in efficiency and utility. More efficient, reproducible, and even automated methods, tools, and standardized interoperable software capabilities for QSP model construction and simulation will enable faster implementation and return on investment to address pragmatic usage challenges. Improving model management/tracking and increasing model sharing across the community engenders more rapid progress and expands use. Health authorities could, where feasible (for example, in the case of disease platforms), even require public disclosure and sharing of "sanitized" models and reproducible code within a reasonable time frame if used in regulatory decision-making.

Finally, the use of similar computational platforms will create the technical base to grow the collaboration and integration among different modeling approaches, including QSP, pharmacometrics, biostatistics, and machine learning. Enhanced collaboration will broaden the utility of all the approaches. This will require training the next generation of QSP modelers to be fluent or at least familiar with multiple modeling approaches and programming languages.

4.2.1 Model Reuse to Accelerate Forward and Reverse Translation

Often, before more complex disease models can be developed, smaller mechanistic models that can inform experimental design or molecular design criteria can be quickly developed and form the foundation for larger translational models. The impact of QSP models during the drug discovery and development process is most tangible when models are applied at the right time. This requires that the modeler anticipate the question(s) and prepare a model in advance—weeks, months, or even years before the questions need to be addressed to impact project decisions. A quicker path is to reuse published QSP models that can be quickly implemented, validated, modified, and reused.

One recent example is an impactful reuse of the original semi-mechanistic three-compartment model of the ocular pharmacokinetics of therapeutic antibodies given by intravitreal injection (Hutton-Smith et al. 2017) in the root cause analysis to identify the mechanistic drivers of immunogenicity against the anti-VEGF biotherapeutic brolucizumab (Kearns et al. 2023). The combination of experimental and modeling approaches identified several factors, including a shared epitope with bacteria and the development of non-native drug species in the eye, that contributed to the immunogenicity of brolucizumab and induced formation of immune complexes.

While there are successful examples of model reuse, often when groups try to re-implement published models, important details are missing and executable code is rarely shared—and neither are the raw data source used to train the models. The reluctance to share code and data is understandable, since model building takes time and effort. However, if we look at the field of AI, the rapid advancement in the development of foundation models (Moor et al. 2023) was only possible because of the fast and transparent sharing of code and publicly available data. Transparent, documented, and efficient sharing of work could even in the future enable automated or even "generative" approaches to new model development.

5 Conclusions

Ultimately, the revolutionary changes in biology, therapeutic modalities, data, and analytics make this not only an exciting time but also a critical time for identifying and investing in the path forward for QSP. To seize the opportunity requires equally brave changes and not incremental progress. While we have outlined pragmatic directions in which the field of QSP can evolve, we encourage the community to think bigger and to be bolder. What if Artificial Intelligence coupled with systems biology helps us identify, analyze, and connect the data we need in a meaningful way, even constructing a QSP model automatically based on causality? Imagine a QSP platform that incorporates various modeling frameworks and/or operates seamlessly with other models and vast amounts of easily accessible data. Consider novel therapeutics being designed and then translated into the clinic using a suite of complex in vitro systems and QSP models linked to PBPK and statistical disease outcome models to predict clinical endpoints before the therapeutic enters the clinic and without ever touching an animal. Imagine this QSP platform leveraging these different modeling approaches *and* multimodal patient data to create equally multi-dimensional multi-*model* digital twins and virtual populations for clinical trial simulation, precision medicine, or even individualized treatment decisions in a mechanistically *explainable* manner. Our vision is thus that QSP partners closely with other quantitative fields linked by the same data. Only then, will we capitalize on the recent and future advances in data generation, computation, and AI/ML to radically increase the efficiency and success rate of creating efficacious and safe treatments for the patients in need.

References

Aghamiri SS, Amin R, Helikar T (2022) Recent applications of quantitative systems pharmacology and machine learning models across diseases. J Pharmacokinet Pharmacodyn 49(1):19–37. https://doi.org/10.1007/s10928-021-09790-9

Allen A, Siefkas A, Pellegrini E, Burdick H, Barnes G, Calvert J, Mao Q, Das R (2021) A digital twins machine learning model for forecasting disease progression in stroke patients. Appl Sci 11(12):5576. https://doi.org/10.3390/app11125576

Alsmadi MM, Abudaqqa AA, Idkaidek N, Qinna NA, Al-Ghazawi A (2024) The effect of inflammatory bowel disease and irritable bowel syndrome on pravastatin oral bioavailability: in vivo and in silico evaluation using bottom-up wbPBPK modeling. AAPS Pharm Sci Tech 25(4):86. https://doi.org/10.1208/s12249-024-02803-z

Avila AM, Bebenek I, Mendrick DL, Peretz J, Yao J, Brown PC (2023) Gaps and challenges in nonclinical assessments of pharmaceuticals: an FDA/CDER perspective on considerations for development of new approach methodologies. Regul Toxicol Pharmacol 139:105345. https://doi.org/10.1016/j.yrtph.2023.105345

Ayyar VS, Song D, Zheng S, Carpenter T, Heald DL (2021) Minimal physiologically based pharmacokinetic-pharmacodynamic (mPBPK-PD) model of N-acetylgalactosamine–conjugated small interfering RNA disposition and gene silencing in preclinical species and humans. J Pharmacol Exp Ther 379(2):134–146. https://doi.org/10.1124/jpet.121.000805

Bai JPF, Earp JC, Florian J, Madabushi R, Strauss DG, Wang Y, Zhu H (2021) Quantitative systems pharmacology: landscape analysis of regulatory submissions to the US Food and Drug Administration. CPT Pharmacometrics Syst Pharmacol 10(12):1479–1484. https://doi.org/10.1002/psp4.12709

Bai JPF, Earp JC, Pillai VC (2019) Translational quantitative systems pharmacology in drug development: from current landscape to good practices. AAPS J 21(4):72. https://doi.org/10.1208/s12248-019-0339-5

Balbas-Martinez V, Asin-Prieto E, Parra-Guillen ZP, Troconiz IF (2020) A quantitative systems pharmacology model for the key interleukins involved in Crohn's disease. J Pharmacol Exp Ther 372(3):299–307. https://doi.org/10.1124/JPET.119.260539

Baran S (2022) Perspectives on the evaluation and adoption of complex in vitro models in drug development: workshop with the FDA and the pharmaceutical industry (IQ MPS affiliate). ALTEX. https://doi.org/10.14573/altex.2112203

Barrett JS, Nicholas T, Azer K, Corrigan BW (2022) Role of disease progression models in drug development. Pharm Res 39(8):1803–1815. https://doi.org/10.1007/s11095-022-03257-3

Bart van der Worp H, Howells DW, Sena ES, Porritt MJ, Rewell S, O'Collins V, Macleod MR (2010) Can animal models of disease reliably inform human studies? PLoS Med 7(3):1–8. https://doi.org/10.1371/journal.pmed.1000245

Békés M, Langley DR, Crews CM (2022) PROTAC targeted protein degraders: the past is prologue. Nat Rev Drug Discov 21(3):181–200. https://doi.org/10.1038/s41573-021-00371-6

Benam KH, Burgess JK, Stewart AG (2022) Editorial: accelerated translation using microphysiological organoid and microfluidic Chip models. In: Frontiers in pharmacology, vol 12. Frontiers Media S.A. https://doi.org/10.3389/fphar.2021.827172

Betts A, Haddish-Berhane N, Shah DK, van der Graaf PH, Barletta F, King L, Clark T, Kamperschroer C, Root A, Hooper A, Chen X (2019) A translational quantitative systems

pharmacology model for CD3 bispecific molecules: application to quantify T cell-mediated tumor cell killing by P-cadherin LP DART®. AAPS J 21(4). https://doi.org/10.1208/s12248-019-0332-z
Betts A, van der Graaf PH (2020) Mechanistic quantitative pharmacology strategies for the early clinical development of bispecific antibodies in oncology. Clinical Pharmacology & Therapeutics 108(3):528–541. https://doi.org/10.1002/cpt.1961
Bloomingdale P, Housand C, Apgar JF, Millard BL, Mager DE, Burke JM, Shah DK (2017) Quantitative systems toxicology. In: Current opinion in toxicology, vol 4. Elsevier B.V, pp 79–87. https://doi.org/10.1016/j.cotox.2017.07.003
Bordukova M, Makarov N, Rodriguez-Esteban R, Schmich F, Menden MP (2024) Generative artificial intelligence empowers digital twins in drug discovery and clinical trials. Expert Opin Drug Discovery 19(1):33–42. https://doi.org/10.1080/17460441.2023.2273839
Bradshaw EL, Spilker ME, Zang R, Bansal L, He H, Jones RDO, Le K, Penney M, Schuck E, Topp B, Tsai A, Xu C, Nijsen MJMA, Chan JR (2019) Applications of quantitative systems pharmacology in model-informed drug discovery: perspective on impact and opportunities. CPT Pharmacometrics Syst Pharmacol 8(11):777–791. https://doi.org/10.1002/psp4.12463
Brubaker DK, Paulo JA, Sheth S, Poulin EJ, Popow O, Joughin BA, Strasser SD, Starchenko A, Gygi SP, Lauffenburger DA, Haigis KM (2019) Proteogenomic network analysis of context-specific KRAS signaling in mouse-to-human cross-species translation. Cell Syst 9(3):258–270.e6. https://doi.org/10.1016/j.cels.2019.07.006
Chan JR, Allen R, Boras B, Cabal A, Damian V, Gibbons FD, Gulati A, Hosseini I, Kearns JD, Saito R, Cucurull-Sanchez L, Selimkhanov J, Stein AM, Umehara K, Wang G, Wang W, Neves-Zaph S (2022) Current practices for QSP model assessment: an IQ consortium survey. J Pharmacokinet Pharmacodyn. https://doi.org/10.1007/s10928-022-09811-1
Chan P, Marchand M, Yoshida K, Vadhavkar S, Wang N, Lin A, Wu B, Ballinger M, Sternheim N, Jin JY, Bruno R (2021) Prediction of overall survival in patients across solid tumors following atezolizumab treatments: a tumor growth inhibition–overall survival modeling framework. CPT Pharmacometrics Syst Pharmacol 10(10):1171–1182. https://doi.org/10.1002/psp4.12686
Chang H-P, Li Z, Shah DK (2022) Development of a physiologically-based pharmacokinetic model for whole-body disposition of MMAE containing antibody-drug conjugate in mice. Pharm Res 39(1):1–24. https://doi.org/10.1007/s11095-021-03162-1
Chen X, Hickling T, Vicini P (2014) A mechanistic, multiscale mathematical model of immunogenicity for therapeutic proteins: part 1 – theoretical model. CPT Pharmacometrics Syst Pharmacol 3(9):1–9. https://doi.org/10.1038/psp.2014.30
Cirit M, Stokes CL (2018) Maximizing the impact of microphysiological systems with: in vitro – in vivo translation. In: Lab on a chip, vol 18. Royal Society of Chemistry, pp 1831–1837. https://doi.org/10.1039/c8lc00039e
Claas A, Bullara D, Gaudet S, Wartchow C, Csibi F, Estrada Diez J, D'Hennezel E, Bonazzi S, Schoeberl B (2023) Modeling targeted protein degradation from cells to the clinic to guide Helios degrader development. ACOP
Claret L, Mercier F, Houk BE, Milligan PA, Bruno R (2015) Modeling and simulations relating overall survival to tumor growth inhibition in renal cell carcinoma patients. Cancer Chemother Pharmacol 76(3):567–573. https://doi.org/10.1007/s00280-015-2820-x
Cucurull-Sanchez L, Chappell MJ, Chelliah V, Amy Cheung SY, Derks G, Penney M, Phipps A, Malik-Sheriff RS, Timmis J, Tindall MJ, van der Graaf PH, Vicini P, Yates JWT (2019) Best practices to maximize the use and reuse of quantitative and systems pharmacology models: recommendations from the United Kingdom quantitative and systems pharmacology network. CPT Pharmacometrics Syst Pharmacol 8(5):259–272. https://doi.org/10.1002/psp4.12381
Cuquerella-Gilabert M, Merino-Sanjuán M, García-Arieta A, Mangas-Sanjuán V, Reig-López J (2024) Regulatory requirements and applications of physiologically based pharmacokinetic models. In: ADME processes in pharmaceutical sciences. Springer Nature, pp 193–219. https://doi.org/10.1007/978-3-031-50419-8_9

Dai W, Rao R, Sher A, Tania N, Musante CJ, Allen R (2021) A prototype QSP model of the immune response to SARS-CoV-2 for community development. CPT Pharmacometrics Syst Pharmacol 10(1):18–29. https://doi.org/10.1002/psp4.12574
Deng J, Wei W, Chen Z, Lin B, Zhao W, Luo Y, Zhang X (2019) Engineered liver-on-a-chip platform to mimic liver functions and its biomedical applications: a review. Micromachines 10. MDPI AG. https://doi.org/10.3390/mi10100676
Desikan R, Padmanabhan P, Kierzek AM, van der Graaf PH (2022) Mechanistic models of COVID-19: insights into disease progression, vaccines, and therapeutics. Int J Antimicrob Agents 60(1):106606. https://doi.org/10.1016/j.ijantimicag.2022.106606
Eddy DM, Schlessinger L (2003) Archimedes. Diabetes Care 26(11):3093–3101. https://doi.org/10.2337/diacare.26.11.3093
Eduati F, Jaaks P, Wappler J, Cramer T, Merten CA, Garnett MJ, Saez-Rodriguez J (2020) Patient-specific logic models of signaling pathways from screenings on cancer biopsies to prioritize personalized combination therapies. Mol Syst Biol 16(2):10.15252/msb.20188664
Eichler GS, Imbert G, Branson J, Balibey R, Laramie JM (2022) Democratizing data at Novartis through clinical trial data access. In: Drug discovery today, vol 27. Elsevier Ltd, pp 1533–1537. https://doi.org/10.1016/j.drudis.2022.02.019
FDA (2023) Real-world evidence. https://www.fda.gov/science-research/science-and-research-special-topics/real-world-evidence
Feder A, Keith KA, Manzoor E, Pryzant R, Sridhar D, Wood-Doughty Z, Eisenstein J, Grimmer J, Reichart R, Roberts ME, Stewart BM, Veitch V, Yang D (2022) Causal inference in natural language processing: estimation, prediction, interpretation and beyond. Trans Assoc Comput Ling 10:1138–1158. https://doi.org/10.1162/tacl_a_00511
Flowers D, Bassen D, Kapitanov GI, Marcantonio D, Burke JM, Apgar JF, Betts A, Hua F (2023) A next generation mathematical model for the in vitro to clinical translation of T-cell engagers. J Pharmacokinet Pharmacodyn 50(3):215–227. https://doi.org/10.1007/s10928-023-09846-y
Friedrich CM (2016) A model qualification method for mechanistic physiological QSP models to support model-informed drug development. CPT Pharmacometrics Syst Pharmacol 2015:43–53. https://doi.org/10.1002/psp4.12056
Go N, Arsène S, Faddeenkov I, Galland T, Martis BS, Lefaudeux D, Wang Y, Etheve L, Jacob E, Monteiro C, Bosley J, Sansone C, Pasquali C, Lehr L, Kulesza A (2024) A quantitative systems pharmacology workflow toward optimal design and biomarker stratification of atopic dermatitis clinical trials. J Allergy Clin Immunol. https://doi.org/10.1016/j.jaci.2023.12.031
Guo Y, Chen X, Gong P, Li G, Yao W, Yang W (2023) The gut–organ-axis concept: advances the application of gut-on-chip technology. Int J Mol Sci 24. MDPI. https://doi.org/10.3390/ijms24044089
Haid RTU, Reichel A (2023) A mechanistic pharmacodynamic modeling framework for the assessment and optimization of proteolysis targeting chimeras (PROTACs). Pharmaceutics 15(1). https://doi.org/10.3390/pharmaceutics15010195
Hardiansyah D, Ng CM (2019) Quantitative systems pharmacology model of chimeric antigen receptor T-cell therapy. Clin Transl Sci 12(4):343–349. https://doi.org/10.1111/cts.12636
Harrison RK (2016) Phase II and phase III failures: 2013-2015. Nat Rev Drug Discov 15(12):817–818. https://doi.org/10.1038/nrd.2016.184
Hasselgren C, Oprea TI (2023) Annual review of pharmacology and toxicology artificial intelligence for drug discovery: are we there yet? https://doi.org/10.1146/annurev-pharmtox-040323
Herland A, Maoz BM, Das D, Somayaji MR, Prantil-Baun R, Novak R, Cronce M, Huffstater T, Jeanty SSF, Ingram M, Chalkiadaki A, Benson Chou D, Marquez S, Delahanty A, Jalili-Firoozinezhad S, Milton Y, Sontheimer-Phelps A, Swenor B, Levy O et al (2020) Quantitative prediction of human pharmacokinetic responses to drugs via fluidically coupled vascularized organ chips. Nat Biomed Eng. https://doi.org/10.1038/s41551-019-0498-9
Hosseini I, Gadkar K, Stefanich E, Li C-C, Sun LL, Chu Y-W, Ramanujan S (2020) Mitigating the risk of cytokine release syndrome in a Phase I trial of CD20/CD3 bispecific antibody

mosunetuzumab in NHL: impact of translational system modeling. Npj Systems Biology and Applications 6(1):28. https://doi.org/10.1038/s41540-020-00145-7

Howard FM, He G, Peterson JR, Pfeiffer JR, Earnest T, Pearson AT, Abe H, Cole JA, Nanda R (2022) Highly accurate response prediction in high-risk early breast cancer patients using a biophysical simulation platform. Breast Cancer Res Treat 196(1):57–66. https://doi.org/10.1007/s10549-022-06722-0

Hutton-Smith LA, Gaffney EA, Byrne HM, Maini PK, Gadkar K, Mazer NA (2017) Ocular pharmacokinetics of therapeutic antibodies given by intravitreal injection: estimation of retinal permeabilities using a 3-compartment semi-mechanistic model. Mol Pharm 14(8):2690–2696. https://doi.org/10.1021/acs.molpharmaceut.7b00164

Ingber DE (2022) Human organs-on-chips for disease modelling, drug development and personalized medicine. Nat Rev Genet 23:467–491. https://doi.org/10.1038/s41576-022-00466-9

Joshi A, Ramanujan S, Jin JY (2023) The convergence of pharmacometrics and quantitative systems pharmacology in pharmaceutical research and development. Eur J Pharm Sci 182: 106380. https://doi.org/10.1016/j.ejps.2023.106380

Joslyn LR, Huang W, Miles D, Hosseini I, Ramanujan S (2024) Digital twins elucidate critical role of Tscm in clinical persistence of TCR-engineered cell therapy. Npj Systems Biology and Applications 10(1):11. https://doi.org/10.1038/s41540-024-00335-7

Junhee Seok H, Warren S, Alex GC, Michael NM, Henry VB, Xu W, Richards DR, McDonald-Smith GP, Gao H, Hennessy L, Finnerty CC, López CM, Honari S, Moore EE, Minei JP, Cuschieri J, Bankey PE, Johnson JL, Sperry J et al (2013) Genomic responses in mouse models poorly mimic human inflammatory diseases. Proc Natl Acad Sci U S A 110(9):3507–3512. https://doi.org/10.1073/pnas.1222878110

Karelina T, Demin O, Demin O, Duvvuri S, Nicholas T (2017) Studying the progression of amyloid pathology and its therapy using translational longitudinal model of accumulation and distribution of amyloid beta. CPT Pharmacometrics Syst Pharmacol 6(10):676–685. https://doi.org/10.1002/psp4.12249

Katsoulakis E, Wang Q, Wu H, Shahriyari L, Fletcher R, Liu J, Achenie L, Liu H, Jackson P, Xiao Y, Syeda-Mahmood T, Tuli R, Deng J (2024) Digital twins for health: a scoping review. Npj Digital Med 7(1):77. https://doi.org/10.1038/s41746-024-01073-0

Kearns JD, Wassmann P, Olgac U, Fichter M, Christen B, Rubic-Schneider T, Koepke S, Cochin de Billy B, Ledieu D, Andre C, Hawtin S, Fischer B, Moretti F, Hug C, Bepperling A, Brannetti B, Mendez-Garcia C, Littlewood-Evans A, Clemens A et al (2023) A root cause analysis to identify the mechanistic drivers of immunogenicity against the anti-VEGF biotherapeutic brolucizumab. Sci Transl Med 15(681). https://doi.org/10.1126/scitranslmed.abq5068

Keuper-Navis M, Walles M, Poller B, Myszczyszyn A, van der Made TK, Donkers J, Eslami Amirabadi H, Wilmer MJ, Aan S, Spee B, Masereeuw R, van de Steeg E (2023) The application of organ-on-chip models for the prediction of human pharmacokinetic profiles during drug development. In: Pharmacological research, vol 195. Academic Press. https://doi.org/10.1016/j.phrs.2023.106853

Kim KS, Ejima K, Iwanami S, Fujita Y, Ohashi H, Koizumi Y, Asai Y, Nakaoka S, Watashi K, Aihara K, Thompson RN, Ke R, Perelson AS, Iwami S (2021) A quantitative model used to compare within-host SARS-CoV-2, MERS-CoV, and SARS-CoV dynamics provides insights into the pathogenesis and treatment of SARS-CoV-2. PLoS Biol 19(3):e3001128. https://doi.org/10.1371/journal.pbio.3001128

Kirouac DC, Schaefer G, Chan J, Merchant M, Orr C, Huang S-MA, Moffat J, Liu L, Gadkar K, Ramanujan S (2017) Clinical responses to ERK inhibition in BRAF V600E-mutant colorectal cancer predicted using a computational model. Npj Syst Biol Appl 3(1):14. https://doi.org/10.1038/s41540-017-0016-1

Kirouac DC, Zmurchok C, Deyati A, Sicherman J, Bond C, Zandstra PW (2023) Deconvolution of clinical variance in CAR-T cell pharmacology and response. Nat Biotechnol 41(11):1606–1617. https://doi.org/10.1038/s41587-023-01687-x

Kirouac DC, Zmurchok C, Morris D (2024) Making drugs from T cells: the quantitative pharmacology of engineered T cell therapeutics. Npj Systems Biology and Applications 10(1):31. https://doi.org/10.1038/s41540-024-00355-3

Kletting P, Thieme A, Eberhardt N, Rinscheid A, D'Alessandria C, Allmann J, Wester H-J, Tauber R, Beer AJ, Glatting G, Eiber M (2019) Modeling and predicting tumor response in radioligand therapy. J Nucl Med 60(1):65–70. https://doi.org/10.2967/jnumed.118.210377

Kuemmel C, Yang Y, Zhang X, Florian J, Zhu H, Tegenge M, Huang S, Wang Y, Morrison T, Zineh I (2020) Consideration of a credibility assessment framework in model-informed drug development: potential application to physiologically-based pharmacokinetic modeling and simulation. CPT Pharmacometrics Syst Pharmacol 9(1):21–28. https://doi.org/10.1002/psp4.12479

Lee J, Yoon W, Kim S, Kim D, Kim S, So CH, Kang J (2020) BioBERT: a pre-trained biomedical language representation model for biomedical text mining. Bioinformatics 36(4):1234–1240. https://doi.org/10.1093/bioinformatics/btz682

Ma H, Wang H, Sove RJ, Jafarnejad M, Tsai C-H, Wang J, Giragossian C, Popel AS (2020) A quantitative systems pharmacology model of T cell engager applied to solid tumor. AAPS J 22(4):85. https://doi.org/10.1208/s12248-020-00450-3

Maass C, Sorensen NB, Himmelfarb J, Kelly EJ, Stokes CL, Cirit M (2019) Translational assessment of drug-induced proximal tubule injury using a kidney microphysiological system. CPT Pharmacometrics Syst Pharmacol 8(5):316–325. https://doi.org/10.1002/psp4.12400

Mcaleer CW, Long CJ, Elbrecht D, Sasserath T, Bridges LR, Rumsey JW, Martin C, Schnepper M, Wang Y, Schuler F, Roth AB, Funk C, Shuler ML, Hickman JJ (2019) Multi-organ system for the evaluation of efficacy and off-target toxicity of anticancer therapeutics. https://www.science.org

Meimetis N, Pullen KM, Zhu DY, Nilsson A, Hoang TN, Magliacane S, Lauffenburger DA (2024) AutoTransOP: translating omics signatures without orthologue requirements using deep learning. Npj Systems Biology and Applications 10(1):13. https://doi.org/10.1038/s41540-024-00341-9

Miyazawa K, Liu Y, Bazzazi H (2024) Development of a minimal PBPK-QSP modeling platform for LNP-mRNA based therapeutics to study tissue disposition and protein expression dynamics. Front Nanotechnol 6. https://doi.org/10.3389/fnano.2024.1330406

Moingeon P (2023) Artificial intelligence-driven drug development against autoimmune diseases. Trends Pharmacol Sci 44(7):411–424. https://doi.org/10.1016/j.tips.2023.04.005

Monine M, Norris D, Wang Y, Nestorov I (2021) A physiologically-based pharmacokinetic model to describe antisense oligonucleotide distribution after intrathecal administration. J Pharmacokinet Pharmacodyn 48(5):639–654. https://doi.org/10.1007/s10928-021-09761-0

Moor M, Banerjee O, Abad ZSH, Krumholz HM, Leskovec J, Topol EJ, Rajpurkar P (2023) Foundation models for generalist medical artificial intelligence. Nature 616(7956):259–265. https://doi.org/10.1038/s41586-023-05881-4

Mougiakakos D, Krönke G, Völkl S, Kretschmann S, Aigner M, Kharboutli S, Böltz S, Manger B, Mackensen A, Schett G (2021) CD19-targeted CAR T cells in refractory systemic lupus erythematosus. N Engl J Med 385(6):567–569. https://doi.org/10.1056/NEJMc2107725

Myers RC, Augustin F, Huard J, Friedrich CM (2023) Using machine learning surrogate modeling for faster QSP VP cohort generation. CPT Pharmacometrics Syst Pharmacol 12(8):1047–1059. https://doi.org/10.1002/psp4.12999

Office of Clinical Pharmacology, F (2023) 2023 annual report. https://www.fda.gov/media/175903/download?attachment

Paczkowska M, Barenboim J, Sintupisut N, Fox NS, Zhu H, Abd-Rabbo D, Mee MW, Boutros PC, Abascal F, Amin SB, Bader GD, Beroukhim R, Bertl J, Boroevich KA, Brunak S, Campbell PJ, Carlevaro-Fita J, Chakravarty D, Chan CWY et al (2020) Integrative pathway enrichment analysis of multivariate omics data. Nat Commun 11(1):735. https://doi.org/10.1038/s41467-019-13983-9

Padmanabhan P, Desikan R, Dixit NM (2022) Modeling how antibody responses may determine the efficacy of COVID-19 vaccines. Nat Comput Sci 2(2):123–131. https://doi.org/10.1038/s43588-022-00198-0

Porto EM, Komor AC, Slaymaker IM, Yeo GW (2020) Base editing: advances and therapeutic opportunities. In: Nature reviews drug discovery, vol 19, pp 839–859. https://doi.org/10.1038/s41573-020-0084-6

Qi T, Liao X, Cao Y (2023) Development of bispecific T cell engagers: harnessing quantitative systems pharmacology. Trends Pharmacol Sci 44(12):880–890. https://doi.org/10.1016/j.tips.2023.09.009

Ram RN, Gadaleta D, Allen TEH (2022) The role of 'big data' and 'in silico' New Approach Methodologies (NAMs) in ending animal use – A commentary on progress Computational Toxicology 23, 100232 https://doi.org/10.1016/j.comtox.2022.100232

Ramanujan S, Chan JR, Friedrich CM, Thalhauser CJ (2019) A flexible approach for context-dependent assessment of quantitative systems pharmacology models. CPT Pharmacometrics Syst Pharmacol 8(6):340–343. https://doi.org/10.1002/psp4.12409

Ramanujan S, Gadkar K, Kadambi A (2016) Quantitative systems pharmacology: applications and adoption in drug development, pp 27–52. https://doi.org/10.1007/978-3-319-44534-2_3

Rao R, Musante CJ, Allen R (2023) A quantitative systems pharmacology model of the pathophysiology and treatment of COVID-19 predicts optimal timing of pharmacological interventions. Npj Systems Biology and Applications 9(1):13. https://doi.org/10.1038/s41540-023-00269-6

Ribba B (2023) Quantitative systems pharmacology in the age of artificial intelligence. CPT Pharmacometrics Syst Pharmacol 12(12):1823–1826. https://doi.org/10.1002/psp4.13047

Rodney S, Banerji U (2024) Optimizing the FDA's project optimus: opportunities and challenges. Nat Rev Clin Oncol 21(3):165–166. https://doi.org/10.1038/s41571-023-00853-z

Rogers KV, Martin SW, Bhattacharya I, Singh RSP, Nayak S (2021) A dynamic quantitative systems pharmacology model of inflammatory bowel disease: part 1 – model framework. Clin Transl Sci 14(1):239–248. https://doi.org/10.1111/cts.12849

Ruiz-Martinez A, Gong C, Wang H, Sové RJ, Mi H, Kimko H, Popel AS (2022) Simulations of tumor growth and response to immunotherapy by coupling a spatial agent-based model with a whole-patient quantitative systems pharmacology model. PLoS Comput Biol 18(7):e1010254. https://doi.org/10.1371/journal.pcbi.1010254

Schäfer PSL, Dimitrov D, Villablanca EJ, Saez-Rodriguez J (2024) Integrating single-cell multi-omics and prior biological knowledge for a functional characterization of the immune system. Nat Immunol 25(3):405–417. https://doi.org/10.1038/s41590-024-01768-2

Selvaggio G, Leonardelli L, Lofano G, Fresnay S, Parolo S, Medini D, Siena E, Marchetti L (2021) A quantitative systems pharmacology approach to support mRNA vaccine development and optimization. CPT Pharmacometrics Syst Pharmacol 10(12):1448–1451. https://doi.org/10.1002/psp4.12721

Shebley M, Sandhu P, Emami Riedmaier A, Jamei M, Narayanan R, Patel A, Peters SA, Reddy VP, Zheng M, de Zwart L, Beneton M, Bouzom F, Chen J, Chen Y, Cleary Y, Collins C, Dickinson GL, Djebli N, Einolf HJ et al (2018) Physiologically based pharmacokinetic model qualification and reporting procedures for regulatory submissions: a consortium perspective. Clinical Pharmacology & Therapeutics 104(1):88–110. https://doi.org/10.1002/cpt.1013

Sorger PK, Allerheiligen SRB, Abernethy DR, Altman RB, Brouwer KLR, Califano A, David Z, Argenio D, Iyengar R, Jusko WJ, Lalonde R, Lauffenburger DA, Shoichet B, Stevens JL, Subramaniam S, Van Graaf P Der WR, Sorger PK, Ma B, . . . Altman RB (2011) Quantitative and systems pharmacology in the post-genomic era: new approaches to discovering drugs and understanding therapeutic mechanisms an NIH White Paper by the QSP Workshop Group – October, 2011. An NIH white paper, pp 0–47

Stein RR, Fouché M, Kearns JD, Roth H-J (2023) A model-informed method to retrieve intrinsic from apparent cooperativity and project cellular target occupancy for ternary complex-forming compounds. RSC Chem Biol 4(7):512–523. https://doi.org/10.1039/D2CB00216G

Susilo ME, Li C, Gadkar K, Hernandez G, Huw L, Jin JY, Yin S, Wei MC, Ramanujan S, Hosseini I (2023) Systems-based digital twins to help characterize clinical dose–response and propose predictive biomarkers in a Phase I study of bispecific antibody, mosunetuzumab, in NHL. Clin Transl Sci 16(7):1134–1148. https://doi.org/10.1111/cts.13501

Taylor DL, Gough A, Schurdak ME, Vernetti L, Chennubhotla CS, Lefever D, Pei F, Faeder JR, Lezon TR, Stern AM, Bahar I (2019) Harnessing human microphysiology systems as key experimental models for quantitative systems pharmacology. In: Handbook of experimental pharmacology, vol 260. Springer, pp 327–367. https://doi.org/10.1007/164_2019_239

Terranova N, Renard D, Shahin MH, Menon S, Cao Y, Hop CECA, Hayes S, Madrasi K, Stodtmann S, Tensfeldt T, Vaddady P, Ellinwood N, Lu J (2023) Artificial intelligence for quantitative modeling in drug discovery and development: an innovation and quality consortium perspective on use cases and best practices. Clin Pharmacol Ther. https://doi.org/10.1002/cpt.3053

Terranova N, Venkatakrishnan K, Benincosa LJ (2021) Application of machine learning in translational medicine: current status and future opportunities. AAPS J 23(4):74. https://doi.org/10.1208/s12248-021-00593-x

Thau H, Neuber S, Emmert MY, Nazari-Shafti TZ (2024) Targeting lipoprotein(a): can RNA therapeutics provide the next step in the prevention of cardiovascular disease? Cardiol Ther 13(1):39–67. https://doi.org/10.1007/s40119-024-00353-w

Thiel C, Smit I, Baier V, Cordes H, Fabry B, Blank LM, Kuepfer L (2018) Using quantitative systems pharmacology to evaluate the drug efficacy of COX-2 and 5-LOX inhibitors in therapeutic situations. Npj Systems Biology and Applications 4(1):28. https://doi.org/10.1038/s41540-018-0062-3

Venkatakrishnan K, van der Graaf PH (2021) Model-informed drug development: connecting the dots with a totality of evidence mindset to advance therapeutics. Clin Pharmacol Ther 110(5):1147–1154. https://doi.org/10.1002/cpt.2422

Venkatapurapu SP, Iwakiri R, Udagawa E, Patidar N, Qi Z, Takayama R, Kumar K, Sato Y, Behar M, Offner P, Dwivedi G, Miyasaka H, Suzuki RK, Hamada AL, D'Alessandro PM, Fernandez J (2022) A computational platform integrating a mechanistic model of Crohn's disease for predicting temporal progression of mucosal damage and healing. Adv Ther 39(7):3225–3247. https://doi.org/10.1007/s12325-022-02144-y

Viceconti M, Pappalardo F, Rodriguez B, Horner M, Bischoff J, Musuamba Tshinanu F (2021) In silico trials: verification, validation and uncertainty quantification of predictive models used in the regulatory evaluation of biomedical products. Methods 185:120–127. https://doi.org/10.1016/j.ymeth.2020.01.011

Wang Y-S, Kumari M, Chen G-H, Hong M-H, Yuan JP-Y, Tsai J-L, Wu H-C (2023) mRNA-based vaccines and therapeutics: an in-depth survey of current and upcoming clinical applications. J Biomed Sci 30(1):84. https://doi.org/10.1186/s12929-023-00977-5

Watkins PB (2020) Quantitative systems toxicology approaches to understand and predict drug-induced liver injury. Clin Liver Dis 24(1):49–60. https://doi.org/10.1016/j.cld.2019.09.003

Wu F, Shah H, Li M, Duan P, Zhao P, Suarez S, Raines K, Zhao Y, Wang M, Lin H, Duan J, Yu L, Seo P (2021) Biopharmaceutics applications of physiologically based pharmacokinetic absorption modeling and simulation in regulatory submissions to the U.S. Food and Drug Administration for New Drugs. AAPS J 23(2):31. https://doi.org/10.1208/s12248-021-00564-2

Zhang S, Gong C, Ruiz-Martinez A, Wang H, Davis-Marcisak E, Deshpande A, Popel AS, Fertig EJ (2021) Integrating single cell sequencing with a spatial quantitative systems pharmacology model spQSP for personalized prediction of triple-negative breast cancer immunotherapy response. ImmunoInformatics 1–2:100002. https://doi.org/10.1016/j.immuno.2021.100002

Zhang T, Androulakis IP, Bonate P, Cheng L, Helikar T, Parikh J, Rackauckas C, Subramanian K, Cho CR, Androulakis IP, Bonate P, Borisov I, Broderick G, Cheng L, Damian V, Dariolli R, Demin O, Ellinwood N, Fey D et al (2022) Two heads are better than one: current landscape of integrating QSP and machine learning. J Pharmacokinet Pharmacodyn 49(1):5–18. https://doi.org/10.1007/s10928-022-09805-z

Zhang X, Yang Y, Grimstein M, Fan J, Grillo JA, Huang S, Zhu H, Wang Y (2020) Application of PBPK modeling and simulation for regulatory decision making and its impact on US prescribing information: an update on the 2018-2019 submissions to the US FDA's Office of Clinical Pharmacology. J Clin Pharmacol 60(S1). https://doi.org/10.1002/jcph.1767

Zhao H, Narjes F (2023) Kinetic modeling of PROTAC-induced protein degradation. ChemMedChem 18(24). https://doi.org/10.1002/cmdc.202300530

Zhao J, Zhu X, Tan S, Chen C, Kaddoumi A, Guo XL, Lin YW, Cheung SYA (2022) Editorial: model-informed drug development and evidence-based translational pharmacology. In: Frontiers in pharmacology, vol 13, Frontiers Media S.A. https://doi.org/10.3389/fphar.2022.1086551

Zheng B, Wille L, Peppel K, Hagen D, Matteson A, Ahlers J, Schaff J, Hua F, Yuraszeck T, Cobbina E, Apgar JF, Burke JM, Roberts J, Das R (2021) A systems pharmacology model for gene therapy in sickle cell disease. CPT Pharmacometrics Syst Pharmacol 10(7):696–708. https://doi.org/10.1002/psp4.12638

Zineh I (2019) Quantitative systems pharmacology: a regulatory perspective on translation. CPT Pharmacometrics Syst Pharmacol 8(6):336–339. https://doi.org/10.1002/psp4.12403

MIX
Papier aus verantwortungsvollen Quellen
Paper from responsible sources
FSC® C105338

If you have any concerns about our products,
you can contact us on
ProductSafety@springernature.com

In case Publisher is established outside the EU,
the EU authorized representative is:
Springer Nature Customer Service Center GmbH
Europaplatz 3, 69115 Heidelberg, Germany

Printed by Libri Plureos GmbH
in Hamburg, Germany